Paris Savant

Paris Savant

Capital of Science in the Age of Enlightenment

BRUNO BELHOSTE

Translated by
SUSAN EMANUEL

Foreword by
DENA GOODMAN

OXFORD
UNIVERSITY PRESS

OXFORD
UNIVERSITY PRESS

Oxford University Press is a department of the University of Oxford. It furthers
the University's objective of excellence in research, scholarship, and education
by publishing worldwide. Oxford is a registered trade mark of Oxford University
Press in the UK and certain other countries.

Published in the United States of America by Oxford University Press
198 Madison Avenue, New York, NY 10016, United States of America.

Library of Congress Cataloging-in-Publication Data
Names: Belhoste, Bruno, author. | Emanuel, Susan, translator.
Title: Paris savant : capital of science in the age of Enlightenment /
Bruno Belhoste ; translated by Susan Emanuel.
Description: New York, NY : Oxford University Press, 2018. | Translation of
the French language edition of: Paris savant : parcours et rencontres au
temps des Lumieres (Paris : Colin, 2011). | Includes bibliographical
references and index.
Identifiers: LCCN 2018045239 | ISBN 9780199382545 (hbk)
Subjects: LCSH: Science—France—Paris—History—18th century. |
Scientists—France—Paris—Biography. | Paris (France)—Intellectual
life—18th century.
Classification: LCC Q127.F8 B4413 2018 |
DDC 509.44/09033—dc23
LC record available at https://lccn.loc.gov/2018045239

1 3 5 7 9 8 6 4 2

Printed by Sheridan Books, Inc., United States of America

Contents

Illustrations

Figures

Maps

Foreword

IN THIS WONDERFUL book, Bruno Belhoste, a noted historian of science at the Sorbonne, gives us a sweeping panorama of Paris in the decades leading up to the French Revolution in which scientists take center stage. For what Belhoste aims to demonstrate is both that Paris was the capital of the sciences at the end of the eighteenth century and that the conduct of science and scientists created a dynamic that pulsed through the city: cutting across its many neighborhoods old and new, cutting through and creating surprising bonds within a hierarchical social order, and offering a new vitality to a monarchy that did not know it was doomed.

We start near the center of the city, in the old palace of the Louvre, which had been vacated by the royal court when Louis XIV moved it to Versailles at the end of the seventeenth century and was now, to the king's glory, the meeting place of his academies, including the Academy of Sciences. We follow the academicians or *savants* about their business into every quarter of Paris and its suburbs: across the Seine to the Paris Observatory and the Royal Botanical Garden (now the Jardin des Plantes) and then to the schools and print shops of the Latin Quarter. We follow them too into the fashionable neighborhoods, mostly on the right bank, where they mingled with *le monde* and *les grands*—the social and political elite of high aristocrats, wealthy financiers, and government officials—in mirrored salons and well-appointed dining rooms. Members of this powerful elite might serve as patrons for the savants, but they validated and supported them in other ways too: by opening their natural history collections to them or attending the public courses they offered in physics, chemistry, and natural history. We then follow the crowd out to the new boulevards on the northern and western edges of the city where science was a form of spectacle and good business, where education met theater, and where commerce, science, and entertainment fueled each other. Returning to the center we spend many happy hours in the Palais Royal, whose arcades and gardens had been turned into a microcosm of this modern Paris by its royal owner, the Duke de Chartres. In the Palais Royal dealers in scientific instruments sold their wares in fashionable shops next door to theaters, cabinets of curiosities, and cafés; above them were private rooms in which members could meet in clubs to pursue shared interests and discuss new ideas; and above them were much smaller rooms to which prostitutes brought the men they met in the gardens after dusk. We are now back almost where we began: just across the rue Saint-Honoré from the Louvre, the "vast caravanserai that housed painters, sculptors, architects, engravers, literati, and scientists, as well as a multitude of rats." But here in the Palais Royal we are not peeking inside the closed doors behind which the privileged producers of culture lived and worked; here, as Belhoste shows us, science was for everyone, and we experience the pleasure and excitement of the men and women from all walks of life who participated in it.

I suggest that you read *Paris Savant* in a comfortable chair but with a laptop or tablet nearby so that you can locate the places to which Belhoste takes you on a map of Paris. Although a few street names have changed and some buildings have disappeared, for the most part you will find Paris Savant on Google Maps: the Louvre, of course, and the Palais Royal, where in 1784 the inventors Quinquet and Lange displayed their new oil lamps for the admiration of the public at Daguerre's fashionable shop. Farther along the rue Saint-Honoré is the townhouse where Madame Geoffrin brought together men of letters, savants, and wealthy amateurs at her Wednesday dinners for more than two decades. You can also follow the path of fresh water flowing into the center of Paris from the new steam pumps at Chaillot on the right bank of the Seine and the Gros-Caillou on the left bank. Best of all, you can imagine the city from above as you climb into the basket of the hydrogen balloon that ascended to the clouds from the Champs de Mars on August 27, 1783, and landed eighteen miles away in the village of Gonesse.

The social reach of Belhoste's *Paris Savant* matches its geographic reach, extending as far beyond academicians as it does beyond the Academy. It includes the inventors of oil lamps and hydrogen balloons, the makers of scientific instruments, and the public of women and men who read books and periodicals, took public courses, and were captivated by the spectacle of science set before them in theaters, balloon launches (the crowd on the Champs de Mars was said to number one hundred thousand), and the tubs of magnetic fluid with which Dr. Mesmer cured all manner of people of all manner of illnesses.

Belhoste shows us a world in which rivalries could certainly be fierce—between inventors of rival oil lamps, for example, or editors of competing dictionaries—but this dynamic world was also driven by public spiritedness and a zeal for improvement. The passion for science and the passion for the public good were, if not the same, closely entangled. And the sense of importance that inflated the egos of the savants was due in part to their belief that the work of science was itself important. Belhoste devotes a whole chapter to the problems of public health that beset Paris at a time when migrants streaming into the city were making its population increasingly dense, but its infrastructure remained largely unchanged since the Middle Ages. In the 1770s and '80s the government turned to the savants to address these problems. Should Paris's oldest hospital, which served both the sick and the poor, simply be rebuilt after it suffered a catastrophic fire in 1772? Or did the tragedy provide an opportunity to implement design reforms that reflected current understandings of disease and contagion? The savants were consulted, but entrenched interests blocked the reforms they proposed. In 1776, the reforming pharmacist Cadet de Vaux issued a report (probably at the behest of the lieutenant of police, according to Belhoste) in which he concluded that the air around the Cemetery of the Innocents, located in the same quarter as the central market, was the least healthy air in Paris. Although an attempt was made to suppress the report, the cemetery was closed in 1780, after reports of "cadaverous gases" that were affecting public health began to circulate. Over the course of six months, and in the dead of night so as not to upset the neighbors, the cemetery's contents were moved to a site outside the city limits that had been dug to accommodate them. These are the catacombs that tourists visit today.

The cadaverous gases were emblematic of the larger problem that confronted the city and its leaders: the unhealthiness of its air and water. Within the city limits, sewers flowed directly into the Seine and noxious gases emanated from cesspits in which human and animal waste was collected from apartment buildings and abattoirs. Belhoste shows us not only how the savants were mobilized to address these problems but how science itself (and the chemistry of gases in particular) was both advanced through and brought to bear on it. Popular science, academic science, invention, and public policy breathed the same air.

What we see here and throughout the book are the ways in which savants, government officials and agencies, inventors and manufacturers, elite artisans, and even charlatans interacted, sometimes with hostility and rivalry, but often collaboratively—if we can talk about collaboration in a social order defined by hierarchy and privilege. Yes, the savants had to take a stand against charlatanry if they were to stand for anything at all, if science was to mean anything, but this was not simply a matter of social elitism. Lavoisier, by then a towering figure in the Academy of Sciences, stood for a new form of science, what Belhoste calls the "severe science," whose experimental method depended upon rigorous calculation and exact measurement. For Lavoisier and the growing number of savants who joined him, unveiling the deceptions practiced by Mesmer and others who, in their view, streamed into Paris to take advantage of the public's gullibility, was not just an assertion of their authority and privilege—although it was that; it was both a public service and a moral crusade. The moral authority they claimed for themselves rested on serving both truth and the public at the same time.

In *Paris Savant* Bruno Belhoste gives us a tour of Paris on the eve of the French Revolution that brings the city to life in a new way. But he also shows us how the modern scientist emerged at the same time as the modern democratic state, or rather how out of the matrix of the Ancien Régime the savant became a scientist and the state became modern and democratic and the two came to work together and depend upon each other. And the character of this partnership reflects the geography and sociology of science in Paris in the late eighteenth century, where the severe science was developed in a laboratory located in the royal arsenal by the outstanding chemist of his day because he was also the head of the Gunpowder Administration. The laboratory itself was just across the courtyard from Lavoisier's living quarters, in which the same savants who collaborated on and observed his experiments also enjoyed the sociability of his wife's salon. Marie-Anne Paulze, too, went back and forth between the living space and the laboratory, assisting her husband with his experiments and drawing the precise figures for the engraved plates that were an essential element of their publication.

Fifty years ago Robert Darnton captured the attention of historians and the public alike with the little book *Mesmerism and the End of the Enlightenment in France* (Harvard, 1968), in which he painted a picture of Paris in the 1780s captivated by science and especially the science of animal magnetism, or Mesmerism. Darnton sought to redirect our attention from the canonical writers of the Enlightenment to Mesmer and his followers, who, he argued, brought the Enlightenment to the masses and thus to an end by fueling the Revolution. It was Mesmerism, not Voltairean reason, which reflected the radical "mood" of pre-revolutionary Paris. Bruno Belhoste gives us a more balanced view of Paris before and during the Revolution. Perhaps today we are more sympathetic to Lavoisier, who led the investigation for the Academy of Science that revealed Mesmer's treatments to be without therapeutic value and his "magnetic fluid" to be imaginary. We know what can happen when science and political power work together, as well as how privilege contributes to the pursuit of knowledge even as it hampers it. But we live in a time when the claims to scientific authority that Lavoisier sought to shore up can no longer be taken for granted, as they were in the 1960s, when the authority of scientists and the power of science were at their peak. These are the contradictions that Bruno Belhoste addresses in his last chapter on Paris Savant in the Revolution.

Surprisingly, perhaps, Belhoste does not want us to conclude simply that the seeds of the magisterial science of the nineteenth century were planted before the Revolution, but neither does he take the tragic view that in closing the Academy of Sciences and guillotining Lavoisier, the Revolution showed itself to be the enemy of science itself. In keeping with our revisionist understanding of the French Revolution, the picture he paints is colored and nuanced rather than black and white; it is a tapestry whose warp is continuity but whose woof is change. Most important, *Paris Savant* guides us through a vibrant city filled with interesting characters and animated by science that, once we come to know it, we can recover in the streets, buildings, and gardens that are still there for us to visit today. Like the shadow of a leaf pressed into the page of a botanist's notebook, the imprint of *Paris Savant* will be visible to anyone who has read and enjoyed this book as much as I have.

Dena Goodman
Ann Arbor, 2018

Preface

"PARIS SAVANT": THE EXPRESSION appeared for the first time in 1841, in a story by Balzac, a satire inspired by the quarrel between two famous naturalists (Geoffroy Saint-Hilaire and Cuvier). Balzac's "An Ass's Guide for the Use of Animals Who Want to Acquire Honors" tells the story of a donkey transformed into a black zebra with yellow stripes that walks like a giraffe, and of its master, the inventor of "instinctology." Balzac uses "Paris savant" to poke fun at nineteenth-century academicians, professors, and scientific popularizers. The "Paris Savant" presented in this book, devoted to the sciences in Paris during the Age of Enlightenment, is quite different. It does include savants (some of whom would later be called scientists or scholars), but also inventors, artists, booksellers, collectors, charlatans—and their audiences. The work and the discoveries of Parisian savants appear in this book as products of the city itself, on a par with luxury goods, works of art, literary creations, and ideas of all kinds that the city produced in abundance. This is why this history is also a history of Paris.

The sciences flourished through encounter and exchange. This scientific exchange, bringing together men and ideas from all over, was concentrated in a few centers open to the world, from which it radiated outward by means of print and education. More than any other metropole, Paris played the role of capital of the sciences in the eighteenth century. At the time the city was witness to intense intellectual activity. Exploring this flurry of activity, the book wanders through the places, small or large, where so many Parisians, through study and invention, contributed to the culture of the Enlightenment.

The book opens with the Académie Royale des Sciences, which dominated and controlled Paris Savant; this was the institution that set the tone for scientific life before the Revolution. While members of the company had premises in the Louvre in the former royal apartments, they were everywhere around the city: working in offices and laboratories, teaching in colleges and giving public lectures, meeting each other at salons and in cafés, frequenting galleries and workshops. The French monarchy placed at their disposal grand establishments like the Paris Observatory, the Royal Botanical Garden, the Mint, and the military laboratory at the Arsenal. The city itself was the theater for their prowess and achievements in civic improvement, from the cleansing of cemeteries and cesspools to public balloon launches and the construction of mathematically designed buildings. From Paris they had dealings with the French provinces, Europe, and the rest of the world.

The savants formed a small universe unto themselves, but they maintained permanent relations with all the representatives of the Republic of Letters in Paris, with scholars, professors, writers, journalists, and printer-booksellers. Around them orbited amateurs (and the curious) who belonged to high society or the financial world. Thus, savants were connected to the highest echelons of government, the court, and princes of the blood, who often played the role of patrons. In another orbit, an astonishing number of all stripes of autodidacts buzzed around them, besieging the savants with their discoveries and inventions, or what they claimed as such.

By means of spectacle, publications, and instruction, Parisian savants and those who emulated them reached a wide audience, including the popular classes. They thereby contributed to the invention and dissemination of the Enlightenment. Moreover, they actively participated in the production of the *Encyclopédie*, that monument to the new age that was conceived and laboriously crafted over many years in Paris. Beginning in the 1770s, public lectures, Masonic lodges, *musées*, and salons burgeoned around the capital. Books and journals publicized their activities, in which the natural sciences occupied a privileged place. This was the time when public opinion became enthusiastic about animal magnetism, but also about ballooning, pneumatic chemistry, and public hygiene.

A spirit of rebellion arose during the final years of the Ancien Régime. Relations between the official world of science and the Parisian public deteriorated. The Académie became increasingly irritated at seeing its authority challenged. At the Arsenal, Lavoisier was inventing a new style of science that was more austere, the very antithesis of the entertaining salon science that flourished in Paris. Its partisans launched a crusade against the ravings of the imagination, systematizing dogmatism, and charlatanism. They had Mesmerism condemned. Victors, they were now detested by all those whose work the Académie had rejected.

This book concludes with the French Revolution and the upheavals it provoked within the world of Parisian savants. The creation of the metric system was certainly a great success for the Académie, but the end of privileges and monopolies over scholarly discourse, and the crumbling of the system of patronage challenged the very existence of academic institutions. Lavoisier's final struggle failed to prevent the closing of the Académie des Sciences in August 1793.

So vibrant and colorful, this Paris Savant seems to be quite forgotten today. For example, what is evoked for visitors by room 33 at the Musée du Louvre, also called the Henri II room? If they crane their necks upwards, they may admire a beautiful ceiling painting by Georges Braque titled *Les Oiseaux* (*The Birds*). But will they know that the Academy of Sciences held their meetings in this room from 1699 to 1793? And what about number 25 Boulevard Bourdon, near the Bastille? On the façade of a soulless building built in the 1970s, where everything has long since been obliterated, appears an improbable commemorative plaque signaling that this was the site of Lavoisier's legendary laboratory.

Strolling through Paris, you can if you choose discover a few heroes of the Enlightenment, sadly petrified, alas, in the form of statues near the sites they once frequented: Diderot in his armchair watches the passing cars on the Boulevard Saint-Germain, a stone's throw from the building in which he lived; Buffon, as if by chance, seems to catch a pigeon in flight at the Jardin des Plantes; Benjamin Franklin, also seated, waits for who knows what in a public square in Passy, quite close to his Paris residence; his friend Condorcet surveys with a pensive air the little square between the Institut de France and the Mint, where he lived; finally, on the façade of the Hôtel de Ville (Paris city hall) stands the besmirched statue of poor Bailly, astronomer and mayor of Paris, suffering the inclement weather in the company of his enemy D'Alembert and many others. I hope that this book will bring to life for readers something of these frozen savants and what they accomplished, as well as the places they lived and labored.

I should like to thank all those without whom the book *Paris Savant* would not have seen the light of day. I am grateful to previous authors who have written about the subjects I tackle; their names and publications appear in the bibliography. The work of researching and writing the text occupied me for several years. I was able to benefit in particular from stays at the Dibner Institute (Cambridge, Massachusetts), the Max Planck Institute for the History of Science (Berlin), and the Institute for the History of Science and Technology at the Beijing Academy of Sciences, without which I might not have finished this project. I have presented some of the material in this book to various audiences (I am thinking in particular of the students and other members of the audiences who came to hear me lecture at the EHESS [École des hautes études en sciences sociales] at the Université de Nanterre and the Université Paris 1 Panthéon-Sorbonne). Nor can I forget the first readers of the manuscript. I take pleasure in recalling the comments I gathered on these occasions, which helped me to fill gaps as well as stimulating me to take the story of French science forward into the nineteenth century.

This book was written in French. Susan Emanuel's precise and elegant translation respects it in both letter and spirit. I have been able to correct a few errors from the French edition. This English edition would never have existed without the efforts of my friend Dena Goodman, a great specialist in the French Enlightenment. I know what this book owes to her, and that she should have wanted to introduce it to Anglophone readers is for me an honor and a pleasure.

Bruno Belhoste
2018

I

The Gentlemen of the Academy of Sciences

ON AUGUST 17, 1793, the members of the Academy of Sciences arrived at the Louvre to gather for the first time as a voluntary society, but they found the doors sealed. Locks had been put on them that morning. After waiting for an hour in the corridor, the gentlemen dispersed. The Revolutionary Convention had voted the preceding week to suppress all the academies and literary societies that were officially recognized and subsidized by the nation. Yet the scientists hoped that they would be spared: in the next few days the deputies were supposed to discuss a provisional existence for the old Academy of Sciences so that its various projects could be pursued, in particular those relating to the establishment of the metric system.

The company had met on August 9 to examine its fate. Thirty-three academicians participated in this final assembly. After reading the minutes of the preceding meeting, the chemist Jean Darcet, presiding, proposed ending the meeting immediately, since, he said, the Academy of Sciences had ceased to exist. A lively discussion arose between those who wanted to continue, since the previous day's decree had not yet been officially promulgated, and those who leaned toward submitting without delay to the Convention's decisions. Finally, after Darcet himself declined to preside, the session was adjourned. Those who so wished could form themselves into a club, a free and fraternal society for the advancement of science.

Over the following days, a contest took place between the chemist Lavoisier, who still hoped to save the Academy, and those like the chemist Fourcroy who wanted its immediate and definitive suppression. The Convention itself seemed to hesitate. But on August 15, it voted to lock up the meeting hall. When the academicians tried to enter the premises two days later, they found the doors locked. For four years the Revolution had gotten rid of whatever it wanted: privileges, intermediate bodies, the monarchy itself. What could the Academy of Sciences do to resist the tide? There would be no more talk at the Convention of a provisional extension of its life. Thus ended the existence of the premier scientific institution of France—and of Europe.

The Louvre, Seat of the Academies

The Royal Academy of Sciences, founded by Louis XIV's minister Colbert in 1666, had met at the Louvre, in the former apartment of the Sun King, since 1699. To reach its premises, one entered by the Clock Pavilion, climbed the superb Henri II staircase, and went down a corridor running along the Navy Hall, which exhibited models of ports, dockyards, and ships for the instruction of students of naval construction. Passing through a vestibule, one entered a former

FIGURE 1.1: View of a pavilion of the Louvre taken from the Jardin de l'Infante. Engraving by Louis-Pierre Baltard, in *Paris et ses monuments mesurés, dessinés et gravés par Baltard (L. P.), architecte, avec des descriptions historiques par le citoyen Amaury-Duval*, vol. 1, Paris, 1803.

antechamber of the royal apartment that now served as a meeting room for the Academy. It was a vast rectangular room that was poorly lit by three windows, one overlooking the courtyard and the two others the square of the Old Louvre (Fig. 1.1).

The academicians would sit around a long table almost completely encircling the parquet floor where "foreigners" (nonmembers) came to present their work. When a member of the Academy was reading a report, he generally sat near the secretary, but he could also speak from his usual seat. The arrangement was slightly modified after the adoption of new internal rules in 1785. The walls were covered with blue damask embellished with fleurs-de-lys. A few paintings and busts of illustrious academicians decorated the hall. Above the chimney hung a painting by Antoine Coypel that represented Minerva holding the portrait of Louis XIV, and opposite the president was a fine pendulum clock with a second hand that kept time for the proceedings. A slate in a carved gilt frame was used for demonstrations. Behind the president, a gallery running along the wall accommodated the ladies who were invited to one of the two public sessions each year. Two stoves heated the room during the cold season.[1]

The Academy of Sciences was not the only academy housed in the Louvre. Starting in 1672—that is, even before the king and his court moved to Versailles— the Académie Française had been holding its sessions there. The Academy of Inscriptions and Belles-Lettres arrived in 1685, and then the Academy of Painting and Sculpture and the Academy of Architecture in 1692. Thus, when the Academy of Sciences moved in seven years later, it followed a general migration. The installation of the royal academies marked the palace's new vocation, after having been abandoned by the sovereigns. The Louvre remained a place of power, but it had passed from the purely political to the cultural sphere. In 1608 Henri IV had granted lodging to artists below the Grand Gallery so they could exercise their talents freely. The Royal Printer and the mint for striking medals had arrived in 1640. Little by little, the palace had been transformed into a vast caravanserai that housed painters, sculptors, architects, engravers, literati, and scientists, as well as a multitude of rats, "the ordinary retinue of talent," according to Louis-Sébastien Mercier. On the eve of the Revolution, the painter Jacques Louis David, who was a member of the Academy of Painting and Sculpture, had both an apartment and two studios there, one in the basement, along the Seine, the other at the northeast corner of the Square Courtyard, just above those he had arranged for his pupils. Many savants occupied apartments in the galleries. It was amidst all these pensioned artists and intellectuals that the various academies met. The Royal Society of Medicine in its turn was established there in 1778.

The colonization of the Louvre Palace by artists and men of letters did not diminish its prestige, however: in the program of absolutism, the control of language, knowledge, history, and taste was an essential prerogative of the state and one of the foundations of its authority. The Louvre academies played a fundamental role in this arrangement. They elaborated the rules of fine speaking, fine feeling, and fine thinking and ensured, at least in principle, their dissemination. Through their publications and their judgments, they would impose norms, orient opinion, and make reputations and careers. In this way they effectively policed the intellectual and cultural life of the kingdom.

Academic Authority and Influence

For intellectuals, particularly for savants, this oversight also represented protection. By founding the Academy of Sciences, royal power had taken control over the production and diffusion of knowledge away from the University of Paris—but also and especially from church authorities—and confided it to a body that was directly dependent on the monarchy. This measure accelerated the formation, under the aegis of the state, of an intellectual space that was independent of the traditional religious authority of the Catholic Church and was thus an important stage in the secularization of intellectual life that had begun in Europe with the Renaissance.

At the same time, the creation of the Academy of Sciences had contributed significantly to defining the kinds of knowledge that would henceforth deserve the name of science: only the study of nature, what was traditionally called natural philosophy and natural history, fell within its purview. And excluded thereby was everything that related to theology, metaphysics, logic, and morality. The study of nature was supposed to rely on investigation, observation, and experimentation. Of course mathematics had a place, but only to the extent that it was relevant, that it was applied (as was said after 1750) to the "study of phenomena." Scientific truths were thus clearly distinguished from revealed or imposed truths. There was no dogma here. Knowledge was increased by the ceaseless labor of savants, without anyone able to see where it might lead. Finally, the emphasis was placed on the utility of knowledge, as much for society as for the state, and this was a major reason to encourage its development. As the historian Roger Hahn has noted, science "was expected to be of immense service to the state, and above all to civilize mankind by offering it a new secular religion."[2]

So this was the vast mission of the Academy of Sciences, and in order to fulfill it, its members were responsible for evaluating the work submitted for its judgment, encouraging discoveries and inventions, appointing men worthy of entering the career of a savant, and advising the government on all matters requiring its expertise. Exercising the privilege it enjoyed to publish its findings, the Academy published annual volumes of *Histoire et Mémoires* edited by its members and several supplementary series: principally the *Savants Étrangers* series, which contained papers presented to the Academy by nonmembers, the *Description des arts et métiers*, and the *Connaissance des temps*, which was devoted to calendars and ephemerides. It gave its approval for the printing of all the works published by its members, and although it had no censorship authority with regard to publishing, for a long time it exercised de facto oversight of the Parisian press and book trade in the sciences.

This is why the bookseller Le Breton and his associates had called upon the mathematician Jean Le Rond d'Alembert to direct the *Encyclopédie* in 1749, along with Denis Diderot. Similarly, Panckoucke engaged Condorcet, Lavoisier, Vicq d'Azyr, and other academicians in the *Encyclopédie méthodique* when he launched that project in 1776. As for Parisian periodicals, they almost always entrusted their scientific rubrics to savants of the company: the astronomer Lalande wrote for the *Journal des savants*, the naturalist Daubenton and the chemist Macquer for the *Mercure de France*. Even the *Journal de physique*, the first independent scientific periodical, was closely linked to the Academy of Sciences. When it was launched at the beginning of the 1770s, Abbé Rozier had benefited from the support of the honorary academician Trudaine de Montigny. The Academy of Sciences looked with favor on an initiative that would enable the rapid publication of work by its members, and Lavoisier, for example, published several of his most important articles in the *Journal du physique*.

Like all royal institutions established in Paris, the Academy exercised its authority over the whole kingdom. It maintained communication with the academies in the main provincial cities. The Royal Society of Sciences in Montpellier, founded in 1706 on the model of the Academy of Sciences, was meant, at least on paper, to be an extension of the Parisian society. After the Naval Academy in Brest became a royal academy in 1771, it was directly affiliated with the Academy of Sciences and included among its corresponding members savants of the Parisian company. Lesser institutional links also existed with the academies of Lyon, Rouen, and Dijon, with which the academicians of Paris would deign to be affiliated. Beyond France, similarly, the company had long since woven tight relations with the main European academic institutions, either directly or by the intermediary of a network of foreign correspondents. Indeed, these connections extended well beyond Europe. From the beginning, in fact, the activities of the Academy of Sciences had a worldwide dimension.

The Academy was thus often consulted about distant diplomatic missions, which always included the quest for information. Above all, it organized its own voyages and expeditions. The most illustrious enterprise of this kind was the measurement of the earth's meridian, which was launched in 1735.

Two expeditions had measured the length of an arc of meridian, one in Lapland near the Arctic Circle and the other (longer and more adventurous) in Peru near the equator. The Academy was thus able to confirm that the Earth, in accordance with Newton's theory, was an oblate spheroid. Apart from their scientific purpose, the two expeditions had diplomatic and commercial implications, which are evident in other journeys sponsored by the Academy as well. Consulted in 1785 on the organization of Lapérouse's voyage, conceived as a French response to those of James Cook, the Academy prepared instructions and put forward names of savants to participate in the expedition.

The Company

The prestige of the Academy, as well as its authority, which extended throughout France, and its influence, which went far beyond national borders, all gave it particular responsibilities. These were, of course, collective responsibilities. The Academy of Sciences was both an institution and a corporate body: the power it exercised in the name of the king was shared among its members. Placed under the direct authority of the Crown, it was required to make a regular account of its decisions and activities to the minister of the King's Household, who as such always became an honorary member. The minister appointed new members, and although he almost always confirmed the nominations proposed by the Academy, he reserved the right (at least theoretically) to reject them. For example, in 1777, Jean-Baptiste Lamarck placed second in the election, but he was appointed adjunct in the botany section thanks to the intervention of Georges-Louis Leclerc, Count de Buffon, head of the Royal Botanical Garden. The academicians might well criticize the power of ministers, but they considered themselves to be the king's savants.

At the time of d'Alembert's death in 1783, the company included ninety-four academicians, of whom sixty-four were required to reside in Paris and regularly attend its sessions. Eleven veteran academicians (meaning retired) were exempt from these obligations, and seven foreign associates participated in the work of the Academy only during their stays in Paris, as was the case with Benjamin Franklin during his famous ambassadorship in France.

During his passage through Paris in 1781, the astronomer Anders Johann Lexell of the St. Petersburg Academy, who was a corresponding member, noted the uniformity that reigned in the company. The dress was that of the king's officers, typical of officials of all kinds: "Lots of tiered wigs and black suits."[3] Ecclesiastical attire, though not forbidden, was not welcome. The few clerics in the company, in short coats, had to be content with a short collar. The mineralogist

René-Just Haüy, who was a priest, consulted a theologian at the Sorbonne before deciding to stop coming to the Louvre in a soutane.[4] D'Alembert, Jacques-Henri Meister tells us, "was almost always dressed like Jean-Jacques [Rousseau], from head to toe in a single color."[5] This uniformity of appearance, at least at first sight, projected a certain uniformity of social conditions. In fact, the savants were mostly without fortune, keeping a modest lifestyle, simple and without luxury. Many remained bachelors, like d'Alembert and Lalande, or married late in life, like Laplace and Condorcet.

Their manners were generally unaffected—and often less than refined. The geologist Nicolas Desmarest, a man of the woods and fields, distinguished himself in high society by his blunt manners and his dragging speech; Madame Roland maliciously called the mathematician Gaspard Monge a kind of bear and a "Pasquino" (from the commedia dell'arte), and the almost friendless botanist Michel Adanson shut himself up in his study and communicated with the world only through books.[6] The tiny astronomer Jérôme Lalande, by contrast, paraded his eccentricities, constantly pushing himself forward with his high-pitched voice; a provocateur, he was known to boast that he was uglier than Socrates. Finally, d'Alembert, as witty a man and friend to the powerful as he was, was not much to look at. Barely taller than Lalande, with a voice like a clarinet, he was careless with his appearance, always badly dressed and badly combed. Frugal, he lived with his nursemaid until the age of almost fifty, in a small, uncomfortable bedroom. He then moved to the home of his friend Julie de Lespinasse in the rue Saint-Dominique at the corner of the rue de Bellechasse. Inconsolable at her death in 1776 and penniless, d'Alembert had to be content with an attic apartment at the Louvre, which he occupied as secretary of the Académie Française.

Most of these savants had only one or two servants and could entertain in their lodgings only exceptionally. Lacking carriages, they went about on foot. In the 1770s, Condorcet walked four leagues every week to visit his mother, who lived in Nogent. D'Alembert loved taking strolls in the Tuileries Garden. Lalande summed things up in his last will and testament: "I have few servants, no horses; I am modest; my clothes are simple; I go about on foot; I sleep wherever I find myself; money is of no use to me."[7] But this equality in simplicity should not fool us. The Academy of Sciences under the Ancien Régime was an institution that was both prestigious and profoundly hierarchical. So we have to distinguish four categories of academicians, different in status and rights: the honorary members, the salaried members, called pensioners, the associates, and the adjuncts, to whom we should add about a hundred correspondents in the provinces, in the colonies, and abroad, each attached to a resident member.[8]

Honorary Members

The dozen honorary members stood apart from the rest. From among them the king appointed the president and the vice president of the Academy each year. All were important figures in the state and the court who were being compensated for their service and their merit. Some were former ministers of state, and some held posts in several academies at the same time. The former director of the book trade Guillaume-Chrétien de Lamoignon de Malesherbes, protector of the Encyclopedists, was also a salaried member of the Académie Française (where there were no honorary members) and an honorary member of the Academy of Inscriptions and Belles-Lettres. As ornaments of the company, honorary members were not supposed to take part in its scientific work, but they could participate in academic elections, and they played an important role as representatives and intermediaries with the government and the court.

The honorary members were given pride of place, with seats directly facing the central area where speakers presented their papers, but most of them attended sessions only rarely. Yet all of them were interested in the sciences, and some even devoted their leisure time to them. The Duke d'Ayen, son of the Duke de Noailles, who had been the first champion in Paris of Linnaeus's system for the classification of plants, devoted his spare time to chemistry experiments. In the 1750s, he had recommended to King Louis XV the gardener Claude Richard and the doctor Le Monnier, who set up the botanical garden at the Trianon in Versailles, and he had protected the naturalist Adanson. The Count de Maillebois was fascinated by physics and took under his wing both Jean-Paul Marat and Anton Mesmer. The more serious La Rochefoucauld, the only son of the Duchess d'Enville, was elected in 1782. He was passionate about mineralogy and used his influence to support Condorcet, who was both secretary of the Academy and his friend. Malesherbes devoted himself to botany, corresponding with the botanist Antoine-Laurent de Jussieu and sharing with Jean-Jacques Rousseau a passion for collecting plants. Finally, Bochart de Saron, president of the Parlement de Paris (the highest court in the realm), was a knowledgeable astronomer and an excellent calculator. He had an observatory in his townhouse, as well as a physics cabinet and a laboratory fitted with the best instruments, which he opened up very generously to his colleagues.

Although they were academicians, and even of the first social rank, the honorary members were not considered to be real savants. Their manners and their dress, when they did come to honor the assembly with their presence, set them apart from their colleagues. The visiting astronomer Lexell raved about the Duke d'Ayen, "a very gracious and very polite lord," about President Saron, who received a few colleagues to dinner every Sunday, "so mild, so considerate, and whose manners are so obliging." He found them much friendlier than the savants of the company, whose vanity and maliciousness he denounced.

Ordinary Academicians

About twenty pensioners formed the elite in this rank. They were the only members who were paid (although the amount was only regularized in 1775) and received in addition tokens that rewarded them for attending meetings regularly. These emoluments, though not considerable, gave them enough financial security to be able to devote themselves entirely to the sciences. But as we shall see, the title of academician opened doors to other posts with better rewards.

All residing in Paris, the pensioners participated in all the activities of the Academy and shared with honorary members the right to vote in all elections. It is from among them that the officers of the company were appointed, the director and vice-director (elected each year, to replace the president and vice president during their absence), the secretary and the treasurer (permanent positions), and the members of standing committees. Most of them were true scientific professionals. Election to the post of pensioner in fact marked the crowning of a savant's career. After the reform of 1785, the average age of the twenty-four pensioners was fifty-six. They had spent on average more than twenty-six years in the Academy, starting out as adjuncts and then advancing to associates. For some, consecration arrived rather late, but others had been elected pensioners when they were still young, like Buffon, Lavoisier, and especially Condorcet, who entered the Academy at age twenty-five and became a pensioner before he was thirty.

Among these precocious ones we find the scientific luminaries, superior to their colleagues in position and wealth as much as—or even more than—in their work. Their appearance alone betrayed their preeminence. Lavoisier, a "farmer-general" (financier and tax collector), enjoyed a considerable fortune; although a commoner, he lived like a great lord. Buffon, a commoner who had been made a count by the grace of Louis XV, reigned over Montbard in Burgundy; as for the Marquis de Condorcet, he had lost his money long before, but he belonged to a family of the old military nobility and had the bearing that came with it. There is no doubt that at meetings of the Academy their superior qualities distinguished them from the other men dressed in sober black.

While Buffon often went back at his provincial estate, Lavoisier regularly entertained his colleagues, and Condorcet frequented high society before opening his own salon, led by his young wife, at the Mint. In short, in their conduct they resembled the honorary members more than their pensioner colleagues. Their responsibilities, too, set them apart. Lavoisier was not only a financier but also a senior civil servant, directing the Régie des poudres (Gunpowder Administration); Buffon was the intendant of the Royal Botanical Garden; Condorcet was the permanent secretary of the Academy. In the system of dependence and protections that characterized society under the Ancien Régime, these high-level savants flew well above most of their colleagues. Although not very grand personages in the social hierarchy, they enjoyed support at court and in the government, which enabled them to deploy resources, to recommend protégés, and to play the role of patrons in the world of science. For example, the young mathematician Pierre-Simon Laplace, arriving in Paris from Caen in 1769 without any money or relations, owed his career to d'Alembert, who quickly recognized his genius.

Seated behind the pensioners in the honorific order but enjoying neither salaries nor rewards for attendance, the twenty-four associates and adjuncts had only limited rights within the company. Of course, they participated in discussions on scientific matters, but associates did not participate in academic elections except for the nomination of adjuncts in their discipline, while the adjuncts were excluded from voting entirely. In 1769, d'Alembert took up an idea suggested by Patrick d'Arcy and proposed, in the spirit of equality, combining the adjuncts and associates into a single group and giving votes to the latter in all elections bearing on their discipline, but his colleagues would not go along. The category of adjunct was finally eliminated in 1785, thanks to the diligence of Lavoisier, but in every respect the gap between associates and pensioners remained.

The inequality among academicians was attenuated, however, by the fact that adjuncts were promoted to associates, and associates to the rank of pensioner. Of course, in this race up the ladder common to all academicians, the length of time it took to go from one rung to the next varied a lot from one savant to another. It depended on talent and publications, but also on chance factors like vacant posts and personal protection. One had to wait an average of fifteen years in the lower categories before becoming a pensioner. But for most, promotion arrived sooner or later: thus, all those who were associates in 1780 (and two-thirds of those who were adjuncts that same year) would become pensioners before the closing of the Academy in 1793. Among those promoted we find many who went on to dominate intellectual life in Paris after the Revolution, such as the chemist Claude-Louis Berthollet and the mathematicians Laplace and Monge.

The Academy of Sciences, like every institution in the Ancien Régime, had a complicated organization. Alongside the dozen ordinary associates there were a dozen "free" associates of a wholly different character and status. These academicians, who were not affiliated with any particular science and did not participate in elections, were elected on merit and for their social standing much more often than for their scientific work. They comprised a sort of second-order honorary body, in which we find all sorts of people.

Perronet and Fourcroy de Ramécourt had been elected ex officio, the former in 1765 because he was the foremost engineer in the Roads and Bridges Corps, whose school he directed, and the latter in 1784 because he was effectively in charge of military engineering. In 1789, the Count de Bougainville, explorer of the South Seas and one of the commanders of the fleet, was elected a free associate. The Marquis de Montalembert, contemptuous of the Engineering Corps and sworn enemy of its director, had been elected free associate back in 1747 only because he was under the protection of the Prince de Conti. As for the advisor to the Parlement de Paris Dionis du Séjour, elected free associate in 1765, his profile as an amateur astronomer and an excellent theorist of comets bears a striking resemblance to that of the honorary member Bochard de Saron. His passage into the class of physicists in 1786 confirmed his status as a true savant.

The Classes

The free associates shared with foreign associates, honorary members, and veterans the privilege of being outside the system of classes, meaning the disciplinary organization of the Academy of Sciences. This was reserved for ordinary academicians, the only ones considered to be full-fledged savants. The company was divided into six classes, each of which comprised (since 1716) six members: two pensioners, two associates, and two adjuncts. The classes of geometry, mechanics, and astronomy were said to be related to mathematics, whereas those of chemistry, botany, and anatomy were related to physics. These divisions and their names, which only remotely correspond to our disciplines today, had been fixed in the seventeenth century and were already considered archaic by the end of the following century.

The reform of 1785 attempted to renovate the system by establishing two new classes, one in general physics, the other in natural history and mineralogy, and by slightly modifying the botany class, which now became botany and agriculture, and that of chemistry, which became the class of chemistry and metallurgy. As it was explained by its authors, the reform aimed to bring into the structure of the Academy sciences that had not yet found their place there. The emphasis was placed on useful knowledge: agriculture, mineralogy, and metallurgy. In fact, the disciplinary organization remained rather artificial: strangely, the class of general physics was associated with mathematics and not physics; that of natural history included neither zoology (implicitly included in the class of anatomy) nor botany, which from the start had been in a class of its own.

In the course of his academic career, a savant almost always remained within the same class, even if he completely changed his field of research. D'Alembert, moving from the class of astronomy to mechanics, then from mechanics to geometry, and Buffon, moving from the class of mechanics to botany, were both exceptional cases in this respect. Within the Academy Alexandre-Théophile Vandermonde was a geometer and Mathurin-Jacques Brisson a botanist, although the former had long since abandoned equations to study the trades and the latter, passing from the protection of the entomologist René Antoine Ferchault de Réaumur to that of the physicist abbé Nollet, soon began to specialize in experimental physics.

The 1785 reform did enable the correction of certain anomalies: Monge, who had abandoned geometry, entered into the new class of general physics, which corresponded better to his interests at the time. Haüy, who was classified as a botanist, became a member of the new class of natural history and mineralogy, much better suited to his discoveries. There he found Desmarest, previously classified under mechanics, although he was principally known for his work on volcanism. But keeping Jean-Charles Borda in the geometry class and Brisson in the botany class, or the presence of Bailly (previously a supernumerary astronomer) in the new class of general physics, still seems to have little justification from a scientific point of view.

Such discrepancies between official academic titles and actual scientific activity remind us that in the eighteenth century savants still had very little specialization and that a sense of disciplinarity was often only secondary in the definition of their identity (Fig. 1.2). Alongside some true specialists, such as the astronomer Charles Messier, whom Louis XV called his "comet hunter," the Academy brought together multitalented savants across a broad spectrum: men such as Lavoisier and Buffon roamed over vast empires; many others, such as Lalande, Laplace, and Monge, had interests that extended well beyond their academic specialties.

The Meetings

Apart from the company's vacations, the academicians met twice a week, on Wednesday and Saturday afternoons. Each meeting lasted about two hours. How did they proceed and what did they actually do? The Academy's main activity was that of a scientific tribunal, which is why the meetings were primarily concerned with delivering papers and examining the research submitted for its examination. The president (or in his absence the director) read the letters and articles addressed by mail to the company, and those authors admitted to a session presented their papers, experiments, or inventions to the floor; and the

FIGURE 1.2: Savants observing through a microscope. Engraving in Buffon, *Histoire naturelle*, vol. 2, *Histoire des animaux*, Paris, Imprimerie royale, 1749, frontispiece vignette.

academicians themselves read aloud to their colleagues their reports and sometimes their own work.

During his visit in 1781 Lexell was struck by the limited attention the members gave to these various presentations. They barely listened to the speaker. Those around the table talked to their neighbors. Some people got up and talked together in small groups by the windows or around the stoves. According to another witness, the Baron de Montalembert, in 1787, "it often happens that most of these gentlemen chat or write letters during the reports." The English botanist James Edward Smith noted in 1786 that one could hear only as much of the papers being read as the "incessant talking" permitted; and even though the president could ring a bell to call for silence, he rang it only when "the general noise prevents him hearing himself or his nearest neighbor."[9]

The physician Mesmer, allowed a few years earlier to present his ideas on animal magnetism to the Academy, had also noted the chaos that reigned in the hall during meetings. "I assumed, reasonably, that when the Assembly was full enough to be thought complete, the gentlemen's attention, previously divided, would now be fixed on a single object. I was mistaken; the private conversations continued; and when M. Le Roy [the director of the Academy] wished to speak, he called futilely for attention and silence, which he was not granted. When he persevered in this request he was even more openly rejected by one of his impatient colleagues, who assured him that they would give him neither, adding that he could just leave the paper he was reading on the desk, where whoever wanted to could peruse it." Dr. Mesmer added philosophically: "Reflecting on the kind of veneration I had always had for the Academy of Sciences of Paris, I concluded it was essential for certain purposes to be seen only from a distance. Revered from afar, they don't amount to much close up."[10]

In reality, the two weekly meetings of the Academy were above all social occasions for its members. For them, the institution was a company, a sort of club. Apart from two annual public sessions, one after Saint Martin's Day, the other after Easter, the meetings were held in closed session. Only distinguished visitors, foreign savants or illustrious princes, and a few authorized persons were allowed to attend. In 1781, before the Count du Nord (the alias of Paul, Grand-Duke of Russia) and his wife, Condorcet praised Peter the Great and the sovereigns who had protected sciences and savants. Two years later, Lavoisier explained the nature of the gases that composed atmospheric air in front of the Count de Falkenstein, alias Joseph II, brother of Marie-Antoinette. On June 2, 1785, the Academy of Sciences received Prince Boudakan, son of the King of Ouaire (the Kingdom of Warri, in modern Nigeria), who had arrived in France the previous year. After having attended the meeting, seated near Lavoisier (then director of the company) the African prince visited the Academy's collections and the Navy Hall, and then the savants accompanied him to the Henri II staircase.[11]

During ordinary sessions, academicians exchanged information and debated their different positions. Despite rivalries and enmities, which were not lacking in this tight milieu where ambitions clashed, the familiarity created by these obligatory rendezvous aroused a strong esprit de corps and, along with it, shared feelings and ideas. Often, moreover, after the meeting ended, the gatherings continued on a social level, on Wednesdays at the home of Bochart de Saron, rue de l'Université, and on Saturdays on the rue Croix-des-Petits-Champs, where Madame Lavoisier served tea.

The public meetings were different. A large but select crowd of both sexes witnessed these worldly events, which gazettes covered fulsomely. On these occasions the Academy of Sciences presented its work to the public. The president opened the meeting and announced the prizes, the secretary delivered eulogies for colleagues who had died, and a few academicians read their papers. A dignified tone had to be set, but without making people yawn: the audience did not want to be bored, and it was liable to show its impatience. On November 13, 1784, a balloon made by Fortin to illustrate the aerostat commission's work delighted the spectators. It was blown up several times before the opening of the session, which raised (according to the anonymous *Mémoires secrets*)[12] "brouhahas and cries of joy." But the gas smelled so bad that the windows had to be opened. In his response, Condorcet, as was his custom, seduced the audience with bits of eloquence. However, those who read papers were less fortunate: people got bored listening to Desmarest talk about geology and Cassini about astronomy. Sabatier had more success with a talk on bites by rabid dogs. But most anticipated was Meusnier talking about balloons. At the following public meeting, on April 6, 1785, the balloon was still on the ceiling, which demonstrated that the envelope had been air-tight, but Meusnier, who had come to present the follow-up for the commission's work, disappointed the spectators: he said nothing new, the author of the *Mémoires secrets* complained.[13]

The academic headquarters scarcely functioned outside these meetings. The former state bedroom of the king was now a library, well stocked with 1,200 volumes that the academicians could borrow. Until 1780, one could also see the Coronelli globes there and a few large skeletons, including that of an elephant. The rest of the anatomy collection was stored in the adjoining former bedroom of the king. On the eve of the Revolution, both rooms were filled with the models of machines that Pajot had bequeathed to the Academy and that had remained in their cases undisturbed for thirty years. As for the king's former office, adjoining the meeting hall, it was used for committee meetings and during elections for counting votes. It was here that the publications committee met about once a month, to prepare for the publication of scientific papers, and that the treasury committee kept the accounts. In addition to these permanent committees, the office could accommodate commissions charged with evaluating candidates for prizes and responding to requests for monopoly privileges for inventions. Tourists passing through who wanted to visit these rooms, usually deserted, had to ask for the concierge, Fatori, who lodged at the entry to the meeting hall.

Most often, in fact, the academic commissions met outside the former royal apartments; the commissioners appointed to evaluate research or an invention prepared their reports at home, and they often visited the site of an experiment when the issue was an invention or a discovery. Similarly, meetings of the major expert commissions formed in the 1780s generally took place in the home of one of their members. The great aerostat commission of 1783 held its first meeting in La Rochefoucauld's townhouse on the rue de Seine, presided over by the duke, before taking up their work in the Arsenal. The 1785 commission on the reform of the Hôtel Dieu (the public hospital) met in the galleries of the Louvre in the apartment of its rapporteur, the astronomer Bailly, and the commission on the ciders of Normandy met at the home of Lavoisier.

Moving activities outside was a major trait of the Academy of Sciences after 1750. Far from being shut up within the walls of the palace, the Academy spread out across the city, where its members were active in offices and schools, salons and cafés; they participated, as we shall see, in everyday life. Considered by the administrative state as a pool of expertise in the domain of the arts and sciences, the Academy furnished it not only with the advice it needed but also and above all with personnel. Academic life thus found itself intimately linked to the major institutions of the state. By means of its competitions, its publications, and its expertise, the influence of the Academy of Sciences extended across Europe, and even to the whole world through the correspondence it maintained and the voyages it sponsored. The contrast was striking between this external brilliance, which attracted to the Academy all kinds of ambitious people, and the relative emptiness of its internal activities.

Maneuvers and Intrigues

The main task of the Academy of Sciences was to evaluate, potentially approve, and occasionally publish and honor the scientific work that was submitted to it. Beyond authorizing publication, academic recognition represented a precious advantage in the search for official support and public approval for authors, inventors, and those with projects they sought to realize. To obtain it, those soliciting the Academy spared no effort: interventions from the government, recommendations from important people, enticements to gain the favor of academicians who might become allies within the company. There was a whole game of approaches and maneuvering played in the wings that prepared the way for the official presentation of research. These practices fostered the creation of networks of protection and influence around the Academy of Sciences, outside of which it was difficult, if not impossible, to get into its meetings and obtain its approval. What was true for judgments was of course also true for elections. Entering into the company and climbing the echelons required formal visits, long preparation, and support from both inside and outside the institution. Ultimate success depended as much on the capacity to build and mobilize alliances and allegiances as on the quality of the candidate's scientific work.

The entry of Condorcet into the secretariat of the Academy illustrates the importance of these intrigues. The permanent secretary had always prepared the minutes of the meetings, from which each year he had to compose a *Histoire de l'Académie*, and he delivered eulogies of colleagues who died. This was heavy labor, but also great power. Bernard de Fontenelle had occupied this position brilliantly for more than forty years, eclipsing his mediocre successors, Jean-Jacques Dortous de Mairan and then Jean-Paul Grandjean de Fouchy. Very critical of the latter, D'Alembert wanted to have a protégé elected. He first thought of the astronomer Bailly, before turning in 1772 to Condorcet, a young mathematician of talent and an ardent partisan of the Encyclopedists. D'Alembert himself had just been elected secretary of the Académie Française. The nomination of Condorcet, he thought, would strengthen the hold of the philosophes on the Republic of Letters. Even more: in the twilight of the long reign of Louis XV, it would seal the alliance between the philosophical clan and the reforming elite of the kingdom. This solution had received prior approval from the minister of the King's Household, the Duke de La Vrillière, who oversaw the academies, and of his powerful brother-in-law, the former prime minister Maurepas, then in disgrace. Turgot, of whom Condorcet claimed to be a disciple, warmly approved; so did Trudaine de Montigny.

Fouchy, exhausted by the job after thirty years of service, was ready to hand things over quickly. While waiting, he agreed to take on Condorcet as his assistant, since he had already helped with some eulogies. With everything worked out, it remained only for the Academy to confirm the new assistant secretary proposed by La Vrillière. The plan was executed on March 6, 1773, but with some hesitation, because d'Alembert's adversaries were aroused. Condorcet's nomination had been imposed from on high. Had not d'Alembert solicited the king's order? And what about his promise to Bailly? The powerful Buffon, treasurer of the Academy and old enemy of d'Alembert, assumed leadership of the opponents. Six votes out of twenty-one were withheld from the official candidate. The cabal began.

After the death of Louis XV on May 10, 1774, Turgot came to power and Condorcet found himself closely associated with Turgot's policies as an advisor. A financial arrangement proposed by Turgot in his new capacity as finance minister, and inspired by d'Alembert, now offered a pension of a thousand livres to Fouchy so that he could retire comfortably, and five thousand livres to Condorcet, who would "inherit" his post.[14] But this would reduce by half the annual endowment of twelve thousand livres promised to the Academy for its experiments and would also provide ammunition to the anti-philosophe cabal within the company.

Launching a counteroffensive, Buffon suggested to the Academy setting up a permanent committee to supervise and censor all writings of the secretariat. The overworked Fouchy had asked for this himself; but by extending this measure to his successor, the Academy was in fact aiming to limit the freedom that Condorcet hoped to enjoy once he had acceded to the post. The goal was probably to make him renounce the position by discouraging him with such conditions. The battle between the two clans continued throughout the year 1775. In the end, despite the support of Turgot, d'Alembert and Condorcet did not manage to get the financial arrangement passed or to have the censorship measure that threatened the future secretary nullified by intervention from the minister. Malesherbes, who had succeeded La Vrillière at the head of the King's Household, had to settle for granting Condorcet three thousand livres and leaving the Academy free to examine the work of its secretary.

But the affair ended in July 1776 with Fouchy's resignation, for which it seems he was granted some financial compensation, and the unanimous election of Condorcet as permanent secretary. Therefore, while the philosophe clan won (after intrigues we know nothing about), symbolic satisfaction had to be given to the defenders of academic freedom: it was by entirely renouncing his right to "inherit" the post that the candidate was presented to his colleagues. In return, there would be no question later of censoring the activities of the new secretary.

Peace had returned to the Academy. Buffon, vanquished, withdrew to the Royal Botanical Garden and attended meetings more and more rarely. In 1782, he suffered the humiliation of a new defeat, this time at the Académie Française, when by a unanimous vote Condorcet, actively supported by d'Alembert, won out over his own protégé Bailly. "I am happier at having won this victory than I would be by having squared the circle," joked the philosopher-mathematician.[15] But as soon as d'Alembert died the following year, the Académie Française proceeded to elect Bailly, who hurried to break with his protector, who was now no longer of any use to him. Affected by such ingratitude, Buffon chose never to reappear at the Louvre.

The heirs of the *Encyclopédie* henceforth dominated almost entirely both the Academy of Sciences and the Académie Française. With d'Alembert dead, there remained those whom he had directly protected, like Condorcet and Laplace. Lavoisier, whom Turgot had appointed to the Gunpowder Administration; the anatomist Félix Vicq d'Azyr, who presided over the Royal Society of Medicine; and Lalande, who hung out with the philosophes, shared their views. While Condorcet performed his task as permanent secretary with authority, Lavoisier stood out as the major figure of the Academy, assembling around a vast research program various chemists, physicists, and mathematicians. The other academicians followed as best they could. Bailly, now at peace with Condorcet and on good terms with the ministers, worked to strengthen the power of the company within the state apparatus. Finally, although the quarrels and enmities had not disappeared, everyone was united to defend the Academy of Sciences in the face of an onslaught of attacks that sprang up around the city.

Anti-academism

For a long time, some people had worried about the excessive power of the Academy of Sciences. Was the domination it exercised over intellectual life compatible with the freedom that ought to reign in the Republic of Letters? Those whose work had been rejected, or simply ignored, were the first to complain. In deciding in 1775 to no longer examine solutions to the problems of squaring the circle, trisecting the angle, doubling the cube, and making a perpetual motion machine, the Academy of Sciences increased resentments and recriminations. For the exalted ones attracted by these insoluble problems, silence was more humiliating than condemnation. The publicist Simon Linguet, known for his virulent attacks on the "philosophical sect" that dominated academic institutions, took hold of this subject too, denouncing the diversion to other purposes of a legacy willed to the Academy to reward anyone who could square the circle.[16]

The anti-Academy movement grew over the following decade. The context was changing rapidly, with more and more sites for the exchange of ideas and public debates such as clubs, intellectual societies, museums, cafés, and newspapers, which ultimately undermined the legitimacy of the official monopoly enjoyed by the Academy of Sciences to evaluate scientific research and inventions in Paris. The *Journal de Paris* had opened up its columns to system builders. The sciences seemed to have escaped the control of official authorities to become everybody's business, including entertainers, visionaries, and charlatans. While the last of these continued to seek the approval of the savants, that did not mean that they intended to submit to their judgment. As Jacques-Pierre Brissot wrote in 1782, "Public opinion alone has the right to crown genius."[17]

From this perspective, the controversy around Mesmerism was a turning point. The Academy of Sciences had long ignored the hare-brained schemes submitted to it. Worried about the public fascination with "fake science," with the support of the authorities it now launched a sort of crusade against "charlatanism" not only in medicine but also in science. The report of the academic commission charged with examining animal magnetism, written by Bailly, reduced the effects of the fluid about which beneficial claims were made to the product of imagination alone; in a secret report addressed to the king, the academician went even farther, calling the treatment "a moral danger" and Mesmer's theory about celestial influences an "old chimera." This condemnation quickly gave rise to a wave of pamphlets denouncing the scientific and medical establishment. "For ten years I have observed your way of doing things," wrote Brissot indignantly in an anonymous pamphlet, "and it has always been the same: humble with your superiors and despotic toward those who depend on you. Throughout the ages you have been the enemies of innovations, you have persecuted people of genius who would not kneel down before you."[18] Even the *Journal de Physique* seemed to distance itself from the Academy after Jean-Claude Delamétherie, a physician versed in the sciences, became its editor in 1785.

Still rather rare before the Revolution, attacks on the savants multiplied after 1789. Jacques-Henri Bernardin de Saint-Pierre, ignored by the Academy of Sciences, ridiculed the arrogance of "Brahmins" in his tale *La Chaumière indienne* (*The Indian Cottage*). Jean-Paul Marat, spewing insults, adopted quite another tone in *Les Charlatans modernes* (*The Modern Charlatans*): "[The Academy] has taken as its symbol a radiant sun and as its motto this modest epigraph—INVENIT ET PERFECIT—not that it has ever made any discovery or ever improved anything; nothing has ever left its belly but a heavy load of aborted papers that may sometimes be used to fill a hole in some great library. On the other hand, it has met 11,409 times, it has published 380 eulogies, and it has handed out 3,956 approvals." For Marat, the academicians were all the same: "false lovers of truth, sincere apostles of falsehood, worshippers of wealth; lazy, ignorant, unruly; but very dissipated, very presumptuous, very stubborn; they are interested in distinctions and passionate about gold"[19]—all of a piece. Although an extreme, almost pathological, case, the revolutionary's fury does testify to a strong anti-academic current. Alongside scientific amateurs who had been turned away, craftsmen and inventors complained just as vigorously about the tyranny of the academies. These recriminations spread widely, finding an echo even in the National Assembly. In 1791, it stripped the Academy of Sciences of its responsibility for evaluating inventions.

Paralyzed and divided, the academicians suffered these events without reacting. A majority had welcomed the Revolution, some even with enthusiasm, but the institution was too linked to the Ancien Régime to be capable of reforming itself. Perhaps by way of compensation, it threw itself into the venture of the metric system, which kept it busy from 1790 on. Its permanent secretary Condorcet had entered into politics on the side of the patriots; he proclaimed himself a republican after the king's flight to Varennes. Elected to the Legislative Assembly, Condorcet sat alongside Brissot, who had been the Academy's bitter enemy. On behalf of the Committee of Public Instruction, he presented a vast project for national education that would include five levels of instruction, going from basic primary schools up to a National Society for the Arts and Sciences at the top. This society, charged with directing the whole system, would replace the old academies. The Legislative Assembly, concerned with more urgent tasks, put it aside to be examined later. Meanwhile, the adversaries of Condorcet's plan were unleashed, denouncing an attempt to place national education under the rule of the academic caste. The project was abandoned even before it was studied.

The attacks against the academies reached their height in the summer of 1793. This time, their suppression pure and simple was demanded. While the Academy of Painting and Sculpture was the prime target, all of them were at risk. Lakanal, charged with the Convention's Committee of Public Instruction, sought in vain to separate the fate of savants from that of artists and men of letters. Others within the committee opposed him. Finally, Abbé Grégoire presented the report to the Convention on August 8: he proposed the suppression of academies but asked for the Academy of Sciences to be provisionally maintained while awaiting the organization of a new society "designed for the advancement of the arts and sciences." The Convention, on the intervention of the painter Jacques-Louis David, refused to negotiate. It voted for the immediate abolition of all the academies, postponing for three days the consideration of exceptional measures concerning the sciences. The next day the Academy of Sciences assembled at the Louvre for the last time. Lavoisier, who had led the struggle with Lakanal, still hoped to win the day, but this turned out to be a false hope, since the Convention never took up Grégoire's proposals. As we have seen, on August 17, the savants were denied access to their meeting hall. On September 11, the Convention created a temporary commission on weights and measures charged with taking up the work on the meter begun by the former Academy, for whom things were now really over.

Notes

1. THIÉRY, 1787, vol. 1, 348–352, MAINDRON, 1888, 33–36, and C. FRÉMONTIER-MURPHY, "La construction monarchique d'un lieu neutre: l'Académie royale des sciences au palais du Louvre," in DEMEULENAERE-DOUYÈRE and BRIAN, 2002, 169–203.

2. HAHN, 1971, 57.

3. BIREMBAUT, 1957, 148–166.

4. Eulogy of Haüy, in CUVIER, 1819–1827, vol. 3, 139.

5. GRIMM, vol. 12, 17 (text written by J. H. Meister in January 1784).

6. Eulogies of Desmarest and Adanson, in CUVIER, 1819–1827, vol. 1, 289, vol. 2, 346. Monge was called by Madame Roland "a kind of original who would make much monkey business like the bears I have seen playing in the pits of the city of Berne."

7. *Éloge historique de M. de La Lande par Mme la Comtesse Constance de S.* (= de Salm), Paris, 1810, extract from *Magasin Encyclopédique*, April 1810, 40.

8. All the information on academicians and their careers used in this book is taken primarily from the *Index biographique de l'Académie des sciences, 1666–1978*.

9. J. C. de MONTALEMBERT, *Lettre de M. le Baron de Montalembert à M. de Keralio, en réponse au compte qu'il a rendu dans le Journal des Savants, du Mémoire sur la Fortification perpendiculaire, publié sous le nom de plusieurs officiers du Corps Royal du Génie, en 1786*, London, undated, 28 (the letter, supposedly written by the Baron de Montalembert, cousin of the Marquis de Montalembert, is dated February 28, 1787), quoted by J. LANGINS, "Un discours prérévolutionnaire à l'Académie des sciences: L'exemple de Montalembert," *Annales historiques de la Révolution française* 72, no. 320 (2000), 169, and J. E. SMITH, *A Sketch of a Tour on the Continent in the Years 1786 and 1787*, London, 1793, vol. 1, 131.

10. F. A. MESMER, *Précis historique des faits relatifs au magnétisme animal*, London, 1781, 30–31.

11. On the reception of Prince Boudakan, see *P.-V.* (1785), fol. 143v, July 2, 1785.

12. The *Mémoires secrets*, attributed to Bachaumont, were published in London between 1777 and 1789. They form a chronicle of the Republic of Letters in France, and principally in Paris, compiled on the basis of handwritten notes that circulated in manuscript form.

13. BACHAUMONT, vol. 27, 4–11 (session of November 13, 1784) and vol. 28, 281–287 (session of April 6, 1785).

14. A *survivance* was the right of succession to a post granted to the title holder for life.

15. According to J.-F. LA HARPE, *Correspondance littéraire adressée au Grand-duc de Russie*, Paris, vol. 3, 1801, 312.

16. S. LINGUET, *Annales politiques, civiles et littéraires du dix-huitième siècle*, London, vol. 6 (1779), 145–160.

17. J.-P. BRISSOT, *De la Vérité*, Paris, 1782, 166.

18. [J.-P. BRISSOT], *Un mot à l'oreille des académiciens de Paris*, pamphlet published anonymously in 1784.

19. J.-P. MARAT, *Les Charlatans modernes ou lettres sur le charlatanisme académique*, Paris, 1791.

2

The Capital of the Sciences

WHEN PILÂTRE DE ROZIER and the Marquis d'Arlandes flew over Paris in their hot-air balloon on November 21, 1783, they scarcely had time to admire the astonishing landscape offered below. Wholly occupied with feeding the fire of their machine, they could only glimpse the miniature monuments that paraded in silence along the Seine. Leaving from La Muette in the west, they landed twenty-five minutes later at the Butte-aux-Cailles. If they had observed the spectacle more attentively, they would have been able to see in detail on the Right Bank the imposing mass of the Tuileries and the Louvre, with the Hôtel de Ville in the distance barely standing out from the massed group of houses, and to the north the multitude of church towers like ears of corn glinting in the sun; across the river, behind the Vert-Galant garden on the tip of the Île de la Cité, the two towers of Notre-Dame Cathedral; and finally on the Left Bank as they approached it, the Military College aligning its façade along the Champ de Mars and the gleaming domes of the churches, those of the Invalides and the Quatre-Nations near the river, and higher on the hill those of the Sorbonne and the Val-de-Grâce, as well as the immense Sainte-Geneviève, still under construction, which overlooked the Latin Quarter.

Of this landscape, the religious edifices claimed the lion's share. However, a few monuments apart from the Louvre indicated the presence of the sciences in Paris. All of them were situated on the Left Bank of the river. The Royal Botanical Garden and its natural history gallery (enlarged under the initiative of Buffon) to the east, and the Paris Observatory, built by Charles Perrault, to the south, both testifying to the longstanding interest of the monarchy in science. Several recent edifices in the neoclassical style also housed savants: the Military College to the west, the Mint along the Seine, the Royal Academy of Surgery, and the new buildings of the Collège de France at the heart of the Latin Quarter. These sumptuous monuments, which were already starting to be promoted as tourist attractions, formed the most visible part of a network of establishments for scholastic, industrial, and administrative use where the King's savants kept busy.

At the time, it was said that Paris was the capital of philosophers and savants, just as London was the capital of merchants and inventors, Rome of cardinals and artists, and Vienna of musicians. The city in fact could lay claim to the rank of capital of the European Enlightenment. Its intellectual prestige went back to medieval times. For several centuries, the University of Paris had been foremost in the world, and although it had declined a lot since then, it continued to attract to its colleges and faculties students from everywhere. The institutional foundations of intellectual life in Paris had been renewed under the Valois and Bourbon monarchies. François I had founded the Collège de France in 1530 to promote humanist education. Over the course of the seventeenth century, the royal establishments had multiplied: the academies, the Royal Botanical Garden, the Paris Observatory, and the Royal Library. Through their mediation the monarchy took control of intellectual production. The effort continued in the following century with the creation of, among other things, the Royal School of Roads and Bridges, the Royal Physics Cabinet, the Royal Academy of Surgery, the Royal Society of Medicine, and a Machine Depository.

Science in the Republic of Letters

In the eighteenth century, the European intellectual world took the form of a cosmopolitan and egalitarian republic. Ostensibly it had neither hierarchies nor borders. Men and ideas were supposed to circulate around it freely. Nevertheless, certain places were preeminent: in the eighteenth century, Paris dominated the Republic of Letters, and within Paris everything gravitated around the former royal palace known as the Louvre. This was where the royal academies assembled, charged with directing intellectual and artistic life. Geography thus made visible the oversight that royal power exercised over literature, sciences, and the arts.

The academies were intended to gather together the best in their respective domains: painters, musicians, architects, writers, physicians, and savants. In addition to its field of knowledge and its bylaws, each academy had apartments in the palace (Fig. 2.1). While the Academy of Sciences held its meetings on the first floor in the former royal apartments, the Académie Française occupied the council hall on the ground floor. The Academy of Inscriptions and Belles-Lettres had its quarters next door. The Academy of Painting and Sculpture, which was also a school, had been assigned vast premises in the extension of the former royal apartment facing the Seine. Its students used individual studios (loges) put into the Apollo Gallery, and its members exhibited their work in the Salon Carré that was connected to it. Each year this event (simply called the "Salon") attracted

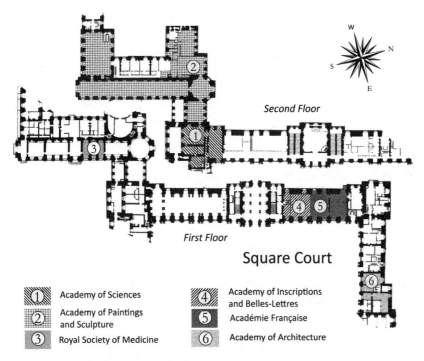

Second Floor

First Floor

Square Court

① Academy of Sciences	④ Academy of Inscriptions and Belles-Lettres	
② Academy of Paintings and Sculpture	⑤ Académie Française	
③ Royal Society of Medicine	⑥ Academy of Architecture	

FIGURE 2.1: The Academies at the Louvre.

during a whole month a crowd of amateurs and art lovers, as well as the art critics who wrote for the gazettes. The Academy of Architecture was housed in the north wing, on the other side of the old Louvre, where, beginning in 1774, it had the use of a meeting hall, a gallery for models, and a study hall on the ground floor. The last academy to arrive, the Society of Medicine, used for its meetings the hearing room of the secretary of state for the King's Household, known as the Hall of Seasons, giving directly onto the garden of the Infanta.

Despite their specialization, the royal academies formed a unified ensemble, placed directly under the supervision of the King's Household (except for the Society of Medicine, which was also placed under the finance ministry). The authority of this protean ministry, which also governed Paris, in fact extended well beyond the service of the king's person and the court and included, among other things, the control of the Republic of Letters, as well as the direction of major Parisian establishments such as the Royal Library, the Paris Observatory, the Royal Botanical Garden, and the Sèvres (porcelain) and Gobelins (tapestries) manufactories. The minister split his time between Versailles, where he reported to the king, and the Louvre Palace, where he had an office and held audiences. Of course, the Academy of Sciences always included this minister among its honorary members. Malesherbes, who occupied this office for a year at the start of Louis XVI's reign, was already in the company by this time. However, his successors had to be elected: Antoine-Jean Amelot and then the Baron de Breteuil, when they were appointed. Breteuil, minister from 1784 to 1788, was closely linked to the astronomers César François Cassini de Thury and Jean Sylvain Bailly and boasted of being a friend of the savants. In 1787 he attracted to Paris Joseph-Marie Lagrange, the foremost mathematician in Europe. After his departure from the ministry, the grateful Academy of Sciences had his bust made by the distinguished sculptor Augustin Pajou. But it did not have time to elect Breteuil's successor, Laurent de Villedeuil, swept away by the Revolution.

Patronage in the domain of the arts related more specifically to the administration of the King's Buildings, attached in principle to the King's Household but actually functioning independently. The Academy of Painting and Sculpture and the Academy of Architecture were placed under its direct supervision, as were the royal manufactories of Gobelins and Sèvres. Named to head this administration in 1751, the Marquis de Marigny, brother of Madame de Pompadour (the king's mistress), gave its work a major boost. In Paris, he continued with the construction of the Military School and launched the building of the church of Sainte-Geneviève, as well as laying out the plan for the Place Louis XV (today the Place de la Concorde) and contributing, along with the architects Ange-Jacques Gabriel and Jacques-Germain Soufflot, to the triumph of the severe neoclassical style of architecture. In the royal manufactories, he encouraged research on porcelain and dyes. After his departure, Count d'Angiviller, an enlightened man who was well placed at court, continued his activity in the domain of the arts. A collector of minerals, he was particularly interested in science and was closely tied to Buffon and Cassini. Aided by his clerk, the mathematician Jean-Étienne Montucla, who was well known as the author of the vast *Histoire des mathématiques* (History of Mathematics), he supported their ambitious projects for expansion at the Royal Botanical Garden and the Paris Observatory in the last years of the Ancien Régime.

Links were especially close between the Academy of Sciences and the literary academies. Tradition had it that the Académie Française always included among its members at least one savant. Fontenelle, permanent secretary of the Academy of Sciences, had been a member of the other company since 1691. The number of "immortal" savants (i.e., members of both academies) grew appreciably after 1740. Among the elect were Buffon and d'Alembert in the 1750s, Condorcet in 1782, and Bailly the following year. This presence expressed an important fact: in the eighteenth century, the world of science fully belonged to the Republic of Letters. Even if savants formed a distinct quarter there, with their own activities, institutions, and publications, they were recognized as full-fledged men of letters.

In fact, at that time there was no strict boundary between literature and the sciences. Literati such as Voltaire and Diderot did not disdain contributing to scientific debates, while savants like Fontenelle and Buffon were also writers. Men of science and men of letters came together in the "philosophe party" that formed around the middle of the century. The initial impulse had been given in the 1740s when the ideas of Isaac Newton and along with them those of John Locke prevailed definitively in Paris at the expense of the Cartesians. The Academy of Sciences had been the epicenter of the battle, but Voltaire played the decisive role, publishing the *Éléments de la philosophie de Newton* (Elements of the Philosophy of Newton) in 1738 and encouraging his intimate friend Madame du Châtelet to translate the *Principia* into French. After 1750, the *Encyclopédie* definitively launched the "philosophes."

Supported in the city and all the way up to the government, this party was that of the Enlightenment; its enemies were the Jesuits and the devout party that was still very powerful at court. The battle was long and strenuous. The publication of the multivolume *Encyclopédie*, twice interrupted by the government (in 1752 and 1759), was relaunched by the audacious tenacity of Diderot and finally completed in 1772. That same year, d'Alembert was elected permanent secretary of the Académie Française, against the advice of the government. Five years later, his protégé Condorcet acceded to the same post at the Academy of Sciences. The party of the philosophes henceforth dominated the major academic institutions of the kingdom.

The World of Parisian Scholars

The sciences had long been confounded with scholarship in general but had begun to be distinct with Galileo and Descartes: while the scholar pored over the writings of men, the physicists and naturalists reserved for themselves the "great book of Nature." However, the distance was not so great between the two intellectual traditions. On the one hand, until the nineteenth century no science was conceived without some reflection on its own history; on the other, history, as a path of secularization, was equally that of nature as of human beings.

Starting in 1663, Colbert had entrusted the task of preparing and writing official history to a few members of the Académie Française. This commission, at first called the "little Academy," gained importance over the following decades. Its reorganization in 1701 made it a real state institution on the model of the Academy of Sciences. Soon thereafter it took the definitive name of the Royal Academy of Inscriptions and Belles-Lettres and began to publish collections of articles. Forty members strong, it was charged with the development of scholarly research throughout the kingdom. Within its ranks were historians of distinction like the Maurist (Benedictine) Bernard de Montfaucon, founder of paleography, and the philologist Nicolas Fréret, its permanent secretary from 1742 until his death in 1749. For almost a century, the Royal Academy of Inscriptions and Belles-Lettres dominated historical scholarship in Europe.

This learned society was closely linked to the Collège de France and especially to the Royal Library, which also answered to the King's Household. In fact, there were old and durable relations between the library and the intellectual world. The Academy of Sciences met in one of its buildings (on the rue Vivienne) from its foundation in 1666 until its installation at the Louvre in 1699. Since that date, savants had continued to frequent its reading rooms regularly. The Encyclopedists collected a good part of their documentation in the Royal Library; d'Alembert, Buffon, and Condorcet were thus among its regular readers.

The origins of the Royal Library went back at least to François I, who had created a royal library at Fontainebleau under the direction of Guillaume Budé. Colbert, entrusted by Louis XIV with his library, gave it a decisive push in 1666 by installing the collection near his townhouse on the rue Vivienne. He tried to improve the deposit of all books printed in the kingdom (which was in principle obligatory), and he encouraged the acquisition of books and manuscripts from distant countries. The establishment was developed further under the direction of Abbé Bignon, named to head it in 1719. He moved the Royal Library to the Hôtel de Nevers, which looked out on the rue de Richelieu, and commissioned a vast expansion. An enlightened and skillful man, he proceeded both to undertake a complete reorganization of services and to enrich the collections by the purchase of many manuscripts. The Royal Library was now divided into five departments, each under the direction of a keeper: printed books, prints, manuscripts, medals, and titles and genealogies.[1]

The direction of the Royal Library remained in the hands of the Bignon family until 1784, when the former lieutenant of police Lenoir was appointed director. Around forty employees took charge of the cataloguing and preservation of the collections. The personnel belonged to the same world as the men of science: the Republic of Letters in its most erudite version. Alongside the keepers and their clerks, a few interpreters of Oriental and European languages had the mission of translating manuscripts. Abbé Barthélémy, the foremost decipherer of the Palmyrian and Phoenician alphabets and author in 1787 of *Voyage du jeune Anarchasis en Grèce* (*Travels of Anarchasis the Younger in Greece*), an Enlightenment best seller, was keeper of the medals. The philologists Joseph de Guignes, secretary of the *Journal des savants* and specialist in Semitic languages and Chinese, and Abraham-Hyacinthe Anquetil-Duperron, the celebrated translator of the Zend Avesta, the sacred text of Zoroastrianism, as well as a Persian version of the Upanishads, were both interpreters at the Library.

At the Academy of Inscriptions and Belles-Lettres, of which they were members, these scholars mingled with great aristocrats and ecclesiastics. This company was very different from the Academy of Sciences in spirit and in appearance. It had been a refuge for Jansenism (an austere form of Catholicism), and remnants of this past lingered on the eve of the Revolution: a certain detachment and austerity that agreed well with the seriousness of erudition. While someone might take an interest in the sciences, it would only be in those of the ancients. Moreover, this particular academy was fundamentally conservative, respectful of religion, and miles away from the philosophical spirit that reigned among the savants of the Academy of Sciences.

Even so, a lot changed in the last years of the Ancien Régime. The creation of a class of resident free associates in 1785 made possible the entry of a young lawyer at the Cour des Monnaies (one of the sovereign courts of France, whose purview was coinage) named Silvestre de Sacy, who was well-versed in Arabic and Hebrew and was destined for a bright future. This new class of members also included two members of the Academy of Sciences: the astronomer Bailly and the physiologist Paul-Joseph Barthez. After publishing several fine articles on astronomy, Bailly had become famous for his theories, inspired by Freemasonry, on the origin of the sciences, which he laid out in his *Histoire de l'astronomie ancienne* (1775) (History of ancient astronomy) and then in published letters addressed to Voltaire. He believed that the astronomical knowledge of the Chinese, Indians, and Chaldeans was the vestige of a common science, invented by a vanished people who had lived before the Flood somewhere in northern Asia. This thesis, though criticized by the philosophes, met with great success. Though Bailly was ignorant of both Greek and Eastern languages, he passed as an expert in ancient astronomy. After his election to the Academy of Inscriptions, he published a *Histoire de l'astronomie indienne* (History of Indian astronomy), in which he pushed the date of the first astronomical tables in India back to the end of the third millennium BCE.

Inspired by the ideas of Antoine Court de Gébelin on comparative mythology, Bailly had interpreted the fables of the ancients in astronomical terms and had drawn from them an argument in favor of a Nordic origin for antediluvian astronomy. Another writer, Charles-François Dupuis, professor of rhetoric at the Collège de Lisieux, in turn advanced an astronomical interpretation of mythology. This protégé of the Duke de la Rochefoucauld d'Enville had assiduously followed Lalande's lectures on astronomy at the Collège de France. In his view, the Egyptians had invented the signs of the zodiac and the constellations fifteen thousand years before Christ. The zodiac was their calendar, and all the fables of mythology were said to be allegories describing the position and movement of constellations. This fundamentally materialist thesis won the support

of Lalande and the philosophes. Bailly, by contrast, fought it vigorously, since it
ran counter to his own theory about the Nordic origin of astronomical sciences.
Despite his strong opinions, Dupuis was elected professor of Latin eloquence at
the Collège de France, then member of the Academy of Inscriptions and Belles-
Lettres in 1787. After the Terror, he would publish the *Origines de tous les cultes
ou la Religion universelle* (The origins of all cults, or universal religion), in which
he treated Christianity with the same radical methods as pagan mythology. The
book was a resounding success.

The Paris Observatory and Astronomy

The interest of savants and scholars in astronomy reminds us of the importance of
this science in early modern Europe. The majesty of celestial phenomena had always attracted the attention of observers. Ancient peoples had filled the heavens
with innumerable divinities. But the astronomers, thanks to their observations
and calculations, had detected remarkable regularities. Copernicus had considerably simplified their models by placing the sun instead of the earth at the center of
the universe. This sixteenth-century revolution, both technical and conceptual,
had marked the point of departure for modern science. Since then, astronomy
had benefited from enormous progress in the means of both observation and calculation, making the data ever more precise. Newton, by a tour de force unique in
the annals of science, had shown that the movements of all astral bodies were the
effect of a single uniform physical law that was expressible mathematically: universal gravitation.

The development of astronomy had contributed to transforming the vision of
the world well beyond savant circles, challenging established dogma both about
the origin of things and about the place of humankind in creation. Newtonianism
had first appeared to support the idea of a watchmaker God who guaranteed the
laws of Nature, but it soon opened a breech to free thinking and the pure and
simple rejection of revealed religion. For in the "crisis of European consciousness" that according to the historian of ideas Paul Hazard characterized the shift
from the classical age to that of the Enlightenment, the role of the sciences was
decisive. From another angle, astronomy offered useful applications, such as the
making of calendars and even more so in the domains of cartography and nav-
igation. This was sufficient reason for governments to protect astronomers and
finance their observations.

In Paris, the Observatory had been founded in 1667 as a necessary complement to the Academy of Sciences, created the preceding year.[2] The establishment
was designed not only for astronomical observations but also for the Academy's
work in the realm of the physical and natural sciences. The original plans called

FIGURE 2.2: The Paris Observatory. Engraving in Jean-Aimar Piganiol de la Force, *Description de Paris, de Versailles, de Meudon, de S. Cloud, de Fontainebleau et de toutes les autres belles maisons et châteaux des environs de Paris*, vol. 5, Paris, Théodore Legras, 1742, 397.

for the installation of all the machines presented to the Academy, the royal natural history collection, and chemistry laboratories, as well as lodgings for savants who would come from all over Europe. These vast projects were quickly abandoned due to a lack of means and because the establishment was too distant from the city center. So it was limited to astronomy. However, from the initial idea there remained a close link between the Paris Observatory and the Academy of Sciences, under whose direction it remained, in theory, until 1771, and also the memory, perhaps, that the establishment had a vocation to observe more than the heavens. The edifice constructed by Claude Perrault on the model of Uraniborg, the famous observatory of the Danish astronomer Tycho Brahe, rose like a castle at the southern edge of Paris, near the "Porte d'Enfer" (now Denfert-Rochereau) at the place called the "Grand Regard," a country site protected from smoke (Fig. 2.2). But this much-admired masterpiece of vaulted architecture based on the science of "stereotomy," or descriptive geometry, was inconvenient and unsuitable for precise observations.

Since 1671, the Cassini family had been in control of the Paris Observatory, and the other Academy astronomers conducted their observations elsewhere. The establishment served principally as the headquarters for research in geodesy and topography, then for executing the map of France, entrusted in 1756 to a private enterprise directed by Cassini de Thury, the grandson of the dynasty's founder. This de facto situation was made official in 1771, when he was given the official title of director general of the Paris Observatory, along with a salary of three thousand livres. Because of a lack of instruments and observers, the astronomical activity of the Paris Observatory was quite limited for a long time. In fact astronomical observations were conducted at many sites scattered across Paris, such as the Military College, the medieval Hôtel de Cluny, the Collège de France, the Collège Mazarin, and many other private observatories, without the Paris Observatory being able to claim it was playing a leading or even a coordinating role.[3] At the end of the 1770s, the monument itself, which by then housed only the Cassini family and the astronomers Guillaume Le Gentil and Edme-Sébastien Jeaurat, threatened to fall into ruin, and the rooms designed for observations were dilapidated.

The son of Cassini de Thury, Jean-Dominique (called Cassini IV), who was effectively in charge of the Paris Observatory after his father fell ill, undertook to restore the building and resume astronomical observations. He had the observation rooms repaired and brought in occasional observers. He also began magnetic observations. But not until after the death of his father in 1784 did he begin an ambitious program of renovating the establishment, with the active support of his friends Baron de Breteuil and Count d'Angiviller. He secured three posts for students to do observations and the creation of a specialized library. He also launched, not without difficulty, the fabrication of new instruments by elite Parisian craftsmen. Since the great enterprise of mapping France was almost over, he proposed a new geodesic project of an international character: the junction between the two meridians of Greenwich and Paris. Finally, the building, which had fallen into a sorry state, had to be restored. Cassini would have liked to lower the upper floor by getting rid of the terrace and vaults, but d'Angiviller opposed the amputation of such a remarkable monument. In the end, the remodeling, begun in 1787, consisted of reconstructing the upper vaults and building a roof. The work was far from complete when the Revolution began, but Cassini had laid the foundation for the new Paris Observatory that would emerge in the following century.[4]

The Naturalists at the Royal Botanical Garden

When he took on the restoration of the Paris Observatory, Cassini was following, rather belatedly, the example of Buffon at the Royal Botanical Garden. When, to general surprise, Buffon was appointed superintendent of the garden in 1732, it was still pretty much as Guy de La Brosse had designed it a century before. Situated in the faubourg of Saint-Victor, along the Bièvre River, it included a little château with an esplanade and two small hills, a labyrinth, and a little butte

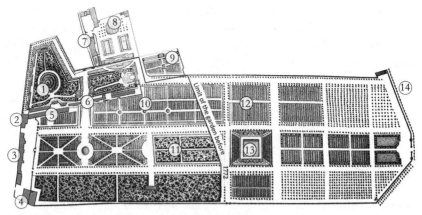

1. Labyrinth. 2. Laboratory of chemistry. 3. Natural history cabinet. 4. Intendance. 5. Orangery.
6. Greenhouses. 7. Hôtel de Magny. 8 Amphitheater of Verniquet. 9. Hot-beds. 10. School of
botany. 11. Plant nursery. 12. New squares of plants. 13. Basin. 14. Quai Saint-Bernard.

FIGURE 2.3: The Royal Botanical Garden under the superintendence of Buffon. After
A.-L. Jussieu.

(Fig. 2.3). Originally conceived as a garden for medicinal plants, it also offered instruction in botany, chemistry, and anatomy for future doctors and apothecaries.
In the years preceding Buffon's arrival, its pharmaceutical collection, housed
in the château, had been converted into a cabinet of natural history where the
collections of Tournefort and Vaillant had been brought together.[5]

Buffon continued the work of his predecessors. A worldly polymath who had
been elected to the Academy for his work in geometry and mechanics, he seemed
to have neither the right nor the competence to serve as superintendent of the
garden. But he quickly proved to be a great naturalist and a remarkable administrator. His *Histoire naturelle* (Natural History) in thirty-six volumes, published
between 1749 and 1788, enjoyed a success comparable to that of the *Encyclopédie*.
The work combined precise descriptions of a multitude of animals and minerals
with general and often audacious general theories about the history of the Earth
and the history of humankind, the origin of life and generation, and the habits
of animals and their domestication. A writer and *philosophe* even more than a
savant, a businessman and a courtier, elected to the Académie Française in 1753,
named a count in 1772, Buffon was incontestably one of the great figures of the
Enlightenment.

The Royal Botanical Garden was renowned for its lectures, which attracted a
wide audience of students and aficionados. Its creation in the seventeenth century
over the strong objections of the Faculty of Medicine had been decisive for the
takeoff of experimental methods in Paris, especially in anatomy and chemistry.
Each of the three subjects taught at the Jardin was endowed with a professorial

chair for theory and another chair for practical demonstration. Buffon had little
interest in the lessons given at the Garden, being content to intervene as superintendent in nominations for vacant posts. If he had little power over the chair
holders, he did exercise powerful patronage over the personnel of the Botanical
Garden.[6]

Even as he was writing his *Histoire naturelle*, Buffon undertook to utterly
transform with the help of his protégés the Botanical Garden of which he was in
charge. He enlarged its cabinets and its collections and extended its footprint by
acquiring land. Little interested in botany, which he left to Antoine de Jussieu,
upon his arrival he concentrated his attention on the Cabinet of Natural History,
at the time limited to two small rooms. He added two large halls taken out of
his own apartment, and then in 1766 he totally abandoned his lodgings in the
château, which allowed him to arrange the collections more methodically in four
large halls: two for animals, one for minerals, and one for plants, plus a room
for the former pharmacy. Since 1745, the Cabinet, open to the public two days
a week, had been under the supervision of the academician Louis-Jean-Marie
Daubenton, a doctor from Montbard like his protector Buffon, with whom he
collaborated on the *Histoire naturelle* (Fig. 2.4).[7]

Buffon began to concern himself with the garden as well in 1771, after he had
almost passed away. First, he had the botany school replanted according to the natural method of Antoine-Laurent Jussieu and doubled its size, then he undertook
to enlarge the property by successive purchases: to the north, parcels belonging
to the Abbey of Saint-Victor, a close called the Patouillet, and municipal lands

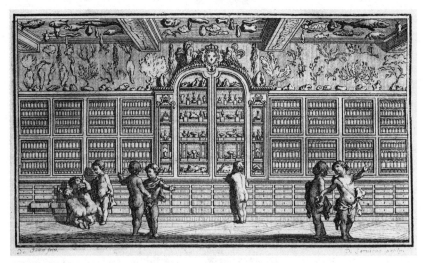

FIGURE 2.4: The king's cabinet at the Royal Botanical Garden. Engraving in Buffon,
Histoire naturelle, vol. 3, Paris, Imprimerie royale, 1749.

near the Seine previously rented to wood merchants; to the south, a whole group of small artisans' houses and a pavilion. The construction began in 1782, under the direction of André Thouin, gardener in chief at the Botanical Garden and Buffon's confidant.

A new street was opened on the Bièvre side, the rue Buffon, and two alleys of lime trees, planted by Buffon's predecessor, were extended to run down to the banks of the Seine; a large square basin was dug, the school of botany grew even larger, and a new greenhouse was built. The acquisition of new land to the north, in particular the Hôtel de Magny in 1787, completed the ensemble. Edme Verniquet, the garden's architect, built a new amphitheater, inaugurated on the eve of the Revolution, for lectures and demonstrations in botany, chemistry, and anatomy. The establishment was becoming one of the finest promenades in Paris and the foremost botanical garden in Europe, with nearly six thousand distinct species of plants (see plate 5). Shortly before his death in 1788, Buffon envisioned a further enlargement of the Cabinet of Natural History, adding a fifth hall on the first floor and a great gallery on the second floor, which would require raising the height of the château.[8]

Although other sites were available for natural history in Paris and its sur-roundings, such as the garden of the College of Pharmacy on the rue de l'Arbalète, the Cabinet of Mineralogy at the Mint, the Veterinary School at Maisons-Alfort, and several private gardens and collections, Buffon had managed to turn the Royal Botanical Garden into a unique center of attraction around which gravitated intellectual activities that extended across the city, the country, and the world. Amateurs came there to attend public lectures, to botanize, to visit the collections, and to meet naturalists. For foreigners passing through, it was an obligatory touristic stop. On a grander scale, Buffon and his protégés had developed a vast network of correspondence, by means of which they funneled information, specimens, minerals, bones, plants, and seeds of all kinds into their collections and their flowerbeds. This movement of capitalization accelerated in the final years of the Ancien Régime. Yet it was just beginning: it was only with the creation of the National Museum of Natural History during the Revolution that all the activity of Parisian naturalists was reorganized around an establish-ment unequaled anywhere in Europe.

The Expertise of Savants

The foundation of the Paris Observatory and the Royal Botanical Garden marked the importance of official science. From its foundation, Colbert had wanted to make the Academy of Sciences contribute to the grandeur and prosperity of the kingdom. It gave its advice on matters of general interest as well as on inventions and particular projects. Everywhere "utility" was the keyword. In fact, the Paris

Observatory and the Royal Botanical Garden had been established primarily
for this purpose. The former had done the geodesic work for the execution of
the map of France. The latter, originally created for the cultivation of medicinal
plants, also accommodated attempts to acclimatize exotic species. More generally, the development of scientific and technical expertise went hand in hand
with that of state bureaucracy during the eighteenth century. Royal power used
the Academy as a breeding ground where it placed experts it could draw upon.
Some savants found themselves entrusted with permanent jobs in the central administrative apparatus; others directed the great scientific establishments that the
monarchy had instituted in the capital. All sectors, both civil and military, were
involved. The need was of course greatest in Paris, capital of the kingdom and seat
of most of the ministries.

Although the main departments of the state dedicated to foreign relations
(war, the Navy, and foreign affairs), were based in Versailles, they maintained regular relations with the savants of the Academy of Sciences. Moreover Versailles
was not so far from Paris, where many departments did remain, such as the
Department of Ordnance at the Arsenal and the Department of Fortifications
on the rue Barbette. The bureaucratic machine set up by Colbert to manage the
colonies was also located along the Seine.[9] Even the Navy's office of charts and
plans, transferred for a while to Versailles, had returned to Paris in 1775, where
it occupied the former Jesuit house on the rue Saint-Antoine. But it was mainly
through educational institutions that soldiers and sailors marked their presence
in the capital.

The Military College (École militaire), established in a prestigious building
in the eighteenth century, had been for a long time an important scientific establishment, especially for mathematics. An observatory had been built there,
and several academicians, including Laplace, had taught and resided at the college. But the place had seriously declined after the closing of the college in 1776
and its replacement by a much smaller school for gentlemen cadets. It was there
that Napoleon Bonaparte studied mathematics in 1785. The shipbuilding school,
which reported to the Navy, was housed in the Louvre, in a hall neighboring
the Academy. Finally, the examiners for the three technical branches of the
military—Engineering, Artillery, and Navy—were required to be drawn from
the Academy of Sciences, and they resided in Paris. In 1768, the mathematician
Charles Bossut took up this post for the Engineering Corps, having succeeded
another mathematician, his protector Charles Étienne Louis Camus. After the
death of the mathematician Étienne Bezout, who served as examiner for both
the Navy and the Artillery, Monge and Laplace shared the posts in 1783. To his
existing responsibilities Laplace added the newly created post of examiner for
students of marine architecture.

Even more than the ministries connected with warfare, the Department of Finance had woven strong ties with the Parisian savants. Initiated by Colbert himself, these bonds were strengthened over the course of the eighteenth century. When Turgot was named finance minister at the start of Louis XVI's reign, he relied on academicians to conduct his financial reforms. He appointed his advisor Condorcet to the Paris Mint and Lavoisier to the Gunpowder Administration; he commissioned Bossut, d'Alembert, and Condorcet to study canals and Vicq d'Azyr to conduct an investigation into bovine epidemics, out of which the Royal Society of Medicine was born. In fact, all subsequent ministers of finance called upon savant expertise, including Jacques Necker, the enemy and successor of Turgot, who was allied with the Buffon clan. The offices of the Department of Finance were just around the corner from the King's Library on the rue des Petits-Champs. The savants came there to advise on matters of commerce, provisioning, and public works, and even on budgets and demography.

The great technical administrations that were under this ministry (the Mint, Mines, Roads and Bridges) were also headquartered in Paris. Their relations with the savant world had been very close for a long time. The minting and regulation of coinage enlisted expertise in chemistry, while the management of mines required knowledge of mineralogy and geology, and the construction of roads and canals needed expertise in mechanics and hydraulics. Over time, these administrations had accumulated great technical and scientific expertise, which they converted into cultural capital in the form of archives and collections of all sorts. To train their inspectors and their engineers as well as to improve their services, they regularly called upon the Academy's savants.

The Mint

On the eve of the Revolution, the Mint was one of the most imposing monuments of the capital. Its long stone façade along the Seine, with sixty-six windows in three straight rows, hid the coining shops and trial laboratories for a vast manufactory and a school, as well as several staff apartments (see plate 4). The building, which opened in 1775, was thus quite new. Since responsibility for coinage had been shifted from the Currency Court to the Finance Ministry, the administration had tried to centralize and rationalize the manufacturing operations. The dilapidation of the old Paris Mint (located near the Louvre) justified the construction of a new building whose location and magnificence would accord with the grandeur of its mission.[10]

The government had at first wanted to reconstruct the building on the new Place Louis XV on the left side of the rue Royale, but the goldsmiths protested to the provost of the merchants. Although they were not restricted to a specific district, for a very long time they had established their studios and workshops between the Louvre Palace and that of the Cité, on the Right Bank of the Seine near the goldsmiths guildhall, as well as on the bridges (Pont-au-change and Pont Notre-Dame) and on the island, and they worried about the inconvenience and lost time if they had to have the quality of their metals inspected so far away. They won, and the government gave up the rue Royale plan and decided on the site of the former Hôtel de Conti, on the Left Bank between the Pont-Neuf and the Collège des Quatre-Nations (now the home of the Institut de France) for the new building. The Mint therefore remained in the heart of Paris, near all the venerable palaces that were major symbols of royal power, and just steps away from the Quai des Orfèvres.

At the end of the Ancien Régime several Academy savants were residing at the new Mint: Condorcet, the Academy's permanent secretary and the inspector-general of Mint, who held a salon there with his wife, but also Mathieu Tillet, royal commissioner for assaying and refining, both appointed by Turgot in 1775, and Balthazar-Georges Sage, professor of docimastic mineralogy, appointed by Necker in 1778. Since there were no more vacant apartments, the chemist Jean Darcet (who replaced Tillet in 1784) and the technician Abbé Rochon (named an inspector in 1786) lodged elsewhere. While the scientific activity of Condorcet and Rochon at the Mint appears to have been almost nil, that of Tillet and Darcet, who had access to a laboratory for their assays, led to several articles presented to the Academy. But the most important activity at the Mint, at least for the sciences, was certainly that of Sage.

An apothecary by training who was thus not valued much by his academician colleagues, Sage conducted research in chemistry and also devoted himself to natural history. In the 1760s he started giving public lectures and little by little built a magnificent mineralogical collection brought together at the Hôtel de Bréant, rue du Sépulcre. Protected by Louis XV, he entered the Academy of Sciences in 1770 and was appointed royal censor for chemistry and natural history, demonstrator for the chemistry course at the College of Pharmacy, commissioner of assays, and finally professor of docimastic mineralogy at the Mint (a post created at his request in 1778). Thanks to his relationships at court Sage had become one of the most powerful and best-paid savants. His research, however, aroused the jeers of his colleagues. He claimed to have found gold in the ashes of plants. When his purported discovery was refuted by the Academy's chemists, he quarreled with them. He was in open conflict with Tillet and loudly rejected Lavoisier's theories. But Sage was too highly appreciated by high society and at court to really suffer from his poor reputation in scientific circles.

The School of Mines and the School of Roads and Bridges

In addition to an apartment at the Mint, Sage had at his disposal an immense and sumptuously appointed salon looking out on the quay, where he installed his laboratory and his mineral collection and gave his lectures. In 1783, despite the hostility of the inspector general of mines, the irascible Antoine Monnet, he also obtained by his personal skills the creation of a School of Mines, also at the Mint, of which he naturally became director. This school, which had only a few students destined to become mining engineers, offered several courses to the public, including Sage's own courses on mineralogy and chemistry and one by the mining inspector Jean Pierre François Guillot-Duhamel, soon elected to the Academy of Sciences, on techniques for the exploitation of mines.[11]

At the same time, Sage continued to give his public lectures in docimastic mineralogy. Profiting from the creation of the School of Mines, he had negotiated on his own terms the purchase by the administration of his mineralogical collections and the remodeling of his classrooms. The lecture hall, ornamented with busts of Louis XVI and Calonne, was now a vast tiered auditorium that could hold up to two hundred people. On its periphery and against the walls, glass-fronted cabinets contained a portion of the collections. Some stairs at the back of the hall led to the ample fireplace where chemical operations were executed. Many other minerals were exhibited in the galleries upstairs, reached by a staircase over which reigned the bust of Savant Sage, an homage by his students to the master of the place.[12]

But due to a lack of job openings, the School of Mines founded by Sage at the Mint stopped functioning in 1788, before being reborn in another form in 1795. In fact, this mining school was far from rivaling the Royal School of Roads and Bridges established by Trudaine in 1747 at the drafting office of the administration of roads and bridges. Directed by Jean-Rodolphe Perronet, the chief engineer of the corps and member of the Academy of Sciences, over the course of the 1780s the school welcomed more than a hundred students, including some who were not enrolled, and all of them lived in the city. Courses were held at the Hôtel Liberal Bruant, situated in the Marais at the corner of the rue de la Perle and the rue Thorigny, and then after 1786 at the headquarters of the Superintendent of Roads and Bridges at the corner of the rue Saint-Lazare and the rue des Trois-Frères.[13]

In truth, the School of Roads and Bridges was barely a school. The training came from Perronet and his assistants, the engineers Lesage and Chézy. Without any appointed professors, the students followed public or private lectures that were given in the capital, like those of Sage in chemistry at the Mint, or of Brisson in physics at the Collège de Navarre. Many students also went to the Oratorian Fathers on the rue Saint-Honoré, which had a chair in hydrodynamics founded by Turgot in 1775 and attached to the Academy of Architecture. The holder of the chair, Abbot Bossut, was replaced first in 1780 by Monge, who presented there (for the first time in Paris) the geometric methods he had been teaching at Mézières for the students of the School of Engineering, and later by an instructor known simply as "Charles the Geometer" to distinguish him from the famous physicist Jacques Charles.[14] On the school's premises, which had a library and room for maps and plans, the students sketched under the supervision of Chézy and Lesage. The more advanced ones helped the younger ones in the theoretical subjects, according to a system of mutual instruction. The summer months were spent on "campaigns" to construction sites arranged through their contacts with engineers.

The Royal Manufactories and the Machine Depository

While the Mint lay at the heart of Paris, the other Parisian industrial establishments that were directly dependent on the Crown were situated on the periphery. The Arsenal was the traditional headquarters of the Artillery and formed a veritable enclave between the Seine and the Bastille, with its own jurisdiction and police. A maze of paths and alleys linked its palaces, houses, and shops. To the south, along the Seine across from the Ile Louviers, was the "Grand Arsenal," with the governor's mansion and his library. To the north, just down the street from the Bastille, the "Petit Arsenal" ran along the ditch that had been dug long before to fill the moat surrounding the fortress. Here was the Royal Gunpowder and Saltpeter Administration. This administrative service was part of the Department of Finance and had been established by Turgot in 1775 to replace the old, notoriously ineffective system of contracting. It was responsible for the harvesting and refinement of saltpeter across the entire kingdom, as well as the production and commercialization of various types of gunpowder. The Arsenal housed its central administration as well as a refinery and gunpowder magazines.[15]

At its head was Lavoisier, appointed by Turgot, who along with three colleagues was installed in the managers' house, situated between the saltpeter courtyard and the Arsenal Ditch. The buildings were burned down during the Paris Commune in 1871 and are gone today. Without neglecting the administrative and financial aspects of his job, Lavoisier was particularly interested in rationalizing and improving production. To this end he launched a serious research program, conducting tests, encouraging innovation, and calling upon expertise in the applied sciences. The effects on gunpowder production began to be felt at the beginning of the 1780s. The demands of war that began with the Revolutionary campaigns of 1792 relaunched research into new procedures, and Lavoisier took part in them until he was arrested.

At the same time as he was improving production, Lavoisier established a system for training gunpowder commissioners that ranged from scientific studies at the Arsenal to internships on the ground. He entrusted to a young protégé named Philippe-Joaquim Gengembre the task of teaching chemistry and mechanics at the Arsenal to students of gunpowder. But the site became famous above all for the private laboratory that Lavoisier fitted out for his own use. In the final years of the Ancien Régime, this place had become the meeting place of partisans of the "chemical revolution," frequented by both the academic elite and foreign savants passing through Paris (Fig. 2.5).

On the opposite bank of the Seine, in the Faubourg Saint-Marcel, another royal manufactory, that of the Gobelins, was of interest to chemists. The establishment, founded by Colbert in 1662, was situated in an industrial neighborhood that traversed the Bièvre. The river was lined with numerous factories, mostly laundries, tanneries, and dye houses. At the Gobelins, independent businessmen made highly prized low-warp and high-warp tapestries under the supervision of the administration of the King's Buildings. Inspectors and painters were attached to the manufactory, as well as a chemist, starting in 1779, who was appointed to

1. House of the Gunpowder Administration   2. Refinery.   3. Presumed location of Lavoisier's laboratory.

FIGURE 2.5: The Gunpowder Administration at the Arsenal.

survey the dyeing workshop, first Claude Melchior Cornette, then Darcet, both of the Academy of Sciences.

The administration was thus inspired by the example of the royal porcelain manufactory established in 1756 in the village of Sèvres two leagues outside of Paris. First attached to the minor ministry of Bertin, responsible for agriculture, mines, manufacturing, and transportation, this establishment in 1780 was transferred to the authority of the King's Buildings. The Sèvres Manufacture was famous for its fine soft porcelain, an imitation of Chinese hard porcelain, whose secret, rediscovered in Germany, remained unknown to the French for a long time. The quality of its production began to improve considerably toward the end of the 1750s, thanks in particular to the research of the chemists attached to the manufactory, all of them members of the Academy of Sciences: Jean Hellot from 1751 to 1766, Pierre-Joseph Macquer from 1766 to 1784, and finally Darcet. In 1769, Macquer succeeded in mastering the technique of hard porcelain by using kaolin, discovered in Saint-Yrieix. He soon presented his results to the Academy of Sciences.

The production of the great royal manufactories like those of Gobelins and Sèvres aimed mainly at encouraging industry. At least since Colbert, administrators systematically supported the improvement of technical processes as well as their dissemination throughout the kingdom. This was the task assigned to the superintendents of the Bureau of Commerce in Paris and to the inspectors of manufactories and mines in the provinces. The Academy of Sciences itself was mobilized on behalf of the arts and trades. The Bureau of Commerce commissioned savants from the company to evaluate new inventions: Hellot, Macquer, and Berthollet were successively commissioners for chemistry; Jacques de Vaucanson, then Alexandre-Théophile Vandermonde and Le Roy for machines. In 1785, two posts as commissioners were also created for inspecting mines, one going to Philippe Friedrich Dietrich, soon elected to the Academy, for metal mines, the other to Faujas de Saint-Fond, a protégé of Buffon's, for coal mines.

A mechanical genius, Vaucanson made himself known to the Paris public in 1737, when he exhibited a superb automaton that played the flute, followed the following year by one that played a tambourine, and above all by a duck that flapped its wings, ate seeds, and defecated, which toured throughout Europe. Finance Minister Orry, impressed by his talents, commissioned Vaucanson to improve the mechanical looms used in the silk industry. Having become an inspector of manufactories and a member of the Academy of Sciences, Vaucanson set up his workshops and assembled his machines in the Hôtel de Mortagne on the rue de Charonne in the faubourg Saint-Antoine.

Upon his death in 1782, Vaucanson's collections reverted to the Crown. They would constitute the kernel of a public depository of machines placed under the authority of the Bureau of Commerce, the management of which was entrusted to the academician Vandermonde. The goal was to stimulate inventors and to "entice capitalists to invest in the products of new machines." New models acquired by purchase or gift and by deposit were added to the sixty-odd machines inherited from Vaucauson. The craftsmen and workmen of the establishment also constructed a good number of machines based on instructions from inventors or at the request of the government.[16] This collection gave rise to the Conservatory of Arts and Trades, created during the Revolution.

Academies, the King's Library, the Paris Observatory, and the Royal Botanical Garden, depositories, manufactories, and great technical schools: these prestigious establishments scattered around the city definitely made eighteenth-century Paris a capital of the sciences that was unrivalled throughout Europe. Mediated through these establishments, the world of science found itself closely integrated into the apparatus of the state, whose interests and purposes it shared. Their presence in urban space, often monumental in scope, also signified the importance that royal power gave to science, which served both its grandeur and its plans. It was around these institutions, finally, that scientific life in Paris was organized. Nevertheless, that life extended well beyond the circle of royal institutions. For it was not just the state and those it employed who made Paris a great home of the Enlightenment but the whole city and its inhabitants.

Notes

1. BALAYÉ, 1988. Description of the library in THIÉRY, 1787, vol. 1, 193–212.

2. The classic study of the Paris Observatory under the Ancien Régime remains WOLF, 1902.

3. On the observation sites in Paris in the eighteenth century, BIGOURDAN, 1930, vol. 2.

4. WOLF, 1902, 229–318.

5. Y. LAISSUS, "Le Jardin du Roi," in TATON, 1964, 287–341.

6. Buffon had hired his own people, Daubenton and Thouin, first, then Faujas de Saint-Fond, Count de Lacépède, and Chevalier de Lamarck, for whom he managed to create posts as correspondents, assistants, and assistant demonstrators in natural history.

7. BOURDIER, 1962, 35–50. Description of the natural history office in THIÉRY, 1787, vol. 2, 172–178.

8. FALLS, 1933. Description of the garden in THIÉRY, 1787, vol. 2, 180–184.

9. REGOURD, 2008.

10. Description of the Mint in THIÉRY, 1787, vol. 2, 473–482.

11. AGUILLON, 1889, and A. BIREMBAUT, "L'enseignement de la minéralogie et des techniques minières," in TATON, 1964, 365–418.

12. Description of the mineralogy office and the courses at the School of Mines in
THIÉRY, 1787, vol. 2, 475–480.

13. DARTEIN, 1906, and PICON, 1992.

14. HAHN, 1964, and THIÉRY, 1787, vol. 1, 334–335.

15. Description in THIÉRY, 1787, vol. 1, 669–673.

16. DOYON, 1963, DOYON and LIAIGRE, 1966, and D. DE PLACE, "Le sort des ateliers
de Vaucanson, 1783–1791, d'après un document nouveau," *History and Technology*
1 (1983), 79–100. Quotation taken from a report dated August 2, 1783, addressed
by the minister of finance to the finance committee and approved by Louis XVI
(DOYON, 1963, 8).

3

Fields of Knowledge in the City

ON JULY 6, 1785, the physicist Martin van Marum, director of the Teyler Museum in Haarlem, one of the most beautiful science museums in Europe, arrived at the Hôtel d'Anjou on the rue Dauphine, close to the Pont Neuf. It was his first visit to Paris. At the Louvre, he attended five meetings of the Academy of Sciences, of which he was a correspondent, as well as a meeting of the Royal Society of Medicine; he visited the King's Library, the Collège de France, the Royal Botanical Garden and its cabinets of natural history, the Mint, and the machine depository at the Hôtel de Mortagne. Many savants invited him to visit them at home: Lavoisier at the Arsenal, Monge on the rue des Petits-Augustins, the anatomist Jacques René Tenon on the rue du Jardinet, Brisson on the rue de Condé, Faujas on the rue de Valois (where he first saw the new Argand lamp), and others as well. He also met Jean-Claude Delamétherie, the editor of the *Journal de physique*, at his home on the rue Saint-Nicaise; he witnessed Ledru's experiments in medical electricity at the Celestine Convent and those of Berthollet on the bleaching power of chlorine. He examined the finest private natural history collections in Paris, Philippe-Laurent de Joubert's and Jean-Baptiste-François Gigot's in the Place Vendôme, Jean-Baptiste Louis Romé de l'Isle's on the rue Neuve des Bons Enfants, and Jacques-Christophe Valmont de Bomare's on the rue de la Verrerie. At the Palais-Royal he discovered the models representing the various crafts that the Périer brothers had made for the children of the Duke d'Orléans; he went to Montmartre to ask quarry workers for fossils, and he visited the Veterinary School in Charenton and the Menagerie at Versailles. Finally, he did not neglect specialized bookshop owners like Didot the Younger on the Quai des Augustins and Cuchet on the rue Serpente or dealers in scientific equipment like Bianchi on the rue de l'Arbre-Sec and Gaillard on the rue de Richelieu.

In a single month Van Marum made the complete tour of Paris Savant. Not one to be easily taken in, he noted in his travel journal the mediocrity of certain scientific cabinets and the antiquated nature of their physics equipment.

Nevertheless, his daily handwritten notes testify to his astonishment that a single city could contain so many scientific resources and activities.[1]

A Geography of Paris Savant

Around 1700, Paris numbered around five hundred thousand inhabitants. At the end of a period of steady growth that accelerated after 1750, the number of Parisians had certainly reached six hundred thousand at the end of the century. Meanwhile the city was enlarged by its faubourgs well beyond the boulevards that had been built where the old ramparts had been. In 1784, the construction of a tax wall known as the Farmers-General Wall (where the "exterior boulevards" are today) marked its new boundaries. Paris would not reach its current scope until 1860 when several neighboring villages, including Belleville, Montmartre, and Passy, were annexed to the city.

Corresponding approximately to today's first twelve *arrondissements*, eighteenth-century Paris was still a city of human scale, which Van Marum could easily cross on foot, if he did not fear either the muddy streets or the jumble of carriages. This did not prevent each quarter, each parish, and almost each street from having its own character. Such diversity was as true for science as for everything else. During his stay in Paris, Van Marum primarily went back and forth between the elegant neighborhoods to the west and the Latin Quarter, but he sometimes also crossed through the center of the city to reach the eastern faubourgs. On occasion, he even ventured outside the city. On his treks about town, the Dutch scientist met many people. His itineraries trace the way in which the sites of the sciences were distributed around the city—at least the ones that interested him.

Van Marum was not content to visit the monuments of official science visible from a bird's-eye view of Paris. Many other less visible places in the area attracted him. Moreover, even these sites were well integrated into their environment. Surrounded by shacks and shops, they were far from being impregnable fortresses; people could wander into them pretty freely. As for the savants of the Academy, they confined themselves neither to the Louvre nor to the townhouses and palaces where they conducted their official duties. Like Van Marum, they crisscrossed the city, its offices, its stores, and its workshops, where they met a thousand merchants, clerks, dilettantes, masters of minor trades, artisans, and workers without whom they would never have been able to complete their tasks. All these places and people represented vital resources for Parisian science. They provided it with raw material, equipment, and processes. The whole ensemble formed a complex system of relationships and interactions, which was itself deeply embedded in its urban environment.

The integration of the savant world into the space of Paris proved a powerful factor of differentiation at the heart of official science. It contributed to the construction of disciplinary boundaries between specialties and to the division of the savant world according to lines of cleavage relating directly to the struggles between various actors with whom the men of science entered into contact. For example, within the medical profession there was a split between physicians and surgeons, and within industry between artisans and manufacturers. At the same time, this immersion gave the savants various means of extending their influence well beyond the tight circle of intellectual and bureaucratic elites. Throughout these relationships were woven networks corresponding to different realms of expertise: mathematics, astronomy, chemistry, botany, mineralogy, and so on. Each one was unique, denser in some quarters, more diffuse in others. These different patterns defined specific zones of scientific activity in Paris.

Thus official science touched all aspects of urban life, either directly or indirectly. It went out to meet worlds of knowledge that were poorly known or unrecognized, kinds of expertise that were in some way invisible. High society was among the most accessible worlds. There the savants found enlightened protectors and collectors, but also charlatans and system makers who had to be reckoned with: their very success meant that they could not be ignored. It was hard to draw a clear line between invention, fantasy, and pure and simple fraud. The Academy of Science thought it had to take charge of these matters in the 1780s because the influence of a kind of spectacular and sometimes sulfurous science had traveled well beyond the salons. Public courses, public dissections, physics demonstrations, and balloon launches: Paris welcomed a multitude of more or less serious demonstrations that took place under the suspicious eyes of the savants.

Professions and trades offered more crucial resources to official science. The expertise of craftsmen was of direct interest to savants, who were striving both to master and to control them, occasionally by using whatever authority they had. Benefiting from their participation in the government, they had access to its offices and departments, where they could meet all kinds of experts and specialists: statisticians from the Finance Ministry, architects and engineers directly or indirectly attached to the King's Household, surgeons and physicians at the Châtelet and the hospitals of Paris, and technicians in the major public and quasi-public departments, such as the Gunpowder Administration, the Post Office, and the Tax Agency. But the principal resource came from artisans, as we shall see.

The ensemble formed a geography of Paris Savant, in which the distribution of scientific activities in urban space, their localization around a few structural poles, and their grouping in clusters corresponding to areas of specialization are all apparent. However, it would seem more difficult to bring to light the interactions that constituted the essence of intellectual labor. Generally speaking, urban space is characterized by the intensity and rapidity of exchanges, which are fostered by the density of buildings and the multiplicity of networks. In cities, not only men and women but also ideas and things are in constant movement. In order to grasp these dynamic relations, we must vary the scale, from that of particular buildings and their subdivisions to the whole city. This requires shifting from the basic spatial unit constituted by, for example, a meeting hall, a lecture hall, or a laboratory to the urban network formed by all the sites of science and the connections between them.

The point of these general reflections is simply to suggest what a systematic geography of Paris Savant in the eighteenth century might look like. Here my aim is more modest: to trace a path that will situate by zones the savants and their activities, as well as to sketch out the dynamics of their interactions with the urban environment. To this end the city must be divided into several parts: the west, with its concentration of wealth and power, which formed Enlightenment Paris properly speaking; the Latin Quarter, home of the University of Paris and heart of intellectual life since the Middle Ages; the overcrowded center, cradle of Paris's arts and crafts; the faubourgs, both rustic and industrial; and the halo of surrounding villages. Finally, we must not forget that Paris, capital and metropole, exercised its influence throughout the kingdom and beyond.

Enlightenment Paris

Since members of the Academy of Sciences were required to live in the capital, where they lived gives us a first indication of the geography of Paris Savant (see map 3.1). Not surprisingly, at the end of the Ancien Régime, their addresses were grouped in the most densely populated part of the city, inside the old walls. Aside from some of the honorary members and certain wealthy free academicians, only those who were housed in public buildings such as the Military College, the Invalides, and the Paris Observatory lived out in the faubourgs. But the savants had deserted the inner city between the central market (Les Halles) and city hall (the Hotel de Ville). A few (but fewer and fewer over time) lived in the Marais, but the vast majority of the addresses (almost three-quarters) were grouped to the west, in twin clusters: one on the Right Bank around the Louvre and the Palais-Royal and creeping northward up to the boulevards; the other on the Left Bank, concentrated on the western flank of the Latin Quarter, from the Seine to the Luxembourg Palace. We are used to contrasting the two banks of the Seine, the productive one to the north, the academic one to the south, but around the Louvre the river was not a formidable boundary: savants residing on the other

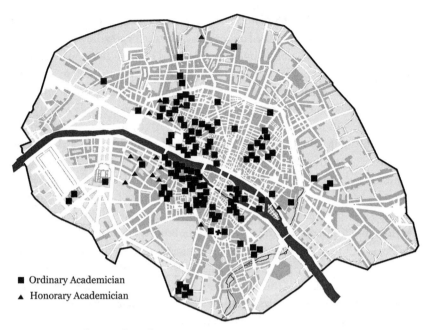

- ■ Ordinary Academician
- ▲ Honorary Academician

MAP 3.1: Residences of Academicians, 1783–1793

side of the river had only to cross one of the bridges to reach the Academy in a few minutes. The quarter at the foot of the Pont-Neuf, behind the Malaquais, Conti, and Les Augustins quays, was basically a sort of extension on the other bank of the residential neighborhoods of the Right Bank.

Let us forget the river for a moment and look at the map: it was indeed around the Louvre, in a vast residential zone developed a century earlier along the ancient walls, that the academicians congregated. There lay the Paris of the Enlightenment with its voluntary societies, public courses, performance-demonstrations, cafés, libraries, journalists, and purveyors of prohibited books. However, the divide between the Right and Left Banks was still significant: to the north (the Right Bank), the side of the Place des Victoires, the Butte Saint-Roch, and the rue Saint-Honoré, savants mingled with the worlds of government, finance, and luxury; on the Left Bank, the side of Saint-Germain and Saint-Michel, they rubbed shoulders with both the high nobility and the university. Half of the honorary members lived in the closest noble faubourg, Saint-Germain; in the other direction, a string of addresses ascended Mont Sainte-Geneviève, extending northward all the way to the islands and southward to the Royal Botanical Garden.

Should we be surprised that the Louvre was at the heart of Enlightenment Paris? After all, weren't all cultural activities placed under the supervision of the academies that resided there? In fact, men of letters and artists seeking support and recommendations converged on the palace. Nevertheless, the presence of the Louvre was only one factor among many that had fostered the takeoff of intellectual life in this part of Paris. The area benefited more generally from an exceptional concentration of power and money. The impetus for this movement lay in the destruction of the ancient ramparts in the seventeenth century.

Beginning in the 1630s, the city expanded north of the Louvre beyond the Palais-Royal built by Cardinal Richelieu. His successor Mazarin had built his own palace along the rue des Petits-Champs on land freed up after the so-called yellow ditches between the rue de Richelieu and the rue Vivienne were filled in. Colbert had built his townhouse on the other side of the latter street. This construction gave birth to a zone of financial and cultural activity that continued to develop up until the Revolution. Louis XIV's brother (Monsieur) had restored the Palais-Royal, which later became the residence of the dukes of Orléans. As for the former palace of Mazarin, divided after the cardinal's death into two townhouses (the Palais-Mazarin and the Hôtel de Nevers), it later housed the Company of the Indies, the Stock Exchange, and the Royal Lottery (in the Palais-Mazarin), as well as the King's Library (at the Hôtel de Nevers). Over the course of the eighteenth century, many government administrations, most of them under the Finance Ministry, would establish their headquarters in the area. At the end of the 1750s, the Finance Ministry itself moved into a building on the rue Neuve-des-Petits-Champs close to the former Palais-Mazarin.

At the same time that it housed offices of the royal administration, the area attracted rich aristocrats and many financiers, who built sumptuous townhouses (*hotels*) nearby. Opulence and the luxury trade that went with it henceforth characterized the neighborhood and its new extensions along the Faubourg Saint-Honoré, the rue Montmartre, and the Chaussée d'Antin. There were shops and boutiques, fashionable cafés, theaters, and other places designed for pleasure. The redesign of the Palais-Royal gardens in 1784 marked the apogee of this period of splendor. The activities of the mind were not absent from this major transformation. The powerful became patrons and assembled mineralogy and natural history collections. The philosophes found a warm welcome in literary salons, such as that of Madame Geoffrin across the street from the Capuchin Monastery on the rue Saint-Honoré, of Farmer-General Helvétius on the corner of the rue Saint-Anne and the rue Thérèse, or of Baron d'Holbach near the Finance Ministry on the rue Royale-Saint-Roch. Toward the end of the century, public courses in the sciences also opened for the upper reaches of the public. Among the most celebrated was the one given by Charles in his physics cabinet on the Place des Victoires, and Pilâtre's Musée on the rue de Valois. Around the same time, Dr. Mesmer was revealing to initiate the secrets of animal magnetism at the Hôtel de Coigny, seat of the Society of Universal Harmony.

One part of Enlightenment Paris also extended onto the Left Bank across from the Louvre. Just as in the north, a new neighborhood had been developed after the destruction of the ramparts. It more or less followed the line of the ditches running from the Collège des Quatre-Nations (built at the wishes of Mazarin along the Seine) up to the church of Saint-Sulpice, built near the fairground of Saint-Germain-des-Prés. Located there was one administrative building, the Mint, and a few lavish townhouses, as well as boutiques, cafés like the famous Procope, and theaters. The opening in 1782 of the Théâtre Français across from the Luxembourg Palace, which had recently become the residence of the Count of Provence (who, as the king's brother, was also known as Monsieur), confirmed the neighborhood's place in high society. The sciences found their place here too, since the quarter adjoined that of the university (see map 3.2).

In the Latin Quarter

For centuries, in fact, people had come from all over to study in the Latin Quarter. The Collège des Quatre-Nations to the west and that of Cardinal-Lemoine near the La Tournelle Quay to the east marked the boundaries of the quarter. To the south, it extended to the summit of Mount Sainte-Geneviève with its abbey. Within this narrow perimeter a dozen prep schools could be found, where the curriculum ranged only from grammar to philosophy, as well as many private schools. In addition to these establishments, whose teachers came from the humanities, were professional schools of law, medicine, and theology, as well as the Collège de France.

With its narrow, winding streets and its antiquated colleges, the "Latin country" evoked the Paris of the Middle Ages more than that of the Enlightenment. "There one observed climbing up and down a horde of Sorbonnists in their cassocks, tutors with their banded robes, law pupils, and students of surgery and medicine: their indigence drives their vocation," noted Louis-Sébastien Mercier with a touch of contempt.[2] In fact, one crossed paths there with more needy clerics than with the philosophes who frequented salons. But intellectual life was lively: in addition to the colleges and libraries, there were lots of courses, both public and private, cabinets of physics and natural history, businesses catering to medicine and the sciences, and all the book trades, which had been connected to the University of Paris since the end of the fifteenth century. The bookshops and printers who were authorized by the university were required by royal regulation to maintain their presses, shops, and storehouses in the Latin Quarter (to which was attached the Palace *quartier* on the Île de la Cité). The print and map dealers were located nearby. Finally, the engravers, bookbinders, gilders, paper marblers, and other specialized crafts had gravitated to their clients and congregated in the

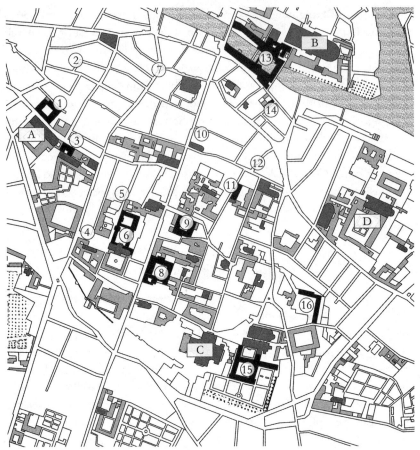

A. Cordeliers convent. B. Notre Dame. C. Church of Ste.-Geneviève. D. Bernardins convent.
1. School of Surgery. 3. Hôtel de Thou (Panckoucke). 3. Royal School of Drawing. 4. Fortin, instrument maker.
5. Favanne de Montcervelle, natural history dealer. 6. Sorbonne (Faculty of Theology). 7. Le Breton (publisher of the Encyclopédie). 8. College Louis-le-Grand. 9. Collège de France. 10. Sigaud's Cabinet of physics. 11. Faculty of Medicine (Jean-de-Beauvais). 12. Desault's amphitheater of surgery. 13. Hôtel-Dieu. 14. Amphitheater of the Faculty of Medicine (rue de la Bûcherie). 15. Abbey of Ste.-Geneviève. 16. College of Navarre (amphitheater of physics).

MAP 3.2: The Latin Quarter

narrow streets of the Saint-Jacques quarter, near the Collège de France and the Sorbonne.

In this trade, the sciences represented a sector that had grown a lot since 1730. Alongside scholarly books destined for a small elite, Parisian publishers published books designed for a wider audience. Above all, the *Encyclopédie* of Diderot and d'Alembert, launched in the middle of the century, was an enormous success. At the end of the Ancien Régime, the market for scientific publishing was dominated by a few specialized publishers belonging to the community's elite, such as Charles-Antoine Jombert, publisher to the king for artillery and engineering, and Charles-Joseph Panckoucke, the "Atlas" of the Parisian book trade. Most of them were located on the west side of the Latin Quarter, between the Quai des Grands Augustins and the Royal Academy of Surgery, that is, very near the Louvre. Jombert, whose shop was on the rue Dauphine near the Pont-Neuf, had held a salon there in the middle of the century, receiving d'Alembert, Lalande, Montucla, and some artists.[3] Later, Panckoucke, publisher and close associate of Buffon and the Encyclopedists, invited his friends from the Republic of Letters to the Hôtel de Thou on the rue des Poitevins, as well as to his country house in the village of Boulogne near Paris. These prominent publishers were the necessary partners of the savants.

However, in the colleges of the Latin Quarter, the sciences were of only secondary importance. The Collège Louis-le Grand on the rue Saint-Jacques had been the exception for a long time. The only Jesuit institution in the capital, it had harbored a small community of priests dedicated to scientific research and publishing, even as they supervised the education of the students who boarded with them. They published the *Mémoires de Trévoux* there, a sort of Jesuit equivalent of the *Journal des Savants*, as well as the *Lettres édifiantes et curieuses* (Edifying and interesting letters) sent back to France by the Jesuit missionaries. But Louis-le-Grand had to close its doors in 1762 when the Jesuits were expelled from France.

In the institutions under the authority of the University of Paris, the sciences were discussed only in the context of the philosophy courses that were attended primarily by students destined for careers in medicine and theology. The physics course was given in the second year (the first year being reserved for metaphysics and ethics). Instruction was somewhat modernized after 1770, as French gradually replaced Latin and Newton's physics that of Descartes; mathematics and (more modestly) experimental physics began to appear in the curriculum; finally, on the eve of the Revolution, the complete separation of physics from philosophy was announced. Despite these changes, however, the sciences continued to play an unheralded auxiliary role in the colleges. Preparation for the mathematical examinations required for admission into the technical branches of the military took place elsewhere: in the provincial military schools and in Paris at a few specialized boarding schools such as Longpré's on the rue de Reuilly and Berthaud's in the Faubourg Saint-Honoré.[4]

Standing out against this forlorn landscape were a few special chairs established by the monarchy that fortunately did raise the scientific level of some of the Parisian colleges. The Collège des Quatre-Nations, which was the largest and most illustrious, had a chair in mathematics, to which was attached a small observatory. After Varignon and La Caille, Abbé Marie taught there until 1785. The Collège de Navarre, located on the hillside close to the new church of Sainte-Geneviève, was distinguished by a chair in experimental physics established especially for Abbé

Nollet, who was succeeded by Brisson of the Academy of Sciences in 1770. The course, which took place in a purpose-built lecture hall with more than four hundred seats, attracted a wide audience from outside the college.

But it was principally at the Collège de France in the Place de Cambrai at the heart of the Latin Quarter that scientific courses at a high level could be found. This institution, which had been independent for a long time, was attached to the University of Paris in 1773. The new financing associated with this loss of independence made possible the renovation of the buildings and the establishment of new chairs, several of which were devoted to the sciences. Now it had a stately hall for formal meetings, six classrooms, an anatomical theater, an observatory, and a modest chemistry laboratory. Out of its nineteen active professors, three occupied chairs in Oriental languages, and nine had scientific chairs.

A diverse audience of amateurs and students came there to attend classes without any equivalent in Paris, such as in Syriac, Arabic, and Persian, but also in differential calculus and mathematical physics. Lalande gathered around his chair the small Parisian world that was passionate about astronomy. In addition to people with a general interest in chemistry, Darcet attracted many medical students. Daubenton gave classes in natural history that complemented his demonstrations at the cabinet of the Royal Botanical Garden. The professors wanted to transform the Collège de France into an academic establishment endowed with a printing press, a library, and an annual prize, which would communicate with all the scholarly societies in Europe. But their plan came to nothing for lack of financial means.[5]

Medical Education

Like the colleges, the Faculty of Medicine was part of the University of Paris. It was actually a completely autonomous professional body comprised of some 150 doctors authorized to practice in Paris, and regent doctors (those qualified to teach) of the faculty. In principle, this established corps regulated the practice of medicine and public health in the capital and lent its expertise on evaluating new remedies and therapeutic procedures; it also organized medical education, whose origin went back to the Middle Ages. Its seat, first located on the rue de la Bûcherie close to the Hôtel-Dieu hospital, had been moved in 1775 into less dilapidated premises on the rue Jean-de-Beauvais recently vacated by the Law Faculty. At the original site there remained only the lecture hall constructed in 1744 for demonstrations.

Instruction from the faculty was in the hands of professors who were elected each year from among the regent doctors. The lectures in medicine were given by the *professor scholarum*, the only one appointed for two years, who covered physiology and hygiene in the first year and pathology and therapeutics in the second year. It also included instruction in anatomy based on dissection of human cadavers. The medical students were also supposed to take classes in surgery, in *materia medica* (botany), and in pharmacy, for which particular professors were designated. Courses in chemistry and ophthalmology were organized on a less regular basis. All this instruction, except in chemistry, was given in Latin. Finally, two faculty courses, taught in French, were addressed to a wider audience, one in surgery for surgeons, and the other in obstetrics for midwives.

Most historians have repeated ad nauseam the harsh criticism of the Faculty of Medicine expressed by its adversaries, in particular by the Royal Society of Medicine.[6] Indeed, the system of instruction was in fact archaic: the professors, who changed from one year to the next, lacked experience for the most part; they lectured attired in robes seated on a raised chair, and their lectures were bookish. The regulations did not call for any clinical teaching. Demonstrations and practical courses were inadequate due to both financial constraints and the lack of a botanical garden and a laboratory. Finally, only a few students actually earned their degrees in Paris.

Nevertheless, this somber picture does not do justice to the importance and variety of medical study in Paris. In anatomy, in botany, and in chemistry, the classes at the Royal Botanical Garden and the Collège de France complemented those of the Faculty of Medicine. Moreover, students could acquire practical and clinical training by following (often for a modest price) one of the many courses that flourished in the Latin Quarter. In this way, they learned anatomy and surgery, scalpel in hand, in the private lecture halls around the Place Maubert and the church of Saint-Séverin, and chemistry and pharmacy in the laboratories and clinical dispensaries in the hospitals, from Jean-Nicolas Corvisart at the Charité or from Pierre-Joseph Desault at the Hôtel-Dieu.

In the hospitals, medical students encountered many other auditors, who came in large numbers and sometimes from far away in order to study medical practice without enrolling at the faculty. Most of them were apprentice surgeons. In fact, Paris had become the center of surgical study in Europe. About eight hundred students were registered at the School of Surgery on the eve of the Revolution, compared to only about a hundred enrolled at the Faculty of Medicine. Paris's corporation of surgeons had broken away from the barbers during the eighteenth century and thanks to the king's protection had been raised to the rank of a liberal profession. The school then changed its name to the College of Surgery. Although it had no formal affiliation with the university, the college (like the Faculty of Medicine) had its school in the Latin Quarter on the rue des Cordeliers.

FIGURE 3.1: The amphitheater of the School of Surgery, Engraving by Claude-René-Gabriel Poulleau, in Jacques Gondouin, *Description des écoles de chirurgie*, Paris, Cellot & Jombert, 1780, plate 29.

In 1774, the School of Surgery abandoned the Saint-Côme lecture hall on the rue des Cordeliers for the new building of the Royal Academy of Surgery across the street. This academy, founded in 1731 on the model of the Academies of Painting and Sculpture, had as its mission the improvement of the art and science of surgery. Presided over by the king's own surgeon, it included as free associates all the master surgeons of Paris, plus associates in the provinces and abroad. In reality, only the forty councilors on its permanent committee, who constituted the elite of the profession in Paris, took part in its work. After 1774, the Royal Academy of Surgery suffered from internal divisions and accomplished little.

Because a master's degree was required for students to enroll in the School of Surgery, the surgeons were considered learned professionals. The curriculum, extending over three years, included five general courses as well as specialized courses, all given in French. Instructors held official appointments, and none of the eight professors who held these posts in 1789 had taught for less than seven years. As for the school's facilities in the Academy of Surgery, they were far superior to those of the Faculty of Medicine. The superb lecture hall could seat 1,400 auditors (Fig. 3.1); the school also included a chemistry laboratory, a special lecture hall for midwives, and a classroom for practical studies reserved for the most advanced students, with four tables for dissections. A hospice with a few beds offered a dozen students clinical training under the direction of Tenon from the Academy of Sciences. All this would serve as a model for the new School of Medicine created under the Revolution.[7]

Parisian apothecaries followed in the footsteps of the surgeons in the final years of the Ancien Régime. Long associated with the greengrocers, they broke away from them in 1777, in the wake of Turgot's edict abolishing the corporations, and took the more prestigious title of "pharmacists." They were now organized into a Royal College of Pharmacy, located on the site of the former apothecary garden. The Parisian pharmacists soon obtained the right to give public courses. In this way, they emancipated themselves definitively from the Faculty of Medicine, which until then had held a monopoly over the teaching of pharmacy. The Royal College of Pharmacy supported research and offered its expertise to the state. It established chairs in chemistry, natural history, and botany and addressed its courses to a wide audience. For their part, pharmacists continued to be trained (as before) on the job, in dispensaries and laboratories scattered throughout the city.[8]

The Paris of the Trades

To go from the Latin Quarter to the heart of Paris on the Right Bank, one had to cross one of the bridges. The Pont Neuf, which traversed the western tip of the Île de la Cité, was the oldest and most beautiful bridge. After using it to cross the smaller arm of the Seine, a pedestrian like Van Marum coming from the rue Dauphine could walk along the Quai des Orfèvres on the southern side of the Île, cut across the courtyard of the Palace of Justice, and then take the next bridge, the Pont-au-Change, to reach the other side of the river. The detour was worth the effort: according to Mercier, the quay boasted "a long line of shops resplendent with the work of silversmiths."[9]

In fact, the whole western part of the Île de la Cité was filled with craftsmen who specialized in precious metals and glass, as well as in luxury goods and books.

Alongside the goldsmiths and jewelers could be found many clockmakers, mirror makers, opticians, and makers of instruments for physics. For these crafts that were at the same time manual and learned, whose clientele belonged to the intellectual and social elites, the location between the two river banks was ideal: to the north, artisans ran into the wealthy neighborhoods around the Louvre and the rue Saint-Honoré with their luxury trades; to the south, they bordered the Latin Quarter with its booksellers and engravers, of which they were an extension.

While thanks to its connections the Île de la Cité represented a major site in the geography of the capital's intellectual activities, it was also an integral part of the Paris of the trades that was the beating heart of the city. Since the Middle Ages, Paris had been home to communities of artisans, and since then it had become a great center of industry. Most of the commercial and artisanal activities were organized into guilds or corporations, placed under royal control. There were 127 of these corporations at the start of the eighteenth century, and even after Turgot's abortive reforms of 1776, there were still forty-four of them. Those known as the Six Corps—the drapers, grocers, mercers, furriers, bonnet makers, and goldsmiths—dominated all the others by their privileges and prestige. By tradition or out of convenience, many trades were concentrated in specific neighborhoods. Many remained in the city center, on the Île de la Cité and around the Hôtel de Ville, between Les Halles and the Marais, the area that had been the cradle of the Paris trades. This is why many of the guilds had their headquarters around the Place de Grève. But some activities that were reliant on the university (especially those related to books) were located on the Left Bank. Others had moved toward the periphery, for example, toward the Faubourg Saint-Antoine and the Faubourg Saint-Marcel.

The few corporations related to the liberal arts had removed themselves from the world of crafts to rise to the rank of prestigious royal institutions. Enjoying great intellectual and moral authority in their domains, they benefited from the right to offer instruction to the public. This was the case above all for artists with the Royal Academy of Painting and Sculpture and for architects with the Royal Academy of Architecture, both housed at the Louvre, as well as the Faculty of Medicine, the Royal College of Surgery, and the Royal College of Pharmacy for the health professions. At a more modest level, in 1779 the Parisian handwriting experts had formed the Academic Bureau of Writing placed under the presidency of Jean-Charles-Pierre Lenoir, the lieutenant of police.

The bureau included eighty members, twenty-four qualifiers (agrégés), twenty-four associates, and an indeterminate number of correspondents. It met four times a month in the rue Coquillière, to study the improvement of handwriting, commercial and financial calculation, and spelling. Valentin Haüy, brother of the mineralogist in the Academy of Sciences, a translator in the Ministry of Foreign

Affairs and a qualifier at the bureau, opened a school of writing and languages there. He developed his celebrated reading method for the blind by means of raised letters there. He presented it to the public at a meeting of the bureau held on November 18, 1784. The Academy of Sciences itself gave its approval a few weeks later. In 1786, the meetings of the Academic Bureau of Writing left the rue Coquillière for the King's Library, while Haüy opened a school for the blind on the rue Notre-Dame des Victoires with the support of the Philanthropic Society.[10]

Apart from these exceptional cases, the Parisian craft guilds had neither the right nor the desire to organize their own public courses to disseminate and improve the knowledge underlying their practices. For that matter, savants and administrators were quick to judge the guilds blinkered and conventional. To be sure, plenty of the descriptions of machines and processes collected by the Academy of Sciences and the *Encyclopédie* were inspired by Parisian industry, like the stocking loom described by Diderot with the aid of the workman Barrat. Yet in the eyes of the savant elite, the sciences alone could truly illuminate and improve the arts.

The Art of Chemistry and the Parisian Trades

The pharmacists Antoine-Augustin Parmentier and Antoine-Alexis Cadet de Vaux tried in this way to bring Enlightenment science to those who made bread. The free baking school, which opened with the support of Lenoir in 1780, was established close to the wheat market on the rue de la Grande Truanderie. The two pharmacists, who had conducted research on bread making, taught the courses. The *Journal de Paris*, which Cadet de Vaux had helped found, gave wide publicity to this initiative, and Mercier himself devoted a chapter to it in the *Tableau de Paris*: "Making bread from wheat is a chemical operation, which should be informed by chemists," his readers learned. Although the school survived until the end of the century, it had almost no success among Parisian bakers, who remained unconvinced of the virtues of chemistry for their art.[11]

The transformation of animal or vegetable matter through the extraction, purification, and mixing of substances, standard operations for grocers, vinegar makers, butchers, and confectioners as well as bakers, was, however, absolutely related to the art of chemistry. And chemistry was relevant to many other occupations outside food making: druggists, paint dealers, varnishers, tanners, dyers, hatters, perfumers, and goldsmiths. It would be impossible to try to cover them all, so the following example will have to suffice.

There were perhaps a thousand artisans and shopkeepers who were devoted to these activities, for the most part small-scale and hidebound. They always used the same formulas and tricks on which their trades were based, but their secrecy (or what was claimed as such) had long since been pierced by the savants. However, among these craftsmen some stood out for their wealth and their success: Maille (on the rue Saint-André-des-Arts), for example, "the Corneille of mustard" according to Grimod de la Reynière, or Fargeon (on the rue de Roulle), the queen's perfumer, or the distiller Lange, who could be found on the rue du Petit-Pont. Often inventive, purveyors to the most demanding clientele in both court and city, these masters sought the support of both the powerful and the savants. From the laboratories behind their shops in the city center or else in suburban workshops came new products and procedures, some of them original, which they would submit for the examination and approval of the academies.

I have already mentioned the jewelers, concentrated on and around the Île de la Cité. Some were true metallurgists, mastering alloys and complex platings. For example, in 1786, the goldsmith Daumy on the rue de la Verrerie, already known for his fabrication of silver-plated tableware, produced with Lavoisier's assistance the silvering for the mirror of the telescope at La Muette. The following year, another Parisian goldsmith, Marc-Antoine Janety, was the first to handle platinum successfully in bulk and to make boxes and other little luxury goods out of it.

Of all the crafts, that of the apothecaries was certainly the most closely associated with chemistry. Of course, the great majority of their laboratories were content to produce either medications from the usual pharmacopeia or specialties that were called at the time "secret remedies." However, some of the workshops were also true research and teaching laboratories, of which the most celebrated belonged to the Rouelle Pharmacy at the corner of the rue Jacob and the rue des Deux-Anges. The pharmacist-chemist Bertrand Pelletier, elected to the Academy of Sciences in 1792, took over this shop and its laboratory in 1784. The laboratories of Demachy and Mitouart, both professors at the College of Pharmacy, were located, respectively, on the rue du Bac and the rue de Beaune. The premises of Cadet de Gassicourt of the Academy of Sciences were on rue Saint-Honoré. The pharmacist Antoine-Louis Brongniart, demonstrator at the Collège of Pharmacy, had his shop on the rue de La Harpe.

On another scale entirely was Baumé's dispensary, located in the rue Coquillière, in the heart of industrial Paris, which included a vast teaching laboratory, a laboratory for the preparation of medications, and above all five laboratories for making chemical products. This veritable complex supplied all sorts of products to apothecaries' dispensaries, as well as to workshops of confectioners, vinegar makers, distillers, and milliners in Paris and beyond. In addition, they supplied chemistry equipment and even scientific books. This very prosperous enterprise had no competitors until the end of the 1770s. But in 1778, a group of capitalists established a chemical factory, known especially for its production of sulfuric acid and nitric acid, on the Seine in Javel at the western edge of the city. The following year, Baumé's former assistant, Quinquet, opened his own wholesale dispensary. Did this competition encourage Baumé, who had already made his fortune, to retire? In any event, in 1780, he sold his pharmacy to Fourcy, who carried on the business.[12]

Urban Machines

Chemistry presents a particularly remarkable case of the integration of savant activity into urban space, a topic to which we will return. Other sciences had formed ties with Parisian trades and crafts, though to a lesser extent. For example, alongside lofty analytical research and its applications to the sciences, mathematics had made a place for the commercial arithmetic of merchants and the practical geometry of builders and artisans. In addition to the writing masters at the Academic Bureau of Writing discussed above, mathematics masters offered Parisian businessmen classes in calculation and accounting. Mathematics was relevant to the arts and trades above all in the realms of geometry and mechanics. At the free Royal School of Drawing, founded by the painter Jean-Jacques Bachelier in 1766 and established since 1775 in the rue des Cordeliers in the former Saint-Côme lecture hall vacated by the surgeons, students received instruction in mathematics, along with architecture, stonecutting, and perspective, as well as practical training in drawing. The school, which was financed by a combination of private subscription and trade communities, trained many of Paris's elite craftsmen, known as "artistes," particularly in mechanics, decoration, and construction.[13]

But in the mechanical industries, Paris was far from being on a par with London, Switzerland, or even the principality of Liège. For the most part the many Parisian clockmakers were content to mount movements imported from Geneva in pretty casings, while a few specialized merchants brought over from England the best scientific instruments. However, Paris did have a small community of elite craftsmen, often of foreign origin, who were capable of producing admirable mechanical devices. A few masters grouped around the Place Dauphine established the reputation of Parisian clockmaking. Supplying the nobility and selling their wares throughout Europe, they also maintained close connections with the savant world. The marine chronometers of Le Roy and Berthoud, which were evaluated by the Academy of Sciences, were equal to the best devices of British clockmakers. Since 1775, the clockmaker Breguet, who had come from Neuchâtel and was trained in mathematics by Abbé Marie, had made multiple technical and aesthetic innovations. He produced exceptional mechanisms in his workshop on the Quai de l'Horloge.

In the locksmith's trade and metalwork, too, while run-of-the-mill work was produced in the provinces, a few Parisian craftsmen specialized in luxury goods and precision pieces, along with a few makers of mathematical instruments who were working in Paris at the time. Around the Faubourg Saint-Antoine worked the mechanics attached to the machine depository of the Hôtel de Mortagne. At the end of the Ancien Régime a handful of makers of English-style mechanical looms could be found there. But it was above all in the trades dedicated to building and decorating the palaces, townhouses, theaters, and public buildings of all kinds, booming in Paris at the time, that elite craftsmen applied themselves to invention. In general, these craftsmen remained in the shadows, however, eclipsed by the powerful contractors who circulated in high society. Yet there was at least one exception made, for the dome covering the Wheat Hall. The press celebrated the technical prowess of its modest builders: the carpenters, ironsmiths, and smelters.

Located at the heart of old Paris, the Wheat Hall was the city's belly, from which the Île-de-France's wheat and flour were sold. This utilitarian building was also a powerful political symbol of a monarchy that was expected to ensure the people's subsistence, that is, to ensure the supply of bread of good quality, in abundance, and at a fair price. Opting for free trade in grain and for technical progress, in the 1760s the royal government had supported a new "economic" milling process and the baking of bread in pans. At the same time, it undertook the construction of a new Wheat Hall that would be covered and modern. Completed in 1767, the building was a majestic ring-shaped monument. Grain and flour were sheltered under twenty-five arcades and in granaries around a circular courtyard. It soon became apparent, however, that the warehouse was too narrow. It was decided to enlarge it by covering the central courtyard with a great dome. Several plans were proposed, of which only two were presented to Lenoir. The architect Bélanger suggested an English-style solution, using iron and glass for the first time. His structure had been designed in collaboration with the ironmaster Deumier. Meanwhile the architects Legrand and Molinos opted for a structure of wood and glass that was lightweight and cheaper. This second plan was selected.[14]

The frame, constructed out of pine boards, was the work of the master joiner André Jacob Roubo, known for his treatise L'Art du Menuisier (The art of the joiner). In consultation with the architects, he had chosen for the dome the technique "à petit bois" (with short timbers), invented by renaissance architect Philibert Delorme, which had fallen into oblivion for two centuries. This type of construction called for no masonry reinforcement. The master carpenter Aldouy constructed suspended scaffolding, for which the same simple, cheap short-timber technique was used. A grand iron lantern created by the master locksmith Raguin (possibly based on an idea from the mechanic Mégnié) was to crown

FIGURE 3.2: View of the Wheat Hall.
Source: Vue de l'extérieur de la Halle au bled. Engraving by Roger after a drawing by Testard

the dome. The master smelter Tournu was charged with the external covering in an alloy of his own invention approved by the Academy of Sciences. Benjamin Franklin was consulted on the installation of a lightning rod.

The work was completed in record time. The day the dome was released from its scaffolding, the *forts des Halles* (porters of the central market) carried Roubo in triumph. When it opened in January 1783, the building was immediately met with unanimous praise. The dome of imposing size covered the courtyard without any apparent support. Twenty-five large skylights were arrayed around the lantern to let in abundant natural light (Fig. 3.2). The *Journal de Paris* devoted a long article to it, paying homage to the builders of this formidable "machine." The Wheat Hall also became a site for entertainment and tourism. "The most superb thing on earth!" exclaimed Thomas Jefferson.[15] For Arthur Young, too, the hall was the most beautiful monument in Paris. But it was also, unfortunately, a fragile monument. The dome structure proved to be very sensitive to variations in temperature. In 1803, a fire reduced it to ashes. After several years of research and investigations, the dome was reconstructed in 1812. This time the plan of the architect Bélanger was selected. The new structure, precisely calculated, was again the work of an entrepreneurial Parisian carpenter, François Brunet, who had already been quite active under the Ancien Régime. It also used the technique of Philibert Delorme that had been revived by Roubo, but as a sign of the times the wood was replaced by Le Creusot cast iron.[16]

On the Outskirts of Paris

There had always been regular communication between the capital and its immediate environs. Like all Parisians, the savants enjoyed the benefits of the fresh air of the countryside. Some of them had country houses where they spent weekends and summers.[17] Others found hospitality in high society, for example, with the d'Envilles at La Roche-Guyon, with Bochart at Nantouillet, or with Trudaine at Montigny. Above all there was Versailles, with its court and ministries, which was for some an obligatory destination.

The environs of the capital also offered many resources for naturalists. His hat jammed in his pocket so he would not lose it, Jean-Jacques Rousseau would go botanizing on the heights of Sèvres and of Romainville, armed with a magnifying glass given to him by the prince of Conti, a tin box, a spade, and a pruning knife fitted to the end of his walking stick. He was not the only one. Lamarck, Jussieu, Haüy, Desfontaines, and many others explored the woods and riverbanks that surrounded Paris in quest of specimens for their natural history collections. Professors from the Royal Botanical Garden took their students there. At the gates of the city, amateurs could find the flower gardens of the botanist Cels and the gardener Cochin near Montrouge and the nurseries of Parisian seed merchants. A little farther away, the privileged few had access to some marvelous private botanical gardens, such as that of Malesherbes in the village of the same name, of Duhamel du Monceau and his nephew Fougeroux in Vrigny, and of Le Monnier in Montreuil, near Versailles. For mineralogical fieldwork, too, the Paris region was fertile terrain. Savants such as Darcet and Jean-Honoré Robert de Paul de Lamanon liked to collect strange fossils dug from the quarries of Montmartre, while the coarse limestone of Grignon and elsewhere furnished collectors with superb shells.

The savants found more than just curiosities of nature when they ventured outside of Paris. The rich agricultural land of the Île-de-France had always supplied the capital. So it was natural that the intendant of Paris, Bertier de Sauvigny, wanted to stimulate progress in livestock raising and agronomy by calling upon Parisian savants, especially after 1780, when he was in charge, within the Finance Ministry, of overseeing the Royal Veterinary School of Alfort and the Royal Society of Medicine. In this enterprise, he relied on Daubenton and his nephew Vicq d'Azyr, the society's secretary, and Auguste Broussonnet, a young physician from Montpellier who was close to the naturalist Joseph Banks, whom he had met in London.

At the Alfort School, established since 1766 in a château near the Charenton Bridge, future blacksmiths and farriers were taught the veterinary art, principally relating to horses. On the advice of Vicq d'Azyr Bertier decided to transform this professional school into an institution of higher education. For this purpose, new chairs were given to members of the Royal Society of Medicine: to Vicq d'Azyr for comparative anatomy and Daubenton for rural economy in 1782, and the next year to Fourcroy for chemistry and botany. These chairs came with deputies and assistants. Broussonnet, under Bertier's protection, became deputy to Daubenton, who got him elected into the Academy of Sciences. The veterinary school was endowed with an experimental farm, a menagerie, cabinets of natural history and anatomy, and laboratories. Now people came from Paris to visit the institution and listen to the savants when they held public meetings. Although the latter did conduct some research at Alfort, they hardly taught there at all. After the fall of the finance minister Calonne in 1787, the government decided to withdraw its financial support. The three scientific chairs were eliminated, and the school reverted to the modest institution it had been before.

Intendant Bertier also undertook to give new life to the Agricultural Society of the Paris Region, which had been inactive since its foundation in 1761. Reformed in 1784, the society included among its members both great landowners and many savants, including Lavoisier, Buffon, and Vicq d'Azyr. Broussonnet, who had just been admitted to the Academy of Sciences in the new class of botany and agriculture, was named its permanent secretary. The society, which met every Thursday in the intendant's offices on the rue de Vendôme, had as its mission to stimulate agronomy and the progress of agriculture. In 1786, the Agricultural Society also succeeded in having an experimental farm built in the royal domain of Rambouillet, to which Abbé Tessier had a few hundred angora goats and merino sheep brought for acclimatization.

Despite all these efforts, the activities of the Agricultural Society of Paris remained limited. For its own purposes, the Finance Ministry preferred to rely on the advice of a Committee of Agriculture appointed from within and whose soul was Lavoisier. In 1788 Bertier finally succeeded in having his society made into a Royal Society whose mission extended to the whole country. Like the Royal Society of Medicine, the Royal Society of Agriculture was supposed to correspond with all the provincial societies of agriculture and establish a solid network of informants throughout the kingdom, but it barely had time to grow. Like all the other academies, it was shut down in August 1793.

The World Comes to Paris

The capital of the kingdom, by the end of the eighteenth century Paris had become a European and world-class metropolis whose role and name went well beyond the city gates. In the Republic of Letters, it represented the cosmopolitan
city par excellence, the chosen homeland of the citizen of the world, as David
Hume wrote to a friend in 1764.[18] Of course, it did not have a monopoly on uni-
versalism: in the eighteenth century, London, Amsterdam, Vienna, Cadiz, Rome,
and Constantinople (to cite just a few European examples) were also global cities
in their own way. What distinguished Paris from the rest was its intellectual vocation, its title as capital of the Enlightenment and hence as capital of science,
which no other metropolis could seriously challenge after 1750. In Paris more than
anywhere else, people, ideas, and goods from all over the world were on display
and into contact with each other. These encounters and exchanges constituted
the strength of Paris Savant.

In the Middle Ages its university had welcomed all the "nations" of Europe.
Even if its prestige had declined, Parisian masters continued to attract foreigners
who came to take public courses, both those offered by official institutions like
the Collège de France, the Royal Botanical Garden, the College of Surgery, or
the Hôtel-Dieu Hospital, and the even more numerous courses offered by private individuals in physics, chemistry, and natural history. Rich foreigners came
too, either just passing through or for a prolonged stay, at first for gambling, fine
dining, and women, but also to become cultivated. Long simply a stage on the
Grand Tour taken by the European aristocracy, the city had thus become by the
end of the eighteenth century the Mecca of cosmopolitans. Travelers stayed in
the many hotels for tourists located between the Palais-Royal and the Chaussée
d'Antin. Tourist guidebooks mentioned scientific sites like the Paris Observatory
and the Royal Botanical Garden among the monuments worth a visit.

Among all these foreigners, some were scholars and savants who had come to
meet their French colleagues. The example of Joseph Priestley is still famous in
the history of chemistry. Arriving in Paris in October 1774, in the company of his
protector and patron Lord Shelburne, he made the acquaintance of Parisian savants, especially of Lavoisier to whom he gave essential information on "dephlogisticated" air, that is to say, oxygen, whose existence he had just discovered. This
encounter was the point of departure for a long dispute between the two chemists
over the priority of the discovery of this gas, which was continued by historians
obsessed with national rivalries. Nine years later, in 1783, Charles Blagden, doctor
and secretary to the English chemist Henry Cavendish, met Lavoisier on a visit
to Paris and informed him of his patron's experiments with the decomposition of
water. Lavoisier quickly undertook his famous experiments on the analysis and
synthesis of water. Once again, a visit to Paris lay at the origin of a controversy

among historians about whether French chemists or English chemists deserved the laurels for this discovery. What is certain is that such visits were an opportunity for fruitful encounters between scientists. Thus when James Watt came to Paris in 1786, he followed with great interest Berthollet's experiments on bleaching with chlorine and, with the full knowledge of the chemist, was inspired upon his return to develop a bleaching industry in Great Britain.

Like all the savants passing through Paris, Martin van Marum received a warm welcome from his colleagues, and thanks to their letters of recommendation he was able to visit the greatest scientific collections in the capital, both public and private. Paris was like a microcosm of the natural world, housing one of the greatest concentrations of natural history objects. Among them, the collections of the Royal Botanical Garden, which had been constantly added to, were distinguished by their richness and diversity. In addition to collections of *naturalia*, Paris offered the treasures of its libraries to visiting scholars: books, manuscripts, drawings, prints, and geographical maps from all points of the globe. So much accumulated knowledge made the city a window open to the world.

Describing the great division between "Western science," universal and cumulative, and knowledge produced elsewhere (feeble, transitory, and local), the anthropologist Bruno Latour has imagined long-distance networks bringing from remote lands to the "centers of calculation" situated at the heart of European metropoles multiple kinds of data and information inscribed in the form of texts, numbers, drawings, and objects.[19] This is a powerful but misleading vision: those interested in other worlds have pointed out that it tends to reduce unfairly everything that is not "the West" to passive peripheries—in short, to terrains simply for the collection of information. Nor does the metaphor of "centers of calculation" scarcely do justice to the diversity of spaces of knowledge that characterize the metropole. In contrast to the open "field" of the anthropologist or the naturalist, the image evokes a kind of fortress in which savants worked between four walls in laboratories and cabinets closed off from the rest of the world. Yet this was by no means the case with Paris, where scientific activity was, as we have seen, deeply overlapping with the urban space.

Moreover, for an observer of Paris like the enlightened freemason Mercier, there were no walls isolating the scientific community. The wind of invention was blowing everywhere. Repeating a commonly-held idea, he thought that the sciences had developed from the observation of the arts and trades: they were born, he believed, out of the spectacle of the street, whose activity aroused curiosity and stimulated intelligence; they were, in short, the natural and necessary product of urban life. In return, these plebeian origins imposed on the savants the duty to get out of their palaces, to serve the public good, to educate the people, and in so doing to bring science back to its birthplace. Such a perspective was not unique.[20] It expressed an ambition that became increasingly strong among men of letters after 1750: to minister intellectually and morally to the public, beyond the control of political and religious authorities, even in opposition to them. Paradoxically, this emancipation took place within the same traditional social formations where new ideas fermented: the nobility, the craft guilds, and the administrative monarchy itself. The savant world, the part of the Republic of Letters that was without doubt best integrated into the absolutist state, was fully involved in this general movement, moving gradually from service to the prince and the patronage of the nobility to driving public opinion. In all innocence, science thus came to have a stake in the political culture of prerevolutionary Paris.

Notes

1. VAN MARUM, 1970.
2. MERCIER, 1782–1788, chap. 80: "Pays latin."
3. C. BOUSQUET-BRESSOLIER, "Charles-Antoine Jombert (1712–1784). Un libraire entre sciences et arts," *Bulletin du bibliophile*, no. 2, 1997, 299–333.
4. BROCKLISS, 1987, 360–390, and COMPÈRE, 2002.
5. J.-J. GARNIER, *Éclaircissements sur le Collège royal de France*, Paris, 1790. Description of the Collège de France in THIÉRY, 1787, vol. 2, 303–308. See GILLISPIE, 1980, 130–143.
6. On criticism of the Royal Society of Medicine, see "Nouveau plan de constitution pour la médecine en France," *HMSM*, 1787–1788, 1–201, also published separately.
7. GELFAND, 1980, and FRIJHOFF, 1990. Description in THIÉRY, 1787, vol. 2, 361–365.
8. G. PLANCHON, "Le Jardin des apothicaires de Paris," *Journal de pharmacie et de chimie*, 5th ser., 28 (1893), 250–258, 289–298, 342–349, and 412–416; 29 (1894), 197–212, 261–276, 326–337; 6th ser., 1 (1896), 254–263, 317–325, 353–362.
9. MERCIER, 1782–1788, chap. 187: "Quartier de la Cité."
10. J. BONZON, *La corporation des maîtres-écrivains et l'expertise en écritures sous l'Ancien régime*, Paris, 1899. On Valentin Haüy, see *Journal de Paris*, November 23, 1784.
11. KAPLAN, 1996, 77–82, and A. BIREMBAUT, "L'École gratuite de boulangerie," in TATON, 1964, 493–509. See also the *Journal de Paris*, June 11, 1780; July 4, 1782; July 23, 1782; August 6, 1782; August 9, 1782; and MERCIER, 1782–1788, chap. 631: "L'école de boulangerie."
12. DAVY, 1955.
13. A. BIREMBAUT, "Les écoles gratuites de dessin" in TATON, 1964, 441–476. On the use of the term *artiste* in the arts and trades, see BERTUCCI, 2017. In general this expression is translated as "elite craftsman" or "craftsman" depending on the context.
14. KAPLAN, 1984, 111–121.
15. Letter from Thomas Jefferson to Mrs. Cosway, October 26, 1786, in *The Works of Thomas Jefferson*, 12 vols., 1904–1905, New York, vol. 4, 203.

16. On the dome of the Hall, see the article "Menuiserie," in the *Encyclopédie méthodique, Arts et métiers mécaniques*, vol. 4, 1785, 673–676, and D. WIEBENSON, "The Two Domes of the Halle au Blé in Paris," *Art Bulletin* 55 (1973), 262–279.

17. For example, the astronomer Bailly had a house in Passy and the chemists Baumé in Ternes, Berthollet in Aulnay-sous-Bois, Sage in Meudon, and Macquer in Gressy-en-France.

18. Letter, dated September 22, 1764, from David Hume, then chargé d'affaires in Paris, to his friend G. Elliott: "I am a citizen of the world; but if I were to adopt any country, it would be that in which I live at present."

19. Bruno LATOUR, *Science in Action: How to Follow Scientists and Engineers through Society*, Harvard University Reprint, 1988, chap. 6.

20. MERCIER, 1782–1788, chap. 1: "Coup d'œil général."

4

The Encyclopédie

ON DECEMBER 14, 1759, the director of the Royal Academy of Sciences and five of his colleagues went to Briasson's scientific bookshop on the rue Saint-Jacques to examine the drawings and proofs of the engravings for the *Encyclopédie*. The architect and draughtsman Pierre Patte had just condemned a plagiarism that amounted to theft in the *Année littéraire* edited by Fréron, the sworn enemy of the philosophes: the associated publishers of the *Encyclopédie* were said to have bribed the Academy of Sciences' engraver to obtain the engraved plates that Réaumur had prepared for the Academy's *Description des arts et métiers*.[1] Patte was in a good position to know about this, since he had been charged with verifying and classifying the preparatory documents for the *Encyclopédie*'s plates. His employers had fired him, so he was taking revenge. For the associated publishers of the *Encyclopédie*—Le Breton, Briasson, David, and Durand—the affair could not have come at a worse time.

Since it was launched, the *Encyclopédie* had been a formidable success—proof, if any was necessary, that the sciences were arousing the public's passionate interest. Even if this interest was not new, the impact of this work marked a turning point, because activities long confined to the scholarly world had now erupted into the public sphere. For science, this phenomenon meant both an expansion of its institutional platform and a shift in its center of gravity. When the monarchy founded the Academy of Sciences, it had not only provided support to the savants but had also placed them at its service. It was in the name of the king that the academicians regulated and evaluated research and inventions, and it was to the various government agencies that they delivered their expertise. If the Academy also contributed to the spread of the sciences to a wider public, this mission remained secondary to that of stimulating scientific progress in the service of the state and industry. The *Encyclopédie*, by contrast, was aimed directly at educated readers, without distinction of rank or expertise. It aimed to enlighten them outside the control of the established powers, by placing at their disposal the tree of knowledge, organized alphabetically and richly illustrated. The sciences were thus no longer the exclusive concern of savants and their powerful protectors, and they became, at least in principle, everyone's concern.

This project displeased some people, however. From the start, the adversaries of the *Encyclopédie* tried in vain to interrupt its publication. The situation was significantly aggravated after the condemnation of the philosophes by the Parlement de Paris, France's highest court, in February 1759. A month later, on March 5, the pope put the *Encyclopédie* on the index of forbidden books, and on March 8 the King's Council revoked the privilege authorizing its printing and sale. By this time, the work had only reached the letter G and was missing its plates. The publishers were still hoping to save the project by publishing the latter, since only the volumes of text had been banned. In fact, the administration granted them a new privilege in September, this time to publish the *Recueil de mille planches* (Collection of a thousand plates). Unfortunately, Pierre Patte's attack was threatening to ruin everything, and this time, it was feared, without any possible recourse.

The Scandal of the Plates

The description of processes in the arts and trades was one of the most original features of the *Encyclopédie*. Yet the idea was not new, since the Academy of Sciences itself had already undertaken the publication of a vast collection bearing on the *Description des arts et métiers* (Description of the arts and trades). This project, whose origin went back to the company's founding, had taken shape in the 1690s; the academician Réamur had directed it since his election to the class of mechanics in 1711. Since then, the work of documentation had advanced considerably, but few volumes had been published. Upon his death in 1757, Réaumur had left many manuscripts and drawings and a few engravings and their copper plates to the Academy. When the scandal over the alleged theft of the plates for the benefit of the *Encyclopédie* erupted, nothing had been done yet about publishing them.

The Academy's commissioners spent three hours in Briasson's shop on December 14, 1759. They saw many drawings and engravings concerning the sciences, but almost nothing on the mechanical arts. They were also shown around forty proofs of Réaumur's plates that bore no relation to those destined for the *Encyclopédie*. The visitors were told that they were only meant to serve as models for the arrangement of the diagrams. Surprised at seeing so few plates on the mechanical arts, in particular on subjects that Réaumur had already treated, the commissioners suspected that some were being hidden from them. Still, the associated publishers assured them that there were no others in their portfolios.

Moreover, they were ready to promise verbally and in writing to submit the plates for the *Encyclopédie* to the Academy for review before their publication, and when the commissioners made their report, the Academy accepted these guarantees. Nollet and Deparcieux were charged with having all of Réaumur's copper plates cleaned and then having three copies printed from each plate for later comparison with those of the *Encyclopédie*.

To calm the concerns of subscribers, the associated publishers quickly invited via the press "those who took a side for or against the *Encyclopédie*" to go to the shops and view almost two hundred plates that had already been printed. Viewers could thus verify that they indeed had nothing to do with those of Réaumur. In fact, only those who wanted to believe were reassured. The enemies of the *Encyclopédie* demanded more specific information: Where were the preparatory documents for the mechanical arts? Why had they not been presented to the commissioners? The associated publishers had every reason to fear new revelations from Pierre Patte. To head off this possibility, they wasted no time having all their plates on the arts and trades inspected by the Academy, this time without exceptions.

At their request, this verification was in fact done very quickly and with the greatest discretion. Unbeknownst to their colleagues and with the agreement of Malesherbes, president of the Academy and the royal official in charge of the book trade administration, Nollet and Deparcieux, accompanied by Morand, went to the shop of Le Breton's, the main publisher, on January 16, 1760. Almost six hundred plates, engraved or drawn, concerning 130 arts, were shown to them. After drawing up a detailed inventory, they certified that they had recognized nothing as having been copied from Réaumur's plates. This was precisely the documentation that the associated publishers of the *Encyclopédie* needed.

Two weeks later, the *Annales littéraires* published a new letter from Pierre Patte, in which the draftsman called the associated publishers' bluff.[2] Unaware of the secret visit on January 16 to Le Breton, he focused on the earlier visit to Briasson on December 14. Patte breezily claimed that anyone with Réaumur's engravings in hand would only have to run through the descriptions in the seven volumes of the *Encyclopédie* that were already published to recognize the same views, the same diagrams, the same processes, and even the same letters in the reference keys. Very well informed, he cited the examples of *Needle Making, Slate Tile Making, Nail Making, Pin Making,* and *Large Forges.* Since the associated publishers had promised to present their plates on the arts and trades before publication, it only remained, he concluded ironically, for them to redraw them and try to make the new drawings accord with the descriptions already made. This argument would have been devastating if Nollet and Deparcieux had not already granted the associated publishers the certificate that officially recognized their innocence.

The new article produced turmoil in the Academy of Sciences. One might imagine that the naturalists Brisson and Guettard, who had not been included in the January 16 visit after having participated in that of December 14, might feel betrayed by their colleagues. One of them (no doubt Brisson) accused them of having compromised the honor and interests of the Academy by hastily granting a certificate to the *Encyclopédie*. The assembly had to be calmed down: on February 20 Malesherbes, who was pulling the strings, proposed yet another visit of all the commissioners to Le Breton's bookshop, which took place two days later and went quite badly. "They are scratching out each other's eyes out at the Academy. Yesterday they called each other by all the insults used in the market," Diderot wrote to Sophie Volland, adding: "I don't know what they could have done today."[3]

But we do know, thanks to the minutes: on February23, Nollet and Deparcieux, protesting the attacks aimed at them and draping themselves in their dignity, reproduced in their report to the Academy the certificate they had granted on January 16. Brisson and Guettard skipped the meeting. The former kept quiet; Guettard alone sent a letter to the president, in which he finally came around to the view of his colleagues. He had compared the bookshops' drawings with Réaumur's engravings that bore on the same arts. From what he had seen, he could confirm that there was no plagiarism. In any case, he added, none of this really mattered. Didn't many of the drawings treat arts that Réaumur had not even tackled? Thus Guettard too exonerated the associated publishers. The company ordered its secretary to communicate these findings to no one and to keep both the letter and the report private. The affair needed to remain internal. The controversy was over and the *Encyclopédie* was saved, once again.[4]

A Project of the Parisian Book Trade

The *Encyclopédie* had been launched ten years earlier, in October 1750. A prospectus written by Diderot had presented the project to subscribers. The objective, he explained, was to "form a general picture of the efforts of the human spirit in all fields and across the centuries." The work would include ten folio volumes, eight of text and two of plates (numbering six hundred in total). Publication would extend from June 1751 to December 1754 at the rate of one text volume every six months; the two volumes of plates were meant to appear with the last volume of text. The subscription price was fixed at 280 livres, of which a deposit of 60 livres was to be paid before May 1, 1751. For those who did not subscribe, the cost would be more expensive by a third, 372 livres. When the subscription closed on April 30, 1,200 subscribers had responded to the appeal. Barely underway, the work was already a success.

In truth, before it became an intellectual adventure, the *Encyclopédie* was a commercial enterprise. The initial idea came from André-François Le Breton, one of the principal printer-booksellers of Paris; through his mother he belonged to the D'Houry dynasty, a powerful family in the Paris book trade. In 1700 his grandfather, Laurent d'Houry, had created the *Almanach Royal*, which was continued by his widow. This administrative annual had made their fortune, while assuring them their entrée into government circles. The young André-François gained his first instruction in the nitty-gritty of his future career under the supervision of his grandmother. In 1733 he was accepted into the guild as a master bookseller, then named printer to the king in 1746. He had taken over the family shop *Le Saint-Esprit*, at the foot of the rue de la Harpe, across from the rue Saint-Séverin. Attached to it was the print shop, where the *Encyclopédie* would be printed from start to finish.

Le Breton, whose time was monopolized by the *Almanach Royal*, had at first published only trifles. Not until 1744, after he had obtained the sole privilege for this valuable periodical, did he truly get into publishing. A first project, a translation by a certain Sellius of a book by the German philosopher Christian Wolff, was quickly abandoned. The same Sellius came back to him in January 1745 with a more attractive proposition: the translation by John Mills, an Englishman living in Paris, of Chambers's *Cyclopaedia*, which had been published in London in 1728. Le Breton quickly sensed a good deal: this encyclopedic dictionary of the arts and sciences in two folio volumes, organized alphabetically with cross references between the articles, had already gone through several editions in England. It could also find a readership France, and so Le Breton signed a contract with Mills and, while he was at it, secured a privilege to print the work. The prospectus announced a volume with 120 plates, or four times the plates of the original English edition. But the agreement between the publisher and the translator did not last long: by August, Le Breton had abruptly broken his contract with Mills, but without giving up the idea of publishing the translation. Looking to share the costs, he turned to other partners, more experienced than he was with this sort of publication: the booksellers Briasson, David, and Durand.

These three colleagues, whose shops were just a short walk from Le Breton's, had been in the book trade for a long time. Antoine Briasson, one of Paris's wealthiest booksellers, had already published over a hundred books, from novels to scholarly works of science and medicine. Michel-Antoine David, called David the Elder, whose bookshop À la Plume d'or was located like Briasson's on the rue Saint-Jacques, was d'Alembert's publisher. In 1743 he published his *Traité de dynamique* (Treatise on dynamics.) An enlightened man, considered to be close to the philosophes, he had set up a cabinet of physics for his own enjoyment. Finally, Laurent Durand, well-known as a scientific publisher, had published Clairaut and Guettard, but also Condillac and Diderot. His bookshop Au Griffon was on the same street. The three booksellers had already collaborated on the publication in 1744 of the French translation of Robert James's medical dictionary in six volumes. They had used Le Breton as their printer.

So all the players in this little world knew each other quite well. It must be said that the Parisian community of booksellers and printers, placed under the aegis of the university, was not huge. After Colbert's reorganization it was closed to newcomers. In addition, its members intermarried so effectively that the shops and workshops remained in the same families. Thus around 1750, the community had barely 250 members, including simple booksellers, booksellers-printers, binders, and type founders, all of whom were required to reside in the Latin Quarter. The royal government itself was primarily responsible for this Malthusianism. Concerned with controlling the production and diffusion of printed works, it imposed a reduction in the number of print shops, limiting them to thirty-six in the capital, and encouraged the concentration of the book trade in a few Parisian hands. These families profited from their proximity to the central government to obtain a de facto monopoly on privileges and prior authorizations. As a consequence, provincial booksellers and printers, once flourishing, gradually declined. In Paris, the syndics of the guild collaborated with the police and the officials controlling the book trade to suppress the trade in illicit or counterfeit books that had come to compete with authorized publications.

Allied with the state and rather conformist, Parisian booksellers and printers for the most part produced works for the clergy, the government, and bourgeois office holders, which is to say, principally books of piety, education, and erudition. However, this kind of output was responding less and less to the aspirations of a reading public that sought novelties, "frivolities," and "works of philosophy" that were often banned by the censors. This was the ever-expanding realm of prohibited books, produced either abroad or clandestinely in France and sold under the counter in Paris in the backrooms of book dealers. Plenty of booksellers in the Latin Quarter had no hesitation about participating discretely in this forbidden but lucrative market.

But even within authorized book production, profitable operations were still possible. Government publications offered opportunities, as we have seen with the *Almanach Royal*. Dictionaries and scientific books were profitable too: the Coignard dynasty of Parisian booksellers-printers had made their fortune by publishing Moreri's historical dictionary for half a century. In a scientific sector that was constantly expanding, there were real bestsellers alongside academic and educational books. Fontenelle's *Entretiens sur la pluralité des mondes* (Conversations on the plurality of worlds), published for the first time in 1686, was regularly reissued by Brunet, a bookseller-printer at the palace. Also in the form of a dialogue, the *Entretiens physiques d'Ariste et d'Eudoxe* (Conversations on physics between Ariste and Eudoxe) by Father Noel Regnault, professor at the College Louis le-Grand, offered worldly readers a sort of encyclopedia of Jesuit science. The book had been published in Paris in 1729 and was reissued several times before it appeared in the catalogues of booksellers David the Elder and Laurent Durand in 1745. In the same genre, *Le Spectacle de la nature* by Abbé Pluche, published in 1732, surpassed all the others, with multiple editions—to the profit of its bookseller, Briasson's mother-in-law, the widow Estienne.[5]

Well versed in these matters, booksellers Briasson, David, and Durand knew when they decided to collaborate with Le Breton on the *Encyclopédie*, that the project would yield major profits. Always on the lookout for a good deal, one of them, Laurent Durand, later became Buffon's publisher for the *Histoire naturelle* (Natural history), another great Enlightenment venture in scientific publishing. The four booksellers signed the contract of association for publishing the *Encyclopédie* on October 18, 1745. Le Breton, who was bearing the burden of the project with his privilege, would keep half the profits and be reimbursed for the expenses he had already committed to the project. He would print the work, and the other associates would share equally the other half of the profits. Briasson, with the most experience, was named the association's director, in charge of the accounts and the stock.

Everyone believed in the success of the project. In the beginning, however, there was no question of financial returns: for the next three years, the associated publishers had to finance the preparation of the encyclopedia at their own expense. By January 1750, expenses had already risen to more than sixty thousand livres, but the four associates remained united to the end, despite the crises. It's easy to understand why: as an Enlightenment best seller, the *Encyclopédie* would prove to be one of the most profitable publishing ventures of the eighteenth-century book trade.[6]

The Editors

By the autumn of 1745, then, the new associates had a royal privilege and the financial and material means, but they were still lacking the essential ingredient: an editorial team for the *Encyclopédie*. They turned first to a professor of Greek and Latin philosophy at the Royal College, who was also an adjunct member of the Academy of Sciences in geometry, Abbé de Gua de Malves. This beanpole of a man, known for his mathematical work, was an eccentric and shady character. He had just had some problems with his colleagues at the Academy, which had led him to request the title of veteran pensioner (emeritus). For the *Encyclopédie*, his work consisted of preparing the publication based on the translation of the *Cyclopaedia* that the associated publishers had acquired from Mills, beefed up with Harris's translation of the *Lexicon Technicum*, purchased from Sellius, as well as some unpublished articles written specially for the French version.

De Gua started work at the end of 1745. In June of the following year, he signed a formal agreement with the associated publishers that laid out exactly what he was supposed to do, but we know little about his actual contribution. He apparently played a role in the metamorphosis of the initial project from a simple improved translation of the *Cyclopaedia* into an entirely new work, but as a result of some kind of disagreement with the associated publishers, he resigned during the summer of 1747. To replace him as chief editors, the booksellers contracted with Denis Diderot and Jean d'Alembert on October 16—and thus the intellectual adventure of the *Encyclopédie* really began.

When Denis Diderot assumed the editorship of the *Encyclopédie*, he had just turned thirty-four. Born in Langres in the province of Champagne, where his father was a master cutler, he went to Paris in 1728 or 1729 to finish his collegiate studies and pursue theology. After three years at the Sorbonne, he had decided that he preferred bohemia first to an ecclesiastical career and then to the law. He frequented public lectures, cafés, and libraries; learned mathematics and English; and was already doing some writing to put food on the table, about which we know nothing. In this way he began to make himself known to Parisian booksellers. In 1740 Briasson commissioned him to do a translation of a *History of Greece* that had been published in London. In 1744, Diderot, who had married in the meantime, again obtained a commission from Briasson (in conjunction with David and Durand), but this time a more important one: a translation of James's medical dictionary, which occupied him until his nomination as head of the *Encyclopédie*.

The young Diderot was more than a paid hack of the publishers. He was passionate about the sciences and philosophy, especially those coming from England. He hung out with Jean-Jacques Rousseau, who had arrived in Paris as an unknown in 1742—for a while the two men were inseparable—and through him he met Étienne Bonnot de Condillac. Anonymously, he translated John Locke's student Shaftesbury, before publishing his first book, *Pensées philosophiques* (Philosophical thoughts) in 1746. The book was published clandestinely with a false imprint of The Hague, but was actually printed by Laurent Durand in Paris. By this time, without having published anything significant, Diderot had become someone who was appreciated by Parisian booksellers as capable of taking on all kinds of publishing jobs. De Gua himself called upon his talents by asking him to check the translation of the *Cyclopaedia*. When the abbé resigned, Diderot naturally appeared to the associated publishers to be the right man for the job. Chancellor d'Aguesseau personally approved his appointment.

The other editor of the *Encyclopédie*, Jean Le Rond d'Alembert, was quite different. A bit younger than Diderot and Parisian by birth, he was the illegitimate son of the well-known *salonnière* Madame de Tencin and the Chevalier Destouches, director general of artillery, who saw to his education. Abandoned on the steps of the church of Saint-Jean-le-Rond near Notre Dame as an infant, he was raised by the wife of a glazier in the Marais. After an excellent education at the Collège Mazarin, he dedicated himself to mathematics. Endowed with exceptional talent, he entered the Academy of Sciences at age twenty-four and did so well that by 1747 he had already made a name for himself in the Republic of Letters. He had written a few articles on mathematics for de Gua the previous year. The associated publishers finally chose him as co-editor because unlike Diderot he represented an incontestable scientific guarantor of the *Encyclopédie*.

The contrast between the two editors did not stop there. Diderot was steadfast, generous, and without airs and graces. Interested in everything, he liked nothing better than to discuss and share ideas. D'Alembert, by contrast, was elusive. A secretive and sensitive man, supremely intelligent, he lived a double life and was distrustful of everything. Nevertheless, the understanding between the two editors was good, at least in the beginning. They complemented each other well. Diderot was responsible for the coordination of the whole project, including relations with all the contributors, writers, artists, and engravers and checking and revising their contributions. But he also wrote innumerable articles, especially everything concerning the arts and trades. D'Alembert was responsible for mathematics and was also supposed to contribute various general articles. In literary and official circles, he was the one who would represent the *Encyclopédie*, for which he wrote and signed the *discours préliminaire* (introduction).

For Diderot, this was the start of a hellish life because very soon, following the Abbé de Gua, the two editors (especially Diderot, if truth be told) imposed on the associated publishers a complete reorientation of the initial project—to which they consented. Gone was the translation of the *Cyclopaedia*, even augmented, as called for in the new privilege granted in 1748. Instead, the *Encyclopédie* would be a work of great scope and entirely new, as the prospectus of 1750 would announce. This would require assembling many authors and giving them assignments. Diderot worked like a demon: writing, calling on people, revising, going back and forth from his new domicile in the rue de la Vieille-Estrapade to the headquarters of the *Encyclopédie* in Le Breton's shop, and rushing around Paris in all directions. This did not prevent him from doing other things, however, for it was during these years of preparation that he became famous.

Already after the publication of some papers on mathematics, in 1748, Diderot found himself on the threshold of the Academy of Sciences: his candidacy for a place as adjunct in February 1749 narrowly failed. But he attracted the attention of the public and the authorities above all because of a scandal. His *Lettre sur les aveugles* (Letter on the blind), published anonymously and distributed clandestinely, had the misfortune of displeasing the authorities: not only did Diderot attack Réaumur for having prevented him from witnessing an operation he had carried out to remove a patient's cataracts, but he also indulged in materialist arguments that were contrary to religion. Interrogated by the police, his bookseller, Durand, must have decided to reveal the author's name, which earned Diderot a three-month stay in Vincennes prison. But persecution of the unfortunate man aroused the sympathy of none other than Voltaire. In a stroke "Socrates-Diderot" became a martyr of philosophy. It also threw the associated publishers into a panic, and they loudly demanded the release of their director, who had become the irreplaceable conductor of the *Encyclopédie*.

Finally, after so many fears and so much effort, the subscription campaign was launched. When the prospectus appeared in October 1750, the work was very far from complete, but this did not prevent Diderot from conjuring up via a little white lie a work that was almost done, with a complete set of manuscripts and drawings. The first volume appeared as planned in June 1751. It opened with a "preliminary discourse by the editors" written by d'Alembert. "The Encyclopedia which we are presenting to the public is, as its title declares, the work of a society of men of letters," it began. The names of Diderot and d'Alembert, with their roles and academic titles, were prominent on the title page: for while d'Alembert already belonged to the Academy of Sciences, both men had also just been named members of the Berlin Academy of Sciences and Letters.

Making the Encyclopédie

The *Encyclopédie* was above all a Parisian venture (see map 4.1). Voltaire reminded Diderot of this when the editor became discouraged and thought of moving to Berlin: "It is only in Paris that you can finish your great project."[7] For this collective work, Diderot and d'Alembert called upon many collaborators. However, they assigned themselves a good share of the articles. Diderot, who often signed his with an asterisk, wrote about five thousand of them, across all fields but especially history and philosophy. Some articles were simple filling, which boiled down to mere compilations (which is to say skillful plagiarisms) of other sources, but others were more developed and original. In addition, Diderot revised innumerable contributions, reworking, simplifying, or adding to them. In this often fastidious work, he greatly benefited from the

A: Royal Academy of Sciences at the Louvre. B: King's Library.
P. Parliament of Paris.
  1 : Diderot's domicile residence, rue de l'Estrapade (1747–1754).
  2 : Diderot's domicile residence, rue Taranne (1754–1784).
  3 : Salon of the baron d'Holbach, rue Saint-Nicaise (1749–1753).
  4 : Salon of the baron d'Holbach, rue Royale-Saint-Roch (1753–1789).
  5 : Salon of Madame de Geoffrin, rue Saint-Honoré, in front of the Capucins (1749–1777).
  6 : Malesherbes' domicile residence, rue Neuve-des-Petits-Champs, near the rue de la Feuillade.
  7 : Le Breton's bookshop and printing house, rue de la Harpe, in front of the rue Saint-Séverin.
  8 : Briasson's bookshop « à la Science », rue Saint-Jacques, near the rue Saint-Séverin.
  9 : Durand's bookshop « au Griffon », rue du Foin in front of the gate of the Mathurins.
10 : David's bookshop « à la Plume d'or », rue des Mathurins in front of the gate of the Mathurins.
11 : Richomme's copper plate printing house, rue du Foin.
12 : Engraving workshop, rue Saint-Thomas.

MAP 4.1: Paris of the *Encyclopédia*

resources of the King's Library thanks to the librarian, his friend Abbé Sallier. Meanwhile, d'Alembert, who signed himself (*O*), concentrated essentially on articles about mathematics and physics. But he also wrote some general articles, like "College" in volume 3 and "Geneva" in volume 7, both of which caused scandals. Other authors were major contributors to the work. The most prolific by far was Chevalier Louis de Jaucourt, who produced more than seventeen thousand articles. Brought in at first to write a few entries that appeared in volume 2, by the end Jaucourt was the main author and compiler, surpassing by a wide margin Diderot himself. He supplied—by himself—more than 40 percent of the articles in the last volumes.

In total, historians have identified with certainty more than 140 contributors to the *Encyclopédie*. Some were famous, like Montesquieu and Voltaire, who collaborated on the article "Taste," and Jean-Jacques Rousseau, who contributed articles on music before he broke with Diderot. But many others were completely unknown or almost so. Diderot and d'Alembert had recruited them from everywhere from all sorts of backgrounds, but the great majority lived in Paris. This was particularly the case with nineteen of the twenty-four authors, who had more than a hundred articles to their credit. In addition to d'Alembert, Venel, who had come to Paris from Montpellier to take the pharmacist Rouelle's courses and was recommended to the editors by Malesherbes, dealt with chemistry; Daubenton, Buffon's collaborator at the Royal Botanical Garden, covered natural history; the Baron d'Holbach, philosophe and friend of Diderot, covered mineralogy; and Louis-Guillaume Le Monnier and Jean-Baptiste Le Roy, both colleagues of d'Alembert at the Academy, covered electricity and mechanics.

Diderot, as we know, kept for himself the description of arts and trades, which occupied him a great deal in the first years. For this he conducted a veritable investigation: he wrote a series of articles himself, but as much as he could he called upon people in the trades for help, usually Parisians. For example, he recruited the clockmakers Berthoud and Romilly for the articles on horology, the jeweler Magimel and the brewer Longchamp for articles on their respective crafts, and the architect Lucotte for the building trades. Diderot enlisted the bookseller Le Breton himself for the article on ink and his compositor Brullé for the article on printing. The cutler Foucou, a friend of his father, with whom Diderot had lodged for a while on the rue Mouffetard, helped him with steel. The workman Barrat advised him on the article "Stocking Loom," in which Diderot gave a very detailed description of a trade that was changing. At first, he had been satisfied with consulting a series of old plates in the King's Library that had been engraved by order of Colbert. But reading a paper especially written for him by Barrat, followed by a visit to his workshop in the Faubourg Saint-Antoine, enabled Diderot to describe the mechanism much more precisely.

In the *Encyclopédie*, the drawings were just as important as the texts, especially when it came to the arts and trades. However, subscribers had to wait ten years before receiving the first volume of plates in 1762. This was not because Diderot had neglected his task. Since his appointment as editor in 1747, he had assembled a mass of drawings and engravings to which the articles published in the first volumes made direct reference. Nor was he held back by scruples, pillaging illustrations from already published books and even (as we have seen) from unpublished documents gathered by Réaumur for the Academy of Sciences' *Description des arts et métiers*. It is true that other entrepreneurs of dictionaries and treatises acted no differently in an era when literary property did not yet exist. But Patte's denunciation and the subsequent scandal in 1760 obliged Diderot to revise completely his plans for publishing the volumes of plates. He relaunched his investigations, this time extending them to the provinces, and commissioned drawings and engraving that were entirely new.

Happily, Diderot was aided in this work by a draftsman of talent, Louis Goussier. Hired in 1747, he worked for the *Encyclopédie* until the end and made a significant contribution to the success of the project. Active and intelligent, he visited factories and workshops, in Paris and in the provinces, and made innumerable on-the-spot sketches. In addition to writing several technical articles, he executed more than nine hundred plates for the *Encyclopédie*.

The engraving of the plates involved many artists and craftsmen. At the time there were almost two hundred engravers in Paris, mostly clustered in the Latin Quarter, where they worked for booksellers and print dealers. The associated publishers started to worry about the production of the volumes of plates in 1757, that is, two years after the date planned for their publication at the time of the subscription campaign. It is true that in the meantime the number of volumes of text had considerably increased. As we have seen, coordination of the work was entrusted to the architect and draftsman Pierre Patte, who would accuse the editors of plagiarism two years later. The great majority of plates were engraved in a workshop on rue Saint-Thomas near the Porte Saint-Jacques.[8]

The *Encyclopédie* was printed in the Latin Quarter too. All the text volumes almost certainly came off Le Breton's presses on rue de la Harpe (Fig. 4.1). Nothing changed when the work was banned in 1759. "It is being printed and I have the proofs here before me. But shhh!" whispered Diderot in 1762 in a letter to Voltaire.[9] Thus, the government seemed to be entirely ignorant of a production that involved the labor of fifty workers, not to mention tons of paper, blank and printed, that was going in and out of the city with impunity. As for the volumes of plates, they were printed entirely legally a little farther away on the rue de Foin by the small copper-plate printer Richomme. We can only suppose that these workshops served as models for the *Encyclopédie* plates on printing.

FIGURE 4.1: Printing workshop, *Encyclopédie*, Plates, vol. 6, 1769, Printing in characters, plate 14, (*The Printing Operation*). Engraving by Robert Bénard after a drawing by Louis Goussier.

The Battle over the Encyclopédie

If the *Encyclopédie* was the publication of the century, it was because it aroused the passions of contemporaries, both partisans and adversaries. No one could have predicted that this would happen at the outset. Basically, the *Encyclopédie* was just a dictionary, no different from so many others. Moreover, the government had found no difficulty in authorizing its publication. When the first volume appeared in 1751, the editors prudently added an overblown dedication to the minister and secretary of state for war, the Count d'Argenson, who was also in charge of the Paris police. In this latter capacity, d'Argenson had sent Diderot to the Vincennes prison for his *Lettre sur les aveugles* (Letter on the blind). But beneath his brusque exterior, the count, who also oversaw the academies, was an intelligent man. As long as he was minister he protected the enterprise discretely, and his disgrace in 1757 was a real loss for the *Encyclopédie*.

The work benefited from an even more effective protector in the person of Malesherbes, director of the Book trade administration (direction de la Librairie). Member of a great Parisian parliamentary family, the Lamoignons, and son of the Keeper of the Seals, he was an enlightened administrator, and even better, a friend of the philosophes and a lover of the sciences with a predilection for botany. He entered the Academy of Sciences as an honorary member in March 1750 and took over the direction of the book trade administration at the end of the same year,

a few weeks after the launch of the *Encyclopédie*. As the royal official charged
with overseeing publishing in the capital, Malesherbes was responsible for censorship and the regulation of the trade. It would be an understatement to say that
he exercised his authority with tact and discernment. Without him, the work
undoubtedly would have been quickly banned, and in any case could not have
been carried out in Paris, where almost immediately the *Encyclopédie* had raised a
chorus of indignation. The devout faction, powerful both at court and in the city,
denounced the impiety of the work and the impudence of its authors. The Jesuits
proved to be the most tenacious adversaries.

The first volume, published in 1751, aroused few hostile reactions. It was the
second volume that lit the powder keg the following January. The archbishop of
Paris having solemnly condemned the book because of an article on theology,
the King's Council banned the two volumes already published. All seemed lost.
A rumor circulated that the Jesuits were going to take over the project. The police were preparing to confiscate the manuscripts. A frightened d'Alembert
considered withdrawing from the editorship of the *Encyclopédie*. Malesherbes
worked to save the project; warning Diderot that the police were coming, he hid
all the documents in his home before they could be seized. At the same time,
he made sure that the *Encyclopédie* held on to its privilege. The king's mistress,
Madame de Pompadour, was able to reassure the editors personally. The prohibi-
tion was quite simply forgotten and they could continue as before. All they had
to do was be a little more prudent in the future with respect to theology.

Far from harming the *Encyclopédie*, of course, the attack gave it a kind of publicity they had not dreamed of. The number of subscribers rose. Meanwhile, a
parti encyclopédique, the party of the philosophes, was organized—though
this should not be taken to mean an organization properly speaking. At most
we can speak of a network of influence, uniting at the heart of the Republic of
Letters those who defended Enlightenment, tolerance, and a critical spirit and
who recognized themselves in the project of the *Encyclopédie*. Its most celebrated
representative, Voltaire, was first in Berlin, at the court of the philosopher-king
Frederick, then at Les Délices, his property near Geneva, but the center of the
party was incontestably on the banks of the Seine. There, despite the hostility of
the king and the court, it could count on powerful support among the high nobility and even within the government.

In the eighteenth century, high society in Paris was organized around a few
salons. Their intellectual role has been much studied, producing an abundant
but uneven literature. A salon host, or more often a hostess, received visitors at
home, bringing together the social elite, artists, and men of letters. Everything
was discussed, including sometimes philosophy and the sciences. Some of these
bureaux d'esprit definitely welcomed the Encyclopedists. D'Alembert himself,

when not in his cubbyhole on the rue Michel-le-Comte, liked to spend his evenings sociably. As early as 1743 he had been introduced by the mathematician Maupertuis into the salon of Madame du Deffand on the rue des Quatre fils, a stone's throw from his apartment. He followed his patron when she moved to the Convent of the Daughters of Saint-Joseph in the Faubourg Saint-Germain, and starting in 1749 he also became a habitué of the salon of Madame Geoffrin on the rue Saint-Honoré.

Until 1750, Diderot, who was still merely an obscure writer, was unknown to high society and had to be content with cafés. If the *Encyclopédie* made him an important man, he was seen hardly more often in the salons. He went neither to the Wednesdays of Madame Geoffrin nor to the evenings of Madame du Deffand, who admitted that she "had nothing in common with him." The only salon he attended—and he did so assiduously—was that of Baron d'Holbach. In truth, rather than a salon for the social elite, the Baron kept an open table on Thursdays and Sundays for his close friends. Born in Germany and settling in Paris, Paul Thiry, known as the Baron d'Holbach, adopted French nationality in 1749. Having studied at Leyden, he was well versed in mineralogy and chemistry, but it was probably music that connected him to the Encyclopedists. Like Rousseau and Diderot, he was truly passionate about Italian opera. He contributed anonymously to the second volume of the *Encyclopédie* and supplied in total more than four hundred articles.

D'Holbach began to receive friends at his home on the rue Saint-Nicaise near the Carrousel around 1750. Having inherited in 1754 the vast fortune of an uncle who had adopted him, he then took the title of baron and in 1759 settled into a magnificent residence on the rue royale Saint-Roch. Whereas d'Alembert preferred to avoid them, Diderot loved the small intimate Thursday gatherings. "This is where you find the true cosmopolitans," he noted.[10] In the summer, d'Holbach invited his closest friends to come to his château de Grandval in Sucy-en-Brie, and Diderot was often included in the party. The "d'Holbach coterie," so-called by Jean-Jacques Rousseau, who quickly distanced himself from it, professed unvarnished atheism and materialism and wholeheartedly supported the *Encyclopédie*. This support was decisive during the difficult years that followed the withdrawal of d'Alembert in 1759, when new collaborators had to be found.

Until 1757, the situation was very favorable to the *Encyclopédie*. In 1753, d'Alembert had written a preface for the third volume in which he was confident enough to attack its adversaries. The following year, at the end of a noisy campaign led by his protector Madame du Deffand, he entered the *Académie Française*. All the philosophes triumphed along with him. Impressed, Voltaire personally bowed before the power of these "gentlemen of the *Encyclopédie*": "These are the lords of the greatest land on earth. I hope they always cultivate it in complete freedom;

they are made to illuminate the world boldly and to crush their enemies."[11] Of course, the project was way behind schedule, but by the end of 1776, six volumes had already appeared and subscribers were flocking to it.

The assassination attempt on January 5, 1757, marked a turning point. A deranged man named Damiens had tried to kill the king. A few weeks later, d'Argenson was dismissed, although we cannot say whether this abrupt downfall was linked to the preceding event. In any case, the king and his government were frightened and sought to find more important plotters behind Damiens's attack. Denouncing "the unbridled license of writings spreading throughout the kingdom that tend to attack religion, upset minds, and threaten authority," the King's Council issued several decrees against the production and dissemination of clandestine books. At the same time, the attacks on the philosophes increased: they were threatened, they were vilified—worse, they were ridiculed by being called *cacouacs*, a supposedly evil and venomous species inhabiting the forty-eighth parallel. When the seventh volume of the *Encyclopédie* appeared in November 1747, the furor increased. The pastors, indignant about d'Alembert's article on Geneva, now joined forces with the priests.

All of this was nothing, though, compared with what was to come. In July 1758, Claude-Adrien Helvétius, a tax farmer turned philosopher, published anonymously with Durand, one of the publishers of the *Encyclopédie*, a book titled *De l'Esprit* (On the mind). In it he calmly professed a radical materialism, criticizing the despotism of religion and advocating happiness through education. The work had passed the censor with no problem, but when an alarmed Malesherbes saw what was in it, he revoked the privilege. Helvétius had to recant, but it was too late to stifle the scandal. The adversaries of the philosophes had found the pretext they were waiting for, and fury ensued.

Not surprisingly, the archbishop of Paris solemnly condemned the book, and then it was the turn of the Parlement de Paris to weigh in. Lumping them all together, the prosecutor general Joly de Fleury added to the indictment seven other philosophical books—as well as the *Encyclopédie*. Little did it matter to him that Helvétius had not written a single article. The Parlement announced the sentence on February 6, 1759: *De l'Esprit* would be burned at the foot of the grand staircase of the Palace of Justice. As for the *Encyclopédie*, its sale was suspended. This measure seemed benign, but it led to the final blow: one month later, the King's Council revoked its privilege, which amounted to an interdiction, pure and simple. In July, the council issued a new decree that required the associated publishers to reimburse each subscriber seventy-two livres in damages, since the work was far from complete. This was really an attempt to sink the project. The condemnation of the *Encyclopédie* by Pope Clément XIII in September capped off the affair.

D'Alembert, beside himself, resigned as editor of the *Encyclopédie* when the privilege was revoked, but Diderot, supported by Baron d'Holbach, held fast and took up his pen to defend his work. As for the associated publishers, threatened with ruin, they could only hope for a miracle. Again, it was Malesherbes who found a way of parrying: a new privilege, this time for the publication by subscription of the volumes of plates. The new subscribers would have to pay seventy-two livres when they signed up and another seventy-two livres for each of the four volumes as they appeared. Existing subscribers to the *Encyclopédie* benefitted from special terms: nothing to pay in the beginning and only twenty-eight livres per volume. Then Patte's attack almost sank the venture once again. Thwarting that stroke, as we have seen, required the goodwill of the Academy of Sciences, fortunately presided over that year by Malesherbes himself.

Victory

The phoenix rose again from its ashes. It is true that from the start the *Encyclopédie*'s eventual success was never in doubt. Remember that when the first volume appeared, there were more than a thousand subscribers. Less than a year later, there were already two thousand, and ultimately their number rose to more than four thousand. Nothing discouraged them, neither delays in publication nor the increase in the number of volumes from eight to seventeen for the text and from two to eleven for the plates. When the new subscription replaced the old one in 1759, nobody asked to be reimbursed. Publication of the volumes of plates, on which everything remained to be done, would take ten more years, from 1762 to 1772. As for the ten volumes of text needed to reach the letter Z, all were published in 1765, under the imprimatur of a Swiss publisher in Neuchâtel. Actually, Le Breton had printed the pages clandestinely over the course of the preceding years and then sent them to a warehouse in the provinces, no doubt in Trévoux near Lyon.

The list of subscribers has been lost, but we know that it was comprised of an educated and well-off elite: some bourgeois, but also aristocrats and clerics. The *Encyclopédie* was a very expensive book. Each folio volume, printed with great care on high-quality paper, cost a great deal, twenty-five livres at the start for those who had not subscribed and up to seventy-two livres for volume 6 of the plates. For the first subscribers, the cost of the complete set, unbound, rose to 850 livres—a considerable sum.

For Le Breton and his associates, the *Encyclopédie* represented a magnificent financial operation. It is estimated that the revenue reached 3.5 million livres for one million livres invested by the end of 1767, which made a return on investment comparable to the most profitable industries, such as printed cottons.

The authors profited much less from this success: some had given their articles without asking for payment; others had been paid by the page. Diderot himself was employed by the associated publishers. After a difficult start, he signed a more favorable agreement, thanks to which he could move with his family to a spacious apartment on the rue Taranne near Saint-Germain-des-Prés. He later negotiated better conditions, which allowed him to enjoy a comfortable existence. But after fifteen years of relentless work, he was far from having made a fortune.

Even more than a commercial and financial success, the *Encyclopédie* represented the triumph of Enlightenment. In its nearly 72,000 articles and 2,500 plates could be found all the knowledge of the era. No other work could rival it for the wealth and diversity of its text and images. The contributions on the sciences and the trades were particularly remarkable. Beyond precise and reliable information, these articles offered the reader a striking portrait of the progress of the human spirit.

Once it was completed, the *Encyclopédie* clearly encouraged others to follow its example. The booksellers Durand and David were already dead when the last volumes of plates appeared under Diderot's supervision in 1772. Le Breton soon sold his interest to Briasson, who remained the sole proprietor of the *Encyclopédie*. But by this time, a new arrival had entered the Parisian book world: Charles Pankoucke, who had already bought the rights to the text and the copper plates from the associated publishers. He continued the *Encyclopédie* with a supplement, but without Diderot, before launching a new *Encyclopédie* under the name *Encyclopédie méthodique*, divided into specialized dictionaries. By this time, the work of Diderot and d'Alembert had ceased being a Parisian affair. Several new editions had come off the presses, some in folio format and others (much less expensive) in quarto or in octavo formats that reached a wider public than the first edition. All of them were published far from Paris, in Switzerland and in Italy.

To conclude, let us return to the Academy of Sciences' *Description des arts et métiers*. Remember that the project, initiated at the end of the seventeenth century, had been taken up by Réaumur in the eighteenth. It had given rise to a few publications before foundering. To prepare the description of the trades in the *Encyclopédie*, Diderot had taken direct inspiration, as we have seen, from the documentation gathered by Réaumur. After the affair with the plates, the Academy decided to launch the project again and entrusted the direction to one of its members, Henri-Louis Duhamel du Monceau. The great folios of the Academy's *Description*, ornamented with superb engraved plates, were printed by the best Parisian booksellers. From the work devoted to the art of the charcoal maker, which appeared in 1761, to the last one, published in 1788, the Academy's *Description* treated a total of eighty-five different trades and crafts. The collection thus surpassed considerably in breadth the description of those arts in the *Encyclopédie* of Diderot and d'Alembert. In addition, when Panckoucke launched the supplement to the *Encyclopédie*, and then the *Encyclopédie méthodique*, he relied heavily on the volumes of the *Description* to fill out the articles.

The savants of the Academy took upon themselves the descriptions of certain arts—like Duhamel, who took on twenty of them, and the astronomer Lalande, who tackled nine of them, including the art of papermaking. For other arts, the task was entrusted to amateurs or craftsmen. The most remarkable of the latter was André Roubo, a simple carpenter living in the Faubourg Saint-Jacques. Between 1769 and 1774 he published an *Art du menuisier* (Art of the joiner) in four volumes, complemented in 1782 by an *Art du layetier* (Art of the box maker) for the Academy's *Description*. He provided all the drawings and engraved most of the plates himself. It was on the strength of his encyclopedic knowledge of carpentry and its history that in 1782 he constructed the frame of the Wheat Hall following the technique of Philibert Delorme, as we saw in the last chapter. This is why his triumph, saluted by the people of Paris, was just as much the triumph of the description of the arts and trades, a majestic and collective work by which the initiative of Diderot was melded with the diligence of the Academy of Sciences. This was indeed proof, if any were needed, that the winds of the *Encyclopédie* blew well beyond the Republic of Letters and the narrow circle of its subscribers.[12]

NOTE: The *Encyclopédie* is available online in French from the University of Chicago at https://encyclopedie.uchicago.edu/ and in an ongoing English translation from the University of Michigan at https://quod.lib.umich.edu/d/did/.

Notes

1. Letter from Patte dated November 23, 1759, *L'Année littéraire* (Fréron) 7 (1759), 341.
2. Letter from Patte dated January 29, 1760, *L'Année littéraire* (Fréron) 1 (1760), 246.
3. DIDEROT, *Correspondance*, ed. G. Roth, vol. 3, Les Éditions de Minuit, 1965, 22. The letter is dated July 1, 1760, certainly incorrectly.
4. *P.-V.* (1760), fol. 94 and following, Saturday, February 23, 1760.
5. S. JURATIC, "Publier les sciences au 18ᵉ siècle: la librairie parisienne et la diffusion des savoirs scientifiques," in PASSERON, 2008, 301–313.
6. On the production and sale of the *Encyclopédie*, see LOUGH, 1971, and DARNTON, 1979.
7. Letter from Voltaire to Diderot, sent from Potsdam, September 5, 1752.
8. F. KAFKER and M. PINAULT-SØRENSEN, "Notices sur les collaborateurs du recueil de planches de l'Encyclopédie," *Recherches sur Diderot et sur l'Encyclopédie*, nos. 18–19, 1995, 200–230.
9. Letter from Diderot to Voltaire, September 29, 1762.
10. DIDEROT, *Salon de 1765*, no. 150.
11. Letter from Voltaire to M. de Gauffecourt, August 30, 1755.
12. On Roubo and his work, see BELHOSTE, 2012.

5

The Court and the Town

ON THE MORNING of April 7, 1778, Voltaire entered the Hôtel de Mézières on the arms of Benjamin Franklin and Antoine Court de Gébelin. He was being received in great pomp into the Masonic lodge known as the Nine Sisters. For four years, the townhouse near the church of Saint-Sulpice had been the headquarters of the Grand Orient of France—after having been, ironically, that of the Jesuit novitiate. Two months earlier Voltaire had left his home in Ferney, near the Swiss border, for the "mud, fracas and incense of Paris," braving at the age of eighty-four both royal punishment and hostility from the devout faction.[1] After twenty years of absence from France, his return was triumphant. At the corner of the Quai des Théatins and the rue de Beaune, the Hôtel de Villette, where he was staying, was constantly full with a throng of visitors. An exhausted Voltaire, spitting and pissing blood, was nonetheless exultant. The exile—the defender of Calas, La Barre, and Sirven against despotism, and the enemy of priests—was taking his revenge. He had come on the pretext of watching the rehearsals for his last tragedy, *Irène*, performed by the Comédie Française, but it was Paris itself that called him, the memory of his youth and the first witness of his glory.

March 30 was the day of his apotheosis: attired in ceremonial dress, he was driven across the Seine in a carriage through an immense crowd to visit the Académie Française and attend a performance of *Irène*. At the Louvre, the academicians lined up to welcome him into the hall of honor and offer him the director's seat. Voltaire listened to d'Alembert read a eulogy of the poet Boileau and went upstairs to his apartment to thank him, before leaving for the Tuileries palace, where the play was being performed. The audience was delirious. The hero of the day, applauded upon his arrival, witnessed the show crowned with laurels. The tragedy was applauded from start to finish. When the curtain finally came down, it rose again to reveal a bust of the poet surrounded by actors who crowned it, hung garlands on it, and covered it with rose petals (Fig. 5.1).

FIGURE 5.1: The coronation of Voltaire at the Théâtre Français. Engraving by Charles Etienne Gaucher after a drawing by Jean-Michel Moreau le Jeune, 1782.

The Nine Sisters

Shortly before this memorable day, on March 25, 1778, a delegation from the Nine Sisters had come to offer homage to Voltaire. The philosopher's host, the Marquis de Villette, was himself a member of the lodge. Voltaire does not seem to have been initiated into Freemasonry, but how could he not have been sympathetic to these friends of Enlightenment? So the idea germinated of receiving him into the lodge: on April 7, after having been prepared, Voltaire was introduced with his face uncovered into the initiation room, the very spot where, as Mercier wittily remarked, the Jesuits had cursed him so often. The hall was vast, decorated with rich blue and white wall coverings, ornamented in gold and silver, with flags and banners of lodges, as well as busts of Louis XVI, Frederick II, the Duke d'Orléans, and Helvétius. A black curtain at the back hid the stage known as the "Orient." The witnesses included all the members of the lodge and 250 visiting brothers. There was a Masonic orchestra.

The ritual had been simplified for the illustrious elder: after a short examination on morality and philosophy, the curtain rose, revealing the Orient: the lighted stage on which the dignitaries sat, among them the astronomer Lalande, a member of the Academy of Sciences and the lodge's master. Voltaire was quickly admitted as an apprentice and given a privileged seat next to the president, from whom he received the laurel crown and the apron of Helvétius, who had been the prime mover behind the Nine Sisters. After speeches, poems, and music, the ceremony ended in a banquet. The apprentice, exhausted by so much pomp and splendor, had already retired. A month later he went to bed and never left it, and on May 30 he died. In secret, his nephews transported the body, embalmed by the pharmacist Mitouard, to Champagne (forty leagues away) in order to avoid the common grave into which the archbishop of Paris had promised to put him, and they gave him a Christian burial.

The admission of Voltaire into the Lodge of the Nine Sisters did not pass unnoticed by either the public or the authorities. His Masonic apotheosis, organized by the same lodge five months after his death, had even more impact. The ceremony, presided over by Lalande, assisted by Franklin and the Russian count Stroganoff and with an audience of about two hundred people, took place on the rue du Pot-de-Fer. Voltaire's niece, Madame Denis, and his "adoptive daughter," the Marquise de Villette, were allowed to attend the spectacle. A vast hall draped in black and illuminated by sepulchral lamps served as a temple. In the middle of the room was a cenotaph crowned by a pyramid and guarded by twenty-seven brothers with swords drawn. Doleful music greeted the visitors. Voltaire's eulogy, delivered by La Dixmerie, was followed by a great roar. As strains of symphonic music arose, the audience in the reilluminated hall saw an immense painting depicting the apotheosis of the departed. The poet Jean-Antoine Roucher closed the ceremony by reading his poem "Les Mois" (The months), in which he denounced the church's last act of persecution of Voltaire: the threat to bury him in a common grave.

The audience applauded and called for encores before going in to the banquet. Reaction was swift: the devout faction was furious, and with the intervention of the government, the Grand Orient imposed sanctions on the lodge. By the end of the year, it had been forbidden to use the Great Temple on the rue du Pot-de-Fer anymore, on the pretext that two women had attended the ceremony in contravention of the Masonic rules. And Lalande was called to appear before the Chamber of Paris of the Grand Orient, where he was told that the archbishop of Paris and the lieutenant of police had received complaints after the reading of scandalous literary works—La Dixmerie's eulogy and Roucher's poem were the targets—and that the ceremony amounted to "a crime with the most dangerous consequences" for Masonry as a whole. The troubles were just beginning for the Nine Sisters and would last for almost two years.

Before describing what finally happened to the lodge, we must return to the time of its foundation. According to its regulations, probably adopted around the time of Voltaire's initiation, "The talents that the Lodge of the Nine Sisters requires of an aspirant in order to justify the name it bears include the sciences and liberal arts; such that any person who shall be proposed to it should be endowed with some kind of talent in either the arts or the sciences, having already given public and sufficient proof of this talent."[2] In short, the lodge aimed to bring together the scientific, literary, and artistic elite in a Masonic framework. Everything suggests that the Nine Sisters as it was constituted in 1776 was in fact merely a continuation or extension of a lodge founded ten years earlier by Helvétius and Lalande as the "Lodge of the Sciences." The adoption of the name Nine Sisters expressed Helvétius's aim before his death to expand its recruitment to include writers and artists.

The creation of the Lodge of the Nine Sisters was part of the revival of freemasonry in France after a long period of crisis and division. After several meetings in which Lalande played an active role, the Parisian lodges had decided in 1771 to create the Grand Orient and had elected the Duke de Chartres (son of the Duke d'Orléans) as grand master. The rules of the Grand Orient, the importance of rituals, and the common ideal of tolerance and fraternity now gave a certain unity to French freemasonry, even though many lodges remained outside its discipline. But each lodge had its own character and composition. While they preached equality among brothers, there was little mingling among Masons of different social conditions. By definition, the Nine Sisters lodge admitted only men of talent. Beyond this particular characteristic of its members, the lodge was distinctive for its liberal spirit: there was little interest in the rituals, and even less in established religion; the leaders did not hide their adherence to the ideas of the philosophes. Its name, taken from Greek mythology, was profane, its master Lalande was a notorious atheist, and its secretary, Court de Gébelin, was a Protestant close to illuminism.

In the beginning, the Nine Sisters had few celebrities apart from Lalande, and even fewer in the sciences. The lodge's orator, Le Changeux, had done some research published in Abbé Rozier's *Journal de physique*. Court de Gébelin made a name for himself with an immense work of erudition, *Le Monde primitif analysé et comparé avec le Monde moderne* (The primitive world analyzed and compared to the modern world), launched by subscription in 1772 as a complement to the *Encyclopédie*. In 1777, the lodge had some sixty members. While the admission of Voltaire and his apotheosis created problems, they also caused its ranks to grow rapidly. By 1778 the number of paying members had already risen to 144. But there were still few men of science. D'Alembert and Condorcet, as well as Diderot, had thought of being initiated but in the end abandoned the idea. Alone among the members of the Academy of Sciences to participate in the work of the Nine Sisters were Lalande, Franklin, and the Count de Milly, all three of whom were elected masters. Later, the naturalist Bernard Germain de Lacépède, the industrialist and hot-air balloon pioneer Montgolfier, the mathematician Gilbert Romme, the physiologist Georges Cabanis, and the mineralogist Jean-Claude Delamétherie also belonged to the lodge, but the vast majority of the brothers were men of letters and artists.

The lodge continued to hold its meetings until the Revolution, first at the Hôtel de Bullion on the rue Coq-Héron, then at the Hôtel de Genlis on the rue Dauphine. If we know little about its work, we do know a bit more about its public activity. In 1779 a so-called lodge of adoption was established for women during a festival organized at the Cirque-Royal on the Boulevard du Montparnasse. The alleged "disorders and indecencies" that occurred on this occasion served as a pretext for bringing the lodge up for charges before the Grand Orient. The real reason, of course, was political, and everything was over by the following year. Other festivals were organized, for example, an "academic festival" held at the Redoute Chinoise, an entertainment venue with a Chinese theme, to celebrate the peace treaty signed at Versailles on May 12, 1783, that ended the American Revolution: Franklin received a medal bearing his likeness. Among other lectures, the Count de Milly presented a paper on the principle of life and of destruction, and Lalande on the new planet discovered by the English astronomer William Herschel. Afterwards, the assembly went out into the grotto and garden, where people dined at little café tables. The evening ended with an oratorio on peace and a ball.[3] If, like other lodges, the Nine Sisters offered a place to meet people and socialize, its intellectual activity, notwithstanding the desire of its founders, was no doubt limited to "august twaddle."[4] This mediocrity can be explained by the fact that the lodge's goals were achieved more through what it did outside the lodge than within it.

Salon Society

A salon lay at the origin of the Nine Sisters. Much ink has been spilled over the role of salons in the literary and artistic life of Enlightenment Paris. However, science has scarcely interested the historians of elite sociability. Of course, the figure of d'Alembert is remembered as a regular of Madame Geoffrin's salon before he became a pillar of Julie de Lespinasse's. But the focus has been less on him as a mathematician than as editor of the *Encyclopédie*, philosophe, and man of letters, the author of the *Essai sur la société des gens de lettres et des grands* (Essay on the society of men of letters and the great) and Julie's unhappy lover. An immense distance separated the austere and dull world of savants working in their cabinets and laboratories from the superficial, frivolous world of gossips and wits.

One of the most recent historians of Parisian high society thus stresses the very minor role of scientific practices in the salons and the difference in the second half of the eighteenth century between true experimental science and the simple curiosity of the salon world.[5] If we remain within the confined space of the salon, then we will find little social activity that could be described as scientific. But society life in the eighteenth century cannot be reduced to salons; it played out on a multitude of stages, public and private, and was pursued behind the scenes in spaces that were discreet or hidden. We should take these sites into account in evaluating the relationship between high society and the world of science.

The space in which the life of the nobility took place (including its imitators and servants) was simultaneously geographical, social, and symbolic. It stretched essentially between Versailles, where the king and his courtiers lived, and the west side of Paris, where the mansions and pleasure venues were concentrated, with annexes in the châteaux and country houses in the region. For Francois de La Michodière, known for his effort to estimate the population of France, it formed a single city, "because so many of the most important people in the state, who have the most domestic servants and other people attached to their service, have residences in both cities and spend a large part of their lives going back and forth between them."[6] This space was called *the world,* or *society,* or *the court and the town.* Throughout his brief final stay in Paris, Voltaire traversed this space to the extent permitted by his health—or at least the city, since to his great regret he was banned from the court until the end. Voltaire the writer wanted above all to be a man of the world. For Voltaire, as for other philosophes and writers, the Republic of Letters was coextensive (so to speak) with the space of high society. Was this also true for the savants? To what extent did they find the resources they needed there? And to what extent could they do without it? In truth, the man of science, like the man of letters, remained dependent on the great, on their power and their protection. To achieve a distinguished career in the sciences, he had to follow the paths of high society, even if it generally sufficed to remain in its shadows.[7]

Yet most savants lacked the qualities that made for worldly success. Many were poor conversationalists and paled in comparison to the men of wit. A taste for solitude and discretion and the very nature of their work often distanced them from the spaces of society life. But not all of them were so uncivilized. Some— and they were the most powerful—did attend the salons and dinners in town. I have already mentioned d'Alembert. Buffon was received in the finest houses and himself held a Sunday salon at the Jardin du Roi; in both Paris and Versailles he was among the close friends of Madame de Marchais, and in his last years was a regular at the after-dinner gatherings of Madame Necker, the wife of the finance minister, whom he called his "divine friend." Lalande, who claimed that studying "the society of intelligent people, and especially of educated women," was his sole form of recreation, was received successively, by his own account, into the "company of Mesdames Geoffrin, du Bocage, du Deffand, de Bourdic, de Beauharnais, [and] de Salm." Bailly, a "pleasant man of society," according to Condorcet, who detested him, often visited the houses of wealthy aristocrats in the village of Chaillot just outside the city; in Paris he frequented the salon of Fanny de Beauharnais as well as that of the "lovely Greek," the mother of the poet André Chénier. In his youth Condorcet was himself a regular at the salon of Julie de Lespinasse, to whom D'Alembert had introduced him. Later, he regularly attended the salon of the Duchess d'Enville as well as the more bourgeois one of the Suards. After his marriage, his wife received guests at the Mint.

Note that all these savants were also among the elite of writers. Marat had them in mind when with his customary outrage he denounced the academicians: "They get up late, their mornings are employed in breakfasting, reading the *Journal de Paris*, receiving and making visits. They dine in town; after dinner they go to the theater, then to an intimate little supper; and if they find themselves with any time to kill, they spend it loading up on a supply of news for their idle chatter. This is how they spend just about every day of the year."[8]

Along with these true high society figures, the salons could also be open to savants of much more modest stature, like Nicolas Desmarest. Arriving in Paris from his native Champagne in 1747 with only a few letters of introduction from his Oratorian teachers, he at first lived very frugally as a mathematics tutor, earning some additional income from minor editing jobs. But after he won a prize from the Academy of Amiens in 1751 for an essay on the ancient land bridge between France and England, the doors of society were suddenly opened to him. Introduced a little later into the society of Madame Geoffrin, he developed a close relationship with the dilettante farmer-general Claude-Henri Watelet and through him with d'Alembert, who became his protector. Now admitted into the society of the Duchess d'Enville, he got to know the royal ministers Trudaine and Turgot, who had him appointed inspector of manufactures in 1762. In 1765, he accompanied the duchess's son, the Duke de la Rochefoucauld, to Italy. In 1770, while Desmarest was working in the provinces, the duke himself alerted him to a vacancy in the Academy of Sciences, exhorting him to come back quickly to attend to the matter. Upon his return to Paris, Inspector Desmarest was duly received into the company.[9]

Do not suppose that sophisticated manners were required to be admitted to noble homes, for aristocrats often affected simplicity and enjoyed intimacy, liking to surround themselves with men of modest airs and without pretensions. Rousseau had condemned the salons after having frequented them a lot, but he continued to have friends and protectors among the aristocracy, where he was valued as much for himself as for his philosophy. When Benjamin Franklin arrived in Paris in 1778, he received the warmest welcome; high society fought over him. In Auteuil, where he was living, he became the darling of Madame Helvétius. The aristocrats loved his familiarity, his broken French, and the simplicity of his attire: he wore glasses and a fur hat, and wore neither wig nor sword, even at Versailles. The person most shocked by this getup was his compatriot John Adams, who was annoyed by the constant praise of Franklin's admirers for his bald pate fringed by unruly locks. Franklin himself was not duped; he was simply playing the parts of the Quaker and Poor Richard that pleased his French friends so much.

Friendship of the High and Mighty

Polite society was not limited to the salons; the life of society was much more fragmented than that. Great nobles received at all hours, their hospitality ranging from the simple visit to an open table. While most of the savants avoided the weekly salons and gala dinners, they still went out into society. The following examples illustrate the encounters between men of science and the social elite. Consider first the friendship of Malesherbes and Turgot with the humble André Thouin, Buffon's chief gardener. Botany brought the three men together. Malesherbes spent hours conversing with the gardener in his kitchen at the Royal Botanical Garden, seated on a bread bin. Starting in 1778, Turgot was also a regular visitor to the Botanical Garden. He took private lessons from Thouin, and took pleasure in having the gardener visit him at his townhouse. "Since I could never have you to dinner if there were ladies present, I warn you, sir, that I am alone here with my sons," he wrote on January 6, 1779. "You are thus invited to dinner either today or tomorrow; have the goodness to let me know if you accept my invitation." A month later he again invited Thouin over, this time for an afternoon chat. These invitations no doubt weighed heavily on the shy Thouin, who hated nothing more than having to leave his garden.[10]

The second example is the genuine friendship between Bochard de Saron and the astronomer Charles Messier. Bochard, who was president of the Parlement de Paris, belonged to one of the principal families of the Paris judiciary. Messier, who, as La Harpe put it, was as "simple as a child," had arrived in Paris with no recommendation other than his clear and readable handwriting and some talent for drawing. Having by chance obtained a position as a copyist for the astronomer Delisle, he became passionate about astronomy and was soon recognized as an exceptional observer. The collaboration between Messier and Bochard lasted thirty years, with one of them identifying comets and the other calculating their trajectories. The two men saw each other regularly in Paris, as well as in Champagne, where Messier joined the judge on his estate at Saron. On a walk one day around the Parc Monceau with Bochard and his children, Messier took a fall on some ice, which left him with a limp.[11]

Bochard and Malesherbes were both honorary members of the Academy of Sciences, and they almost certainly contributed to the elections of their respective protégés, Messier in 1770 and Thouin in 1786. It was said that Buffon was unhappy to "see his inferior, a kind of domestic, become his equal." As we have seen, the honorary members represented both an ornament and an advantage for the company. Their presence also marked the place occupied by high society at the very heart of the premier scholarly institution in the kingdom. Voltaire, during his final Parisian tour in 1778, did not fail to attend the public meeting of the Academy of Sciences on April 29. Attracted by the spectacle, Bachaumont tells us, "the most seductive beauties among the fair sex, the most frivolous and amiable men drawn from the court, and, the most elegant and sought after writers from the literary world took over the hall." Voltaire kissed Franklin to the applause of the crowd before being invited to join the savants.[12]

The Academy of Sciences thus found itself annexed in a way to the literary world. The public believed in the power of the aristocracy over the savants and even exaggerated it. "The duchess d'Enville and her son, the great friends of Monsieur Turgot, were said to have great influence with the Royal Academy of Sciences, to make members at pleasure, and the perpetual secretary Mr. D'Alembert, was said to have been of their Creation, as was M. Condorcet afterwards," noted John Adams in his diary.[13] Some years later, in his eulogy of Cassini, Condorcet himself warned his colleagues about getting too close to the great nobles: "In fact, the kind of domination they like to exercise over the business (even over the feelings) of those they call their friends seems incompatible with the freedom and independence whose loss strips talent of half its strength and its resources. The more we are convinced by reason of the natural equality among men, the more it becomes a law for us to avoid the intimacy of those whom opinion has placed above us."[14] This lesson was not new. In 1753, d'Alembert had denounced the subordination of men of letters to the great, the lowliness of the former, the arrogance of the latter. Nevertheless, he believed that a few great lords (like the Marquis d'Argenson) were capable of "familiarity without pride" and imagined that a man of letters might treat and view them "with full confidence as his equals and friends."[15] Condorcet was much more radical. In the name of the freedom of talent, he called for men of science to break off all close intercourse with the powerful; this was strange advice on the part of a man who had been a regular of the salons and a very close friend of La Rochefoucauld, but it foreshadowed in a nutshell his stance during the Revolution.

Savants as Courtiers

Were the men of science so dependent and subjugated that the permanent secretary had to take on their powerful protectors publicly? D'Alembert considered the courtier to be the lowliest role that a man of letters could play. Yet in the eighteenth century many savants were still in the service of princes, part of their households, that is, living in their entourages. Certainly (with a few exceptions) they did not have to play the part of a flattering courtier, since it was understood that the man of science was first and foremost in the service of truth and that in this capacity he was equal to the person he served. But until the end of the Ancien Régime they could still be found among the clients of the great, and some at least formed part of their household staff.

Few savants, it must be said, frequented Versailles. Each year, the officers of the Academy of Sciences went there to make a presentation to the king about the company's achievements, but that was the extent of the personal relations between the monarch and his Academy. A few savants, such as Buffon and Cassini, were admitted to court. But among men of science, it was above all the physicians and surgeons in the service of the royal family who went to Versailles and lived there for at least part of the year. During his stays at court, François de Lassone, chief physician to Louis XVI and Marie-Antoinette, a veteran pensioner of the Academy of Sciences and president of the Royal Society of Medicine, lived at the château. He also maintained a chemistry laboratory at another royal palace at Marly. The physician Cornette, Lassone's trusted friend and collaborator, lived in his apartments and followed him to court. Elected in 1778 to the Academy of Sciences as a chemist thanks to his protectors and chosen over Pinel to be physician to Mesdames Adélaïde and Victoire (the king's aunts) in 1784, Cornette had a small laboratory in Versailles and had many patients there, the most illustrious of whom was the king's brother, the Count d'Artois. During the Revolution he would emigrate with his protectors.[16]

Louis-Guillaume Le Monnier was undoubtedly the most remarkable of these court physicians. The son and brother of astronomers, he had been a member of the Academy of Sciences as a botanist since 1735. In collaboration with the gardener Claude Richard he developed the cultivation of plants in greenhouses when he was a young physician attached to the royal infirmary of Saint-Germain-en-Laye. Presented to Louis XV, who was interested in what he was growing, he gained favor, advising the king on the arrangement of the new Trianon gardens and having Bernard de Jussieu appointed as scientific director of the botanical gardens there. In 1759, while he was in Germany with the army during the Seven Years' War, Le Monnier was appointed professor of botany at the Jardin du Roi. In 1762 he purchased the succession to the post of chief physician at Versailles (presumably from François Quesnay, the founder of physiocracy). When he finally acceded to that post in 1770, he arranged to have Jussieu's nephew Antoine-Laurent take over his chair at the Jardin du Roi. Introduced into the inner circle of Louis XV, whom he treated during his final illness, he used his influence to import plants and seeds from all over and to send his students abroad in search of more, including André Michaux to Persia and René Desfontaines to the Atlas Mountains.

Le Monnier was connected to Rohan and to the devout faction, and especially to the Countess de Marsan, governess of the royal children. He lived with her at her house in Montreuil until his marriage in 1773 and created a remarkable garden there that made it both pleasant and interesting. Modest and almost self-effacing, having written and published almost nothing, he was no less powerful thanks to all the relationships he had formed, while affecting to remain outside court intrigues. At the death of Lassone in 1788, he finally obtained the post of chief physician to the king.[17]

The households of the princes of the blood also offered protection to savants. The Count de Provence (the king's eldest brother, known as "Monsieur") seems to have taken little interest in the sciences. In the 1780s, though, he supported with his name the popular scientific and educational institution founded by Pilâtre de Rozier, known then as the Musée de Monsieur, which he then purchased after the death of the aeronaut in 1785. Provence's brother, the Count d'Artois, played a more active role, indirectly at least, in supporting savants; not that he demonstrated more interest in science, but he had children, the Duke de Berry and the Duke d'Angoulême. Their education had been entrusted to the Marquis de Sérent, who in 1775 called upon Abbé Marie, professor of mathematics at the Collège des Quatre-Nations, to teach them the sciences. A cabinet of physics and natural history was installed at Versailles for the two young princes. Having entered the court, Abbé Marie soon promoted one of his own young protégés, the Swiss clockmaker Abraham-Louis Breguet, introducing him to a princely clientele. Later he played a decisive role in bringing the mathematician Lagrange to Paris. Like other princely households, that of the Count d'Artois included many positions in the medical fields. The purchase of one of these offices permitted a physician to practice in Paris outside the corporate framework. This is the only reason, no doubt, why Jean-Paul Marat, with a medical degree from the University of Edinburgh, purchased the title of physician to the Count d'Artois's bodyguards in 1777. We do not know if this is how he made the acquaintance of Breguet. In any case, the two men, who had similar backgrounds, became close before the Revolution.

The Orléans engaged in scientific patronage even more than their cousins the Bourbons. The tradition went back to the regent, whose son Louis "the Pious" had retired to the Abbaye de Sainte-Geneviève and was a prince-savant. Passionate about physics, chemistry, and natural history, he had conducted

experiments with Jean-Étienne Guettard, the keeper of his natural history cabinet and member of the Academy of sciences. Without sharing this interest to the
same degree, Louis Philippe, known as "the Fat," inherited from his father a sincere commitment to literary and scientific patronage. In 1774, on the recommendation of his chief physician, Théodore Tronchin, he hired Claude Berthollet to
work for his morganatic wife, Madame de Montesson. The duke soon placed a
laboratory at Berthollet's disposal, where the chemist-physician undertook his
first research in the new field of pneumatic chemistry.

The duke was also interested in the Périer brothers, who had been noticed
by Abbé Nollet and whom he made his mechanics. Installed at the Chaussée
d'Antin next door to the mansion of their protector, the brothers constructed
novel machines for his châteaux at Raincy and Neuilly. The duke also financed
their manufacture at Chaillot where they produced Watt's steam engines. The
duke's son, Philippe, Duke de Chartres, called on the brothers to set up an atmospheric pump in his garden at Monceau as well. In 1783, the elder brother,
Jacques Constantin, was elected adjunct mechanic at the Academy of Sciences.
That same year, the Périers, at the request of Madame de Genlis, Chartres's mistress and the tutor of his children, assisted by the mechanic Calla, constructed for
the children's education relief models of all the objects used in the arts and trades,
which were exhibited to the public in the galleries on the second floor of the
Palais-Royal. Around this time, the Duke de Chartres became passionate about
hot-air balloons. Between trips to London, he followed the flight of Charles and
Robert, whom he welcomed when they landed, and then wrote up his own account of the event for the *Journal de Paris*. When he became Duke d'Orléans at
his father's death in 1785, he enlarged his scientific patronage by extending his
protection to savants like Darcet and Laplace.[18]

These are just a few examples of the importance of the patronage of the great
for those who engaged in the sciences. Until the Revolution, recommendations
remained indispensable for building a career and securing posts. Only with the
personal support of an august personage could one enter into networks of power
and obtain offices and emoluments. In this respect, the situation of savants did
not differ from that of other members of the Republic of Letters. Within the
Academy, a sprinkling of savants with high connections, such as d'Alembert and
Buffon, were themselves in a position to play the role of patrons within their own
spheres, but the high nobility, close to the court and the government, remained
the best recourse for the ambitious.

We could provide many more examples, including the effect of patronage
on elections to the academies. In 1772, the Duke de La Rochefoucauld (who
was later accused of having created savants himself) complained in a letter to
Desmarest: "There is something untoward about a protector in an Academy; he

excludes Rouelle and Darcet, but gives a place to M. Sage; he should be pulled up short; do you have some sort of trick for doing that?"[19] The protector in this case was the monarch himself. Having encountered Louis XV in the gardens at Versailles, the pharmacist Sage had in fact won his confidence and thanks to his direct intervention had gained admission to the Academy as an ordinary associate in 1770. As the criticism raised against these abuses suggests, the patronage of the court and nobility was not well received by the savants; but before the Revolution nobody could escape it.

The case of Gaspard Monge provides another example. Since the beginning of his career, when he taught at the engineering school of Mézières, the geometer had as his immediate protector Abbé Bossut, which placed him in d'Alembert's sphere of influence. When he was elected to the Academy of Sciences in 1780, he was obliged to live in Paris. For several years, he tried with difficulty to reconcile his stays in Paris with his post as professor in Mézières. His appointment as an examiner for the Navy in 1784 finally enabled him to leave the Ardennes region, but it cost him the friendship of Bossut, who was aiming for the same post. Monge would not have been able to obtain such a position without powerful support. In 1778, Monge had made the acquaintance of the Marquis de Castries, who was passing through Mézières, and on this occasion became friendly with the tutor of this important person's son, Jean-Nicolas Pache. Two years later, Castries had become minister of the Navy and made the faithful Pache his secretary general. When a replacement for the mathematician Étienne Bezout had to be named, Monge was not forgotten. From then on he was part of the marquis's entourage: in 1786, the new examiner came to stay at his château at Bruyères, near Étampes, to write his treatise on statics.

A Worldly Science?

There is no doubt that in the eighteenth century the French nobility supported the sciences, just as they supported the arts and letters, or that they contributed to the progress and dissemination of science. We could ask, however, if this system of patronage exercised any influence over how science was practiced and what subjects were chosen. In fact, is there not a natural incompatibility between the worldly culture of the aristocracy (superficial and sparkling) and scientific research (austere and profound)? Aren't the pleasures and illusions of a world of ostentation inconsistent with the austerity of science oriented toward truth? In fact, this question goes back to the seventeenth century, and has for a long time fed into the analyses of historians of science. Charles Gillispie, in his great study of science and society in France at the end of the Ancien Régime, dismisses the very possibility of a worldly science. He maintains that the development of the great scientific institutions under the aegis of the state reduced the scientific activities offered to the aristocratic public to simple amusements. The scornful condemnation of Mesmerism by the Academy of Sciences in 1784 sealed the definitive rupture between high society and the world of science.

More recent historiography has tended to call into question, or rather to contextualize, this supposed separation between illusory and futile forms of knowledge on the one hand and serious (that is, professional) science on the other. While the academic elite in the final years of the Ancien Regime did clearly keep their distance from activities they judged to be incompatible with the character of science, that very position raises questions that are worth asking. Was it simply a matter of an official scientific institution pushing aside impostures and superstitions, or did the campaign actually aim to delegitimize knowledge produced and approved outside its control and inside high society specifically? In other words, was the criticism raised by savants scientific or political or both at the same time? To respond to this question we must first consider what the cognitive activities specific to high society were in the eighteenth century. That is, we must identify and define the practices, values, and actors associated with the type of science that could be called "polite."

Historians have long noted the aristocracy's interest in the most spectacular and recreational aspects of experimental science. In France, Abbé Nollet was the self-proclaimed spokesman for a physics based on demonstrations that took him all the way to Versailles. At the end of Louis XV's reign, electricity was in fashion: in both the court and the town, people made their hair stand on end and submitted themselves to electric shocks. But the court quickly tired of spectacles that only mildly interested the royal family. In Paris, by contrast, people continued to have a passion for physics as entertainment. The magic tricks of a performer who went by the name of Comus were very successful. In 1773, the Duke de Chartres asked to be initiated into his secrets: every Tuesday he attended the master's lessons and practiced in his new house at Monceau. Edme-Gilles Guyot, who had explained these tricks in his book on recreations, sold the props for executing them to rich amateurs wanting to make a splash in the salons. This is perhaps how the scientific career of the physicist Jacques Charles began: employed in a sinecure in the administration of stud farms, he amused his friends with his physics experiments in the 1770s.[20]

Comus's reputation was controversial. Many considered him a charlatan, although he laid claim to no supernatural power. White magic, in vogue since 1770, left behind a sulfurous reputation. Those who engaged in this form of entertainment made sure not to disillusion those who believed in their marvels. Two (presumed) physicians in particular were targeted by those who denounced charlatans: Mesmer and Cagliostro. Both belonged to a category of itinerant adventurers who went from town to town and court to court demonstrating their talents.

Arriving in Paris from Vienna in 1778 with a reputation that preceded him, Mesmer was soon welcoming patients into his home on the rue des Quatre-Fils in the Marais, where he treated them by means of animal magnetism. The Viennese doctor and his universal fluid were not well received by the savants or the physicians from the medical faculty (with the notable exception of Charles Deslon). He was given a warm welcome in the court and the town, however, where he actively sought patronage. Among his patients were the Duchess de Chaulnes and the Princess de Lamballe. Marie-Antoinette herself, without ever having met Mesmer, took an interest in his fate. In 1783, several great nobles, such as the Marquis de Lafayette, the Puységur brothers, and the Marquis de Montesquiou helped to launch the Society of Universal Harmony founded by the lawyer Nicolas Bergasse, to spread Mesmer's therapeutic methods. The society's headquarters were in the former Hôtel de Coigny on the rue Coq-Héron. After the condemnation of Mesmer's theories by two official commissions in 1784, he left Paris and France for good. But society's interest in animal magnetism remained strong right up to the Revolution. The Marquis de Puységur, who had discovered the phenomenon of hypnosis, was performing experiments in the salons, in particular at the home of his friend the Duchess de Bourbon, which led to cures and predictions.

The case of the Count de Cagliostro, alias Giuseppe Balsamo, although different, resembled Mesmer's in some respects. Without doubt a con man, he was nevertheless also a skillful physician. After some long and shady adventures, in 1780 he arrived in Strasbourg, where he rapidly won the trust of respectable society by his medical consultations, and in particular his treatment of the archbishop, Cardinal de Rohan, who had a private laboratory built for Cagliostro in his Saverne Palace for his research into the occult. The cardinal's privy counsellor Louis Ramond de Carbonnières, who was also a protégé of the Duchess d'Enville, served there as (in his own words) a "laboratory boy." A first trip to Paris in 1781, to care for the Maréchal de Soubise at the cardinal's request, introduced Cagliostro into Parisian society. But he did not move to the capital with his wife, the enigmatic Séraphina, until January 1785. His reputation, too, had preceded him.

Settling in style in the Marais, two steps away from the Hôtel de Strasbourg, where the cardinal lived, Cagliostro continued his experiments in alchemy aided by Ramond and practiced necromancy, all the while giving medical consultations. After having sought in vain to take control of the Philalethes, who practiced a mystical version of freemasonry, he organized a Masonic lodge according to the Egyptian rite he had established in Lyon the previous year and of which he claimed to be the grand Copt. The father-in-law of Puységur, the financier Baudart de Saint-James, was its grand chancellor. Everything seemed to be going well for Cagliostro; he was in fashion and many people took him seriously: Houdon sculpted a bust of him, and Fragonard painted his portrait. But he was suddenly brought down with the cardinal in 1785 in the wake of the Diamond Necklace Affair. The magician was accused of being the instigator of the plot and thrown into the Bastille. Acquitted by the Parlement in June 1786, he left France for new adventures in London, only to come to a miserable end in Italy in an Inquisition prison. His disciple Ramond de Carbonnières made a career in politics and science, and in 1802 would be elected to the Institut de France as a naturalist.

All these examples have one thing in common: success in polite society, in science as in everything else, required spectacle and surprise. People had to be entertained and not bored. This is why polite science was above all a pleasure for the senses, especially the eyes (Fig. 5.2). And so physics demonstrations belonged to the realm of the theater and concerts. In society one observed: What could be better for a man of the world than to demonstrate an experiment, like performing chamber music or mounting the boards? Recreational physics became both prestigious and prestidigitation—a magic show. In this sense, it can be seen as one of the society games that were all the rage in the salons. For example, the Duke de Chartres was an aficionado of magic tricks, card games, and gambling.

Moreover, playfulness was not incompatible with a desire to know and to understand. Guyot, in his *Recreations*, written for high society, noted that "those who try to vary the effects of electricity by applying it to things that are entertaining will be able to gain new insights from experiments that they perform every day, from which physicists more initiated than they into the secrets of nature will not fail to profit."[21] He thus stressed that the ambition of polite science was not to construct theories and systems but to reveal phenomena. And who better to conduct such experiments, or at least to judge their success, than those select and sensitive persons who were able to grasp slight and fleeting impressions?? These privileged witnesses would confirm the existence of effects, without necessarily claiming to explain their causes.

This arrangement, however, had several defects. First, it did not take into account seduction or fraud. As we have seen, savants constantly denounced the impudence of imposters and the credulity of the public. It also supposed a surprising lack of intellectual curiosity. If these repeated apparitions of sparks, glimmers, or specters aroused surprise and pleasure, they also demanded explanations. This is why, although it was good form at the time to condemn the Cartesian "spirit of system," polite science was not content with merely exhibiting marvels; it also invited a host of explanations. Behind the "facts" that it claimed to reveal lay a hidden reality that it explored of invisible forces and mysterious fluids. But it

FIGURE 5.2: Visiting a cabinet of natural history. Engraving as frontispiece in Pierre Rémy, *Catalogue raisonné des tableaux, estampes, coquilles et autres curiosités; après le décès de feu Monsieur Dezalier d'Argenville*, Paris, Didot, 1766.

believed it could do so through the exercise of an exquisite sensibility rather than by using reason, which was judged to be blind and often deceptive.

Thus for Rousseau, who was read and appreciated by a broad public, sensibility was the source of all action. Physically, it connects us to the external world and thereby provides motivation for action. Morally speaking, it connects us through feelings to other beings, to whom we would otherwise be indifferent. This is why sensibility was considered to be both the source of our curiosity and the very condition of all social bonds.[22] In accord with this sentimental epistemology, polite science required a movement of the heart toward the world and toward others. It assumed a harmony between man and nature, which both served as its precondition and determined the goals of human activity: to discover the analogies and affinities among beings, to grasp and evaluate the forces that unite and separate them, and to describe the material and immaterial circulations that are established between them.

In this way, science was not merely physical but also social, simultaneously offering the Parisian elite the basis for an art of healing, resources for a collective practice (whether in physics or in natural history), and the building blocks of a political critique. After Rousseau, Bernardin de Saint-Pierre was able to capture this movement of sensibility in the 1780s. This context which Robert Darnton has discussed, also explains the social success of a Mesmer or a Cagliostro during the prerevolutionary decade.[23]

Amateurs and Collectors

If the main goal of science for the idle rich was the spectacle of nature (the title of the best-selling book by Abbé Pluche), it accommodated all kinds of sociability. In the world of the aristocracy, there was room for more solitary and private activities alongside the collective practices of presentation and exchange. Even more than in the theater of high society, this is where the selfish pleasures of contemplation and knowledge blossomed. For example, the spectacle of the heavens attracted some amateurs, but regular observation of the stars was too demanding to be truly amusing. The practice of chemistry attracted a few members of the public who were fascinated by its mysteries, like Trudaine de Montigny and La Rochefoucauld d'Enville (see plate 3). But it was above all natural history that attracted sustained interest. There are at least two reasons for this predilection: first, the aristocratic taste for outings in forests and fields—if they could not hunt, they took long walks and botanized; second, the passion for collections, which gave wealthy owners the pleasure (or the illusion) of possessing an entire world. Alongside painting and other works of art, there was a place for *naturalia*.

Of course, Paris had its great collectors.[24] The foremost was the king himself, and after him the princes of the blood. The natural history collection of the Duke d'Orléans at the Palais-Royal was in the care of Jean-Étienne Guettard of the Academy of Sciences until the latter's death in 1786. In 1785 the Count d'Artois purchased the library of the Marquis de Paulmy at the Arsenal, which also included a fine natural history collection. Noble amateurs also had their collections: the Duke de La Rochefoucauld on the rue de Seine, the Duke de Chaulnes (rue de Bondy), the Duke de Montmorency (rue de Saint-Marc), Calonne (rue du Bac), Turgot (Ile Saint-Louis), and Joubert (Place Vendôme). Many of these collections, like Baudart de Saint-James's on the ground floor of his luxurious Place Vendôme mansion, or Bertin's in his townhouse in the Chaussée d'Antin, were simply curiosity cabinets, mingling shells and porcelain figurines in the "Chinese" style (see map 5.1). After he visited the cabinet of the keeper of the royal stables Aubert, the Dutch traveler Van Marum noted that the mineralogical samples were spread out unsystematically; with the finest pieces placed in front, just "as porcelain is arranged."

But alongside these prestigious cabinets that were open to educated visitors, rich amateurs also possessed working cabinets that were accessible only to savants. The Duke de Croÿ, for example, welcomed visitors to the ground floor of his mansion on the rue du Regard, where there was a painting gallery and a museum fitted with mahogany cases that contained his mineralogy collection, but he received the real connoisseurs in his cabinet upstairs.[25] The growing popularity of herbaria after 1770 testifies to a desire among the aristocracy for intimacy as much as a passion for nature and its charms. Hidden within bound volumes, dried and pressed

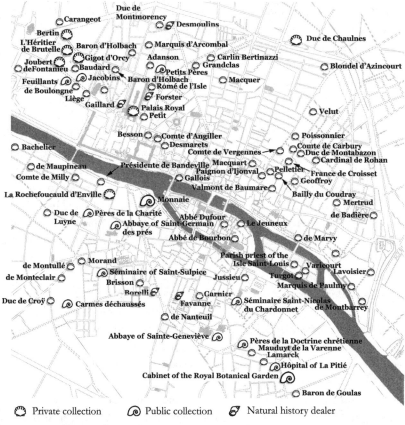

MAP 5.1: Cabinets of natural history at the end of the Ancien Régime

plants were offered to the gaze of only the privileged few who were permitted
to look at them. Malesherbes had in his library a herbarium that contained six
thousand plants arranged in fifty-six portfolios. On the Place Vendôme his friend
L'Héritier de Brutelle, a lawyer at the tax court, had assembled one of the finest
botanical libraries in Europe and an herbarium of almost eight thousand species.

On the model of Bochart and Dionis in astronomy, these great magistrates
collaborated with botanists and served as their patrons. In this way, L'Héritier's
house became the rendezvous of everyone who was passionate about natural
history in Paris, and especially of foreigners passing through. After publishing
at his own expense the first installment of his *Stirpes novae*, in 1786 L'Héritier
started on a description of the plants Joseph Dombey had collected on his expedition to Peru and Chile. For the illustrations, he hired the as yet unknown
Pierre-Joseph Redouté and then introduced him at court, where the artist became the painter of the queen's cabinet. Another great enlightened collector, the
receiver general of finances Gigot d'Orcy, whose superb cabinet was also in the
Place Vendôme, first financed the publication of *Papillons d'Europe* (Butterflies
of Europe), with descriptions by Father Engramelle and drawings by Jean-Jacques
Ernst, and later the publication of an immense *Entomology*, a project he entrusted
to the young physician and naturalist Guillaume-Antoine Olivier, who had been
recommended by Daubenton. Finally, there was the cabinet of Jacques de France
de Croisset on the rue de Chaume in the Marais, generously opened to Haüy,
who took up the study of crystallography there, and Lamarck, who made considerable use of Croisset's collection of fossilized shells.[26]

If the finest cabinets belonged to the court and judicial nobility and the great
financiers, there were many other private collectors in Paris. Some were enlightened amateurs, some were specialized dealers, and still others were professionals,
painters, physicians, apothecaries, and naturalists, for whom a collection was primarily a practical tool for the work they did. The objects collected by the most
prominent members of society thus entered into an intellectual and commercial
circulation in which amateurs, savants, and dealers encountered one another. The
activities of Romé de l'Isle provide the best illustration. Having devoted himself
to mineralogy under the wing of Sage, in order to earn a living Romé drew up
auction catalogues for natural history collections. The famous collector of medals
Michelet d'Ennery befriended him and offered him room and board in his townhouse on the rue Neuve des Bons Enfants. While no doubt helping his protector
to enrich his collection, Romé created his own collection of crystals. Regularly
augmented by specimens that his students and admirers sought out and brought
back to him as homage, it became very large.[27] Like Romé, other savants, such
as Sage and Faujas de Saint-Fond in mineralogy and Lamarck in conchology
(the study of mollusk shells) assembled important collections in their fields of

specialization. Despite their connections with wealthy collectors, these savants made it clear they were to be distinguished from simple amateurs. For example, once he was named the king's botanist in charge of herbaria in the cabinet of natural history, Lamarck became very critical of curiosity cabinets whose "aim was, as it were, to create a visual spectacle and perhaps to offer an idea of the wealth or the luxury of the owner"—which he carefully distinguished from true cabinets of natural history.[28]

A New Master: The Public

The science of high society participated in a symbolic economy in which the enjoyment of knowledge, the arts, and precious objects, whether ostentatious or discrete, directly correlated with a consideration of social rank. For the high nobility, the primary purpose of understanding nature was neither utility nor speculation. Rather, they were looking primarily for a spectacle or a refuge for their own pleasure and curiosity. But this activity, whether hedonistic or ascetic, at the same time performed an important function in the social system: it extended to phenomena and things the domination exerted by those who were powerful in the political sphere by domesticating them. Deployed in the narrow confines of the cabinet or the laboratory, as well as in the more convivial spaces of the salon and the garden, polite science appeared to be at bottom a campaign to annex the natural world to the social order.

Nevertheless, spectacular physics could be appreciated from other perspectives than that of high society. Starting out in a world that was almost exclusively aristocratic, polite science gradually moved outward toward more diverse publics. The practices of naturalists also began to spread to the Parisian bourgeoisie, who increasingly enjoyed collecting specimens and assembling collections of plants, insects, shells, and minerals. This shift was part of a broad transformation of social and cultural practices during this period. Far from Versailles and its court, a new power, public opinion, began to gain ground in Paris in the second half of the century. More than the emergence of a particular content, this was a change of structure: a network of discussions and exchanges carried in speech, spectacle, and print—in other words, a public sphere that served as an intermediary between the private sphere and that of the state. Geographically, this space under construction was the city itself, vibrant and rebellious, illuminated by its cafés, theaters, and streetlights; economically, it constituted a great market in which ideas circulated along with goods; socially, it brought together all the intermediaries, smugglers, traders, and publicists; and ideologically, in a word, it constituted the Enlightenment (see map 5.2). If the public sphere so defined extended across the country and throughout Europe, Paris had the privilege of

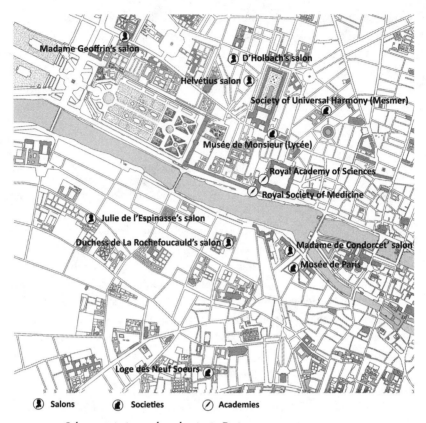

MAP 5.2: Salons, societies, and academies in Paris

being situated in one of its centers: there more than anywhere else public opinion was made, fashions were invented, new ideas were formed. The movement accelerated after 1780, deeply transforming the Republic of Letters.

The Lodge of the Nine Sisters illustrates this evolution. As we saw, it emerged from the salon of Helvétius and his wife. In 1766, the "philosophical dinners" given every Tuesday on the rue Sainte-Anne and limited to a small circle around Madame Helvétius were extended to a larger group of exclusively male invitees. The transformation of this "lodge of the sciences" into the Lodge of the Nine Sisters after the closure of the salon in the rue Sainte-Anne, its move to the rue du Pot-de-Fer, and its takeoff after the initiation of Voltaire, marked a new stage; the lodge metamorphosed into a sort of club, whose activities reverberated with the public. Going even farther, in November 1780 a few members of the Nine Sisters founded the Apollonian Society, rechristened a few months later as the Musée de Paris. This establishment, presided over by Court de Gébelin, was a literary and scientific academy that held meetings weekly, then monthly, at the

Hôtel de Genlis on the rue Dauphine, which also housed the Nine Sisters. But weakened by divisions, the Musée de Paris was soon transformed into an educational institution and in 1786 moved to a hall in the Cordeliers convent on the rue de l'Observance. In 1790, the Nine Sisters was itself transformed into a national society that welcomed women as well as men, organized weekly meetings as well as public sessions, and periodically published its work. Languishing after the storming of the Tuileries on August 10, 1792, the society disappeared by the end of the following year.

The evolution of the Lodge of the Nine Sisters was not an isolated case. Other establishments like the Musée de Paris appeared during these years. The model of polite society associated with private spaces like salons and practices of hospitality and conversation, was giving way to a model of public sociability associated with assembly rooms and the practice of trade and publicity. In 1778, even before Court de Gébelin founded the Apollonian Society, a young adventurer named Mammès Claude Pahin de La Blancherie created a "Salon of Correspondence" that met every month, at first on the rue de la Harpe, then on the rue de Tournon, at the office of the newspaper he had founded on the model of the British *Spectator*.

The basic idea was to bring high society together with artists and men of letters by establishing a means of international communication for the Republic of Arts and Letters. "The general agent of correspondence for the sciences and the arts," as he pompously called himself, secured for this venture the support of both the Academy of Sciences and the Royal Society of Medicine, as well as the financial support of several great nobles, including the king's brothers, the Count de Provence and the Count d'Artois. In both his weekly salon and his newspaper, commercial publicity was closely intertwined with information on the arts, letters, and sciences. Exhibitions for the sale of works of art and objects of curiosity could be viewed there, occasionally including scientific instruments. In 1781, Pahin moved his salon to the Hôtel Villayer on the rue Saint-André-des-Arts, where he continued his activities until 1787, when, stricken by debt, he suspended his salon and his newspaper, before fleeing his creditors by going to London.[29]

From Musée de Monsieur to the Lycée

Pahin's venture, like that of Court de Gébelin, seems to have been, at least in part, a victim of the vogue for the Musée de Monsieur, created by a new arrival on the scene, Jean-François Pilâtre de Rozier. The son of an innkeeper from Metz, Pilâtre had attracted the patronage of Dolomieu and of La Rochefoucauld d'Enville, even though he was still only an apothecary's apprentice. In 1775 he came to the capital to pursue his studies. After spending a year with the apothecary Mitouart, he had abandoned pharmacy for the natural sciences. The physician François

Weiss, who had made a fortune through selling secret remedies, developed an interest in him and took him under his wing. After the doctor's death in 1777, Pilâtre benefited from the support of his widow. Not only did she give him her deceased husband's cabinet of natural history on the rue de Saint-Avoye in the Marais, but she bought him an ennobling office as valet to Madame (the wife of Monsieur, the Count de Provence), a title that Pilâtre, now fashioning himself de Rozier, hastened to convert into the more prestigious post of secretary of the cabinet of Madame.

The young apprentice apothecary had given his first courses in the cabinet in the Marais. In 1780, he also presented his first paper (on the commercial use of dyes) to the Academy, which he later published in the *Journal de physique*. The following year he launched the idea of a "royal musée under the protection of Monsieur and Madame," an establishment open to both women and men that would bring under one roof laboratories, a cabinet of natural history, and a range of courses in the sciences and foreign languages. Pilâtre, who now claimed he was steward of Monsieur's cabinets of physics, chemistry, and natural history, as well as apothecary to the Prince of Limbourg (another made-up title), secured for this project the patronage of the authorities and of powerful aristocrats. He also solicited (although in vain) a seal of approval from the Academy of Sciences and the Royal Society of Medicine.

The Musée de Monsieur opened its doors on the rue Saint-Avoye in December 1781. Pilâtre himself gave the courses in physics, while Jean-Guillaume Wallot taught mathematics, Pierre Flandrin the anatomy of horses, and Jean-Joseph Süe human anatomy. Other professors who had experience with public courses were called upon to teach there. Thanks to subscriptions and loans, Pilâtre was able to buy a lot of equipment, with which he was able to give entertaining physics demonstrations. His flight over Paris in a balloon on November 21, 1783, in the company of the Marquis d'Arlandes, made him famous and helped him to re-launch his musée. A year later, he moved it to the rue de Valois close to the Palais-Royal, at the corner of rue Saint-Honoré, the ideal location for a fashionable establishment.

The opening of these new premises met with great success. At that time the musée had 726 members: 126 founders (men and women), each of whom paid seventy-two livres a year; 133 members of academic institutions (including the Musée de Paris), who paid no dues at all; and 467 subscribers. It was run by a board composed of officers drawn from among the founders. Moreau de Saint-Méry, who was a member of the Nine Sisters, and Bontemps, the secretary to the Count de Provence, were the managers. Pilâtre, crowned by his aerostatic adventure, was merely the treasurer. He no longer taught, but he chose the professors. If the musée's public was larger than that of the salons, it was still limited to the wealthy, mainly aristocratic elite of the capital. Women were a significant minority. The musée also attracted men of letters in search of contacts and patrons.

At its new location on the rue de Valois, the Musée de Monsieur had an enormous laboratory equipped for physics and chemistry, two small laboratories, a large lecture hall, a meeting room, an administrative office, a library, and an exhibition gallery decorated with busts of Étienne Montgolfier and Buffon sculpted by Houdon. The laboratories, which were serviced by three young assistants, were equipped with three hundred scientific instruments, valued at approximately seventeen thousand livres in 1785. The library contained close to five hundred volumes, including the publications of various academies, and subscriptions to fifteen periodicals. Dozens of chairs decorated with balloon motifs furnished the meeting and lecture halls, which were lit by the latest model of Quinquet oil lamps.

The establishment should have foundered after the sudden death of Pilâtre on June 16, 1785, the first casualty of ballooning. The physicist, who had spent money like crazy, left behind substantial debts. Bailed out by the Count de Provence and the Count d'Artois, followed by other lofty personages, the Musée de Monsieur then adopted the name of Lycée. Charged with drawing up the program of the new establishment, the Marquis de Montesquiou gave it a more academic orientation. The Lycée became, at least on paper, a sort of extension of the official academies toward the elite public. Two members of the Académie française gave lectures in history and literature.

Several members of the Academy of Sciences were also invited to serve as professors. Mathematics was given to Condorcet, physics to Monge, and chemistry to Fourcroy. Condorcet gave the opening lecture of his course, but then turned the rest over to a young protégé of Monge's, the mathematician Sylvestre-François Lacroix. Monge himself was content to serve as Deparcieux's patron. Only Fourcroy actually gave his own course, on top of the three he was already teaching, at the Royal Botanical Garden, at the veterinary school of Alfort, and in his laboratory on the rue des Bourdonnais. Over the following years, the number of subscribers remained at a high level despite some decline: 650 in 1786, 600 in 1787, 500 in 1788, and 400 in 1789. On the eve of the Revolution, the Lycée was rich and well attended. Despite serious financial and political difficulties, it managed to traverse the Revolution and, as the Athénée, survive until 1849.[30]

The Journal de Paris

Newspapers played a decisive role in the launching of the new musées. Both a vector of information and a space of relative freedom, the Parisian press had long contributed to the formation of public opinion, despite censorship and the finicky control of royal power. It is striking how much interest the press showed in the sciences, in fact much more than it does today. The gazettes announced public courses in physics and medicine and published reviews of scholarly works and accounts of the public sessions of the various academies. They also discussed new theories and discoveries.

Traditional journals, like the *Mercure de France* and the *Année littéraire* (both published by Panckoucke) and the *Journal des Savants*, controlled by the academies, affected a learned and serious tone. By contrast, the manuscript newsletters and correspondences known as *nouvelles à la main* and written for foreign readers, like those of Grimm and Bachaumont (the *Mémoires secrets*), created a racket that reverberated across Europe from the court and the town. A special place must be given to periodicals that purveyed practical information and classified ads, such as the *Petites affiches* of Paris, which appeared twice a week, the *Avant-coureur*, a weekly that ceased publication in 1773, and above all the *Journal de Paris*, the first Parisian daily, founded in 1777 with the government's blessing. Read in both town and court, including by the royal family, the *Journal de Paris* enjoyed an immediate success and made the fortune of its proprietors. Any allusion to politics was prohibited, but its thousands of subscribers appreciated the precise information it provided on the weather, the theater, new books, and the arts and sciences.

There were close ties between the *Journal de Paris* and the Lodge of the Nine Sisters, to which two of its founders belonged: the publicist Louis d'Ussieux and the pharmacist Antoine Cadet de Vaux. The two other founders were the clockmaker Jean Romilly, author of many articles on clockmaking in the *Encyclopédie*, who provided the meteorological observations published each morning as the lead item in the *Journal*, and his son-in-law, the lawyer Guillaume Olivier de Corancez, a friend of Jean-Jacques Rousseau. Cadet de Vaux, who seems to have been the prime mover, benefited from the patronage of Lieutenant of Police Lenoir and Director General of Finance Necker. A student of Parmentier's, brother of the chemist Cadet de Gassicourt of the Academy of Sciences, and active himself in philanthropy and passionate about the sciences, Cadet was well known for his research on bread making and public health. Of course, the *Journal* regularly mentioned these issues in its columns, an interest that is even less surprising since it was Cadet himself who took charge anonymously of the sections of the paper devoted to scientific matters. He also covered the return of Voltaire to Paris and his admission to the Lodge of the Nine Sisters in 1778. For questions of astronomy, the *Journal* generally called on the academician Lalande, the master of the Nine Sisters.

Despite its relations with the savant milieu, the *Journal de Paris* demonstrated a fine eclecticism, mixing publicity for public courses and plays performed at the fairs with serious announcements about the Academy of Sciences. Information was presented under different rubrics, including the sciences, physics, chemistry, natural history, botany, and medicine. Loyal to the government that was its patron, the *Journal de Paris* praised the action of the authorities and stressed the utility of the sciences and the arts. However, it cannot be reduced to a simple organ of the state. It organized debate and discussion, amplifying and publicizing it, thus contributing to the formation of public science in Paris. From its early years, it campaigned for the development of a scientific policy of public hygiene to address a variety of problems that plagued Paris: emptying cesspools and sewers, improving cemeteries, hospitals, and the water supply all regularly filled its columns. Mercier, Cadet's brother in the Lodge of the Nine Sisters, took up all these themes in his *Tableau de Paris*. The *Journal de Paris* was also very attentive to medical questions such as the use of electricity, obstetrics and childbirth, and animal magnetism.

The *Journal de Paris* was above all a commercial enterprise. Concerned with pleasing readers, it gave publicity to all discoveries (at least those claimed as such) that might astonish readers and attract their attention: Mesmerism, of course, but also Barthélemy Bléton's divining rod, which it covered over several issues. It contributed to the promotion of Mesmer, Thouvenel, Marat, Pahin de la Blancherie, and Pilâtre de Rozier. Its role was even more decisive in the launching of balloon experiments. While it made little distinction between charlatans and serious innovators, it made sure never to cross the savants. Interest in science fell off after 1785, when political passions took over, but the *Journal de Paris*, apolitical and governmental, was not swept up in this movement.[31]

Notes

1. The quotation is from S. LINGUET, "Arrivée de M. de Voltaire à Paris," *Annales politiques, civiles, et littéraires du dix-huitième siècle* 3 (1777), 387.
2. AMIABLE, 1989, 32.
3. AMIABLE, 1989, and Ch. PORSET, "Matériaux inédits relatifs à la loge des Neuf sœurs," in Ch. PORSET, ed., *Studia Latomorum et historica: Mélanges offerts à Daniel Ligou*, Paris, Honoré Champion, 1998, 347–373; on the academic festival for the peace signed at Versailles, see *Journal de Paris*, May 18, 1783.
4. Helvétius reproached the lodges for "neglecting the sciences and arts in order to occupy themselves exclusively with august twaddle" (according to Ch. PORSET, critical commentary in AMIABLE, 1989, 15).
5. LILTI, 2015, 260–272.
6. LA MICHODIÈRE, "Essai pour connaître la population du Royaume et le nombre des habitants de la campagne," *HMAS*, 1783, 703–718, 708.

7. On "the court and the town", see LILTI, 2015, 73–80; for a "literary" definition, see D. GOODMAN, 1994, 90–135. On men of letters and the world, see LILTI, 2015, 169–222, and particularly on Voltaire, 187–188.

8. J.-P. MARAT, *Les Charlatans modernes ou lettres sur le charlatanisme académique*, Paris, 1791, 13, note.

9. Letter from La Rochefoucauld d'Enville to Desmarest, May 8, 1770. Copy in the Desmarest file, Archive of the Academy of Sciences.

10. LETOUZEY, 1989, 128 and 139.

11. On Messier and Bochard de Saron, see J.-D. CASSINI, *Éloge de M. de Saron, Premier président du parlement et membre honoraire de l'Académie royale des sciences de Paris*, Paris, 1810, as well as the manuscript biography of Messier in the Delambre papers, Bertrand Archive, Bibl. Institut, ms. 2041.

12. The statement by Buffon about Thouin is in BACHAUMONT, 1777–1789, vol. 31, 190 (March 25, 1786); the encounter between Voltaire and Franklin is reported in BACHAUMONT, 1777–1789, vol. 11 (April 29, 1778).

13. Diary of John Adams, April 20, 1778.

14. CONDORCET, "Éloge de Cassini de Thury," *HMAS* (1784), 61–62.

15. D'ALEMBERT, "Essai sur la société des gens de lettres et des grands, sur la réputation, sur les mécènes, et sur les récompenses littéraires," *Œuvres complètes*, 1822, vol. 4, 359.

16. F. VICQ D'AZYR, *Éloge de M. de Lassone*, Paris, 1789; A. DESORMONTS, *Contribution à l'étude du XVIIIᵉ siècle médical, Claude-Melchior Cornette, apothicaire, chimiste, hygiéniste, médecin de la cour (1744–1794)*, Paris, 1933.

17. Eulogy of Le Monnier, in CUVIER, 1819–1827, vol. 1, 83–107.

18. BRITSCH, 1926, 124–125, and PAYEN, 1969, 48–51. On the Duke d'Orléans's financing of Darcet's research, see A. PILLAS and A. BALLAND, *Le Chimiste Dizé: sa vie, ses travaux, 1764–1852*, Paris, 1906, 18–22; on the annuity of eight hundred livres granted by the Duke d'Orléans to Laplace in 1786, see HAHN, 2004, 72.

19. Letter from La Rochefoucauld d'Enville to Desmarest, October, 14, 1772. Copy in the La Rochefoucauld file, Archive of the Academy of Sciences.

20. On Comus's classes for the Duke de Chartres, see BACHAUMONT, 1777–1789, vol. 7, 14 (June 21, 1773). The *Nouvelles récréations physiques et mathématiques* by Edme-Gilles Guyot went through three successive editions between 1769 and 1786; the list of material sold by the author, with his price, is found at the end of each volume. See B. BELHOSTE and D. HAZEBROUCK, "Récréations et mathématiques mondaines au XVIIIᵉ siècle: le cas de Guyot," *Historia mathematica*, 4 (2014), 490–505.

21. E. G. GUYOT, *Nouvelles récréations physiques et mathématiques*, 3rd ed., vol. 1, 1786, 218.

22. J.-J. ROUSSEAU, *Rousseau juge de Jean-Jacques*, second dialogue, text written between 1772 and 1776 and published after his death in 1782.

23. DARNTON, 1968.

24. Forty-five cabinets of natural history are described in THIÉRY, 1787.

25. M.-P. DION, *Emmanuel de Croy (1718–1784), Itinéraire intellectuel et réussite nobiliaire au siècle des Lumières*, Brussels, Editions de l'Université de Bruxelles, 1987, 139–140.

26. THIÉRY, 1787, vol. 1, 126–128 (Gigot d'Orcy) and 575–576 (France de Croisset).

27. J.-C. DELAMÉTHERIE, "Notice sur la vie et les ouvrages de M. Romé de l'Isle," *Journal de physique* 36 (1790), 315–323.

28. J. LAMARCK, *Mémoire sur les cabinets d'histoire naturelle et particulièrement sur celui du Jardin des plantes*, Paris, 1790, 2.

29. On the evolution of Parisian musées, see GUÉNOT, 1986; GOODMAN, 1994, 233–280; and LYNN, 2006, 72–75; on the Salon de la Correspondance, see LYNN, 2006, 76–80, and L. AURICHIO, "Pahin de la Blancherie's Commercial Cabinet of Curiosity (1779–87)," *Eighteenth-Century Studies* 36 (2002), 47–61.

30. On the musée of Pilâtre and the Lycée, P. DORVEAUX, "Pilatre de Rozier, apothicaire (1754–1785)," *Bulletin de la Société d'histoire de la Pharmacie*, 1920, 209–220 and 249–258, C. CABANNES, "Histoire du premier musée autorisé par le gouvernement," *Nature* 65 (1937), 577–583, and LYNN, 2006, 82–90.

31. N. BRONDEL, *Journal de Paris* (notice no. 682) in J. SGARD, *Dictionnaire des journaux (1600–1789)*, 2 vols., Paris, Universitas, 1991, vol. 2, 615–627.

6

Spectacles and Marvels

AT THE END of 1783, an anonymous clockmaker who lived a hundred leagues from the capital announced a new kind of "experiment" in the *Journal de Paris*. He proposed to cross the Seine near the Louvre by walking on water, and to do so more rapidly than a horse crossing at a fast trot over the Pont Neuf. He launched a subscription that would guarantee him a hundred louis if he succeeded. The date was fixed for January 1, 1784. His means would be "elastic clogs" attached to each other by a bar "like a two-headed bullet." In a few days, the subscription was almost filled. Among the subscribers listed in the *Journal de Paris* were, it was said, the royal family itself for 1,080 livres (although hidden behind the veil of a "Versailles society"). A huge crowd was anticipated (Fig. 6.1).

The administration was preparing to install grandstands for the subscribers when the revelation came: this was nothing but a prank mounted by a joker named Charles-Jean de Combles, honorary councilor at the Lyon Mint. This was apparently not the first prank he had pulled: being "brazen and loving to play tricks," he had published under a pseudonym a scatological parody of Voltaire's tragedy *Zaïre*. Although in better taste, the hoax was altogether too outrageous this time, for important people had been deceived and made to look ridiculous. A frightened Combles revealed the truth to the intendant of Lyon, who quickly warned the authorities in Paris. The affair went right up to the king, who laughed about it and kidded his brother Monsieur, who had taken part in the subscription. Yet the perpetrator of the prank does not seem to have been subjected to an investigation.[1]

Combles's mischief did not occur by chance. He was inspired by the infatuation of Parisians for balloons, for he thought, "From now on, they could be made to believe in any marvel you choose." All he had to do was to parody the methods of those who had organized the first flights over Paris: announcements in the press and a public subscription. Combles had found unwitting accomplices to his

VUE DU TRAJET DE LA RIVIERE DE SEINE À PIED SEC,
au deſſous du Pont neuf au moyen des *SABOTS* élaſtiques.
Dedié aux Souscripteurs.

Gravé d'après le Deſſin du Redacteur du Journal de Paris Janv.ʳ 1784.

De la part de M. de Comblas Magiſtrat de Lyon, Inventeur.

FIGURE 6.1: View of crossing the Seine with dry feet, Engraving by François Sellier after the drawing by the editor of the *Journal de Paris* (*Journal de Paris*, January 1784). Courtesy of the Musée national du château de Pau, photo Jean-Yves Chermeux.

hoax in journalists, who were as zealous as they were clueless—but they were also his first victims. After having been played, they had to reimburse the subscribers.

While the affair revealed the credulousness of the Parisian public and the influence of newspapers, it also made evident a general passion for science and its achievements. In fact, the hoax posed a serious problem in a burlesque style: how to judge inventions that were handed over to the speculations of commerce and publicity. What confidence could one have in the demonstrators of such phenomena for whom public favor seemed to count more than verifiable facts? What kind of science could one promote in a public sphere in which the tribunal of public opinion had replaced the tribunal of savants and men of refined taste? Ultimately it was the very legitimacy of the spectacle of science that was being challenged.

A Subscription at the Café du Caveau

The first venture that had inspired the parody of the "elastic slippers" had been launched a few months earlier in the garden of the Palais-Royal, the

Parisian residence of the Dukes d'Orléans. On July 28, 1783, Buffon's protégé Barthélémy Faujas de Saint-Fond, a naturalist-assistant at the Royal Botanical Garden, opened a subscription in the Café du Caveau, located in the Galerie de Beaujolais, which looked onto the garden, to pay for the experiments of Jacques Charles. The rumor had just made its way to Paris that the Montgolfier brothers, paper manufacturers in the Auvergne, had tested "an aerostatic ma- chine" on June 4 that had risen to an altitude of more than five thousand feet. The challenge was to reproduce this feat.

In fact, in Paris the idea of such a machine was already in the air. The pre- vious year, the mechanic Jean-Pierre Blanchard claimed to have taken off in a "flying boat." With the financial backing of the Count d'Artois and the Duke de Chartres, he had announced that his flight would take place on May 5, 1782. Spectators crammed into the townhouse of the Abbé Vienney on the rue Taranne, at the intersection of the rue du Sépulcre (today the rue du Dragon), where the machine was located. It was a complete failure. But this did not discourage the inventor, who now proposed to lift anchor in the northeastern suburb of Pantin and head east toward the gardens of Raincy. The Montgolfier brothers' success, conveyed to the Finance Minister Calonne by the provincial authorities and quickly communicated to the Academy, had reached a public that was already primed for it.

While Paris was ignorant of the procedures being used by the provincial paper manufacturers, the savants immediately thought of inflammable air (hy- drogen), whose properties were well known. Weren't people already amusing themselves by watching soap bubbles filled with this light gas rise into the air? Jacques Charles, who had a physics cabinet on the Place des Victoires, where the Italian savant Alessandro Volta had made just such a demonstra- tion, quickly decided (with the support of Faujas) to attempt a flight. From a commercial and publicity standpoint, the affair was an immediate success; its contribution to science was another matter, to which we will return below. Thanks to word of mouth, the subscription filled up in a few days. According to Faujas, "the most illustrious names" contributed to it. For the author of the *Mémoires secrets*, Charles was merely an "experiment maker" who was trying to profit from public curiosity in order to make money. Even if his enthusiasm was totally sincere and disinterested, the venture no doubt also represented a golden opportunity to promote his physics cabinet. Dubuisson, the proprietor of the Café du Caveau, took charge of collecting the money, which reached a hundred louis. Each of the eight hundred subscribers received a ticket that would allow them to witness the experiment when the day came. A month later Faujas made a deal with the Montgolfiers to launch a second subscription at the café for the striking of a medal commemorating the first flight in the Auvergne.[2]

When the subscription for Charles was opened, the Café du Caveau had just reopened its doors, with magnificent decor, in a Palais-Royal that was bubbling with activity. A few years earlier the Duke de Chartres, who lived there, had undertaken a vast real estate operation to refloat his finances. The scheme consisted of subdividing a strip of land around the edge of the garden into lots and building shops on it that would cater to the fashion and luxury trades. The idea had originated with the Marquis de Ducrest, Madame de Genlis's brother, who was the duke's chancellor and a great planner of projects, both technical and financial. This one was entrusted to a young architect, Victor Louis. The duke had secured the consent of the king and had overcome vigorous opposition from neighbors and habitués of the garden.

Work began in the summer of 1781 and was already well advanced two years later. On three sides of the garden rose the uniform and richly decorated façades of new pavilions, with their columns and arcades serving as a public promenade. The interiors had barely been sketched out when the first tenants, shop owners, and restaurateurs moved in. At first public response was lukewarm, but the wind changed quickly as promenading visitors started to come back to the garden and marvel at its splendor. In the summer of 1784, the new Palais-Royal was definitively launched; the replanted trees provided shade and the crowds were squeezed together under the arcades, which were illuminated in the evening by 180 lamps.[3]

The place remained fashionable until the Restoration, but nothing would equal the decade of the man who would later style himself "Philippe *Égalité*," between 1784 and 1794. As Mercier wrote in 1788, the Palais-Royal was at the time "the capital of Paris," "a small luxurious city enclosed within a big one." What struck observers above all was the mixture of social groups and types. Duchesses, tourists, and military veterans crossed paths anonymously with prostitutes, lackeys, and wig makers. "There you see prelates jostled by pedants, princes by street urchins. Anyone who has not seen this chaos, this bizarre mixture, this strange confusion, cannot boast of really knowing Paris," sang Guigoud-Pigale in his comedy *Le Baquet magnétique* (The magnetic tub).[4]

A commonplace of descriptions of the Palais-Royal, this supposed encounter among people of both sexes and all social conditions did reflect an incontestable fact: the diversity of the commercial offerings. The new Palais-Royal was supposed to have been dedicated to luxury alone, but out of financial necessity, the duke had arranged for a series of inexpensive boutiques in the wooden galleries in 1784. This was the famous "Tartar camp" which contributed greatly to the success of the venture. A steady of parade of gawkers and curiosity-seekers came there to be amazed and to window-shop. The boutiques under the arcades of the pavilions were more expensive and more luxurious. There were found the cafés, restaurants, theaters, and gaming halls for which the Palais-Royal was famous.

Gardens and Boulevards

All this was nothing new. The Palais-Royal accommodated and extended the recreational and commercial activities of the fairs and boulevards. The two great Parisian fairs had been held since time immemorial, one in the Faubourg Saint-Germain (between the rue du Four and the rue de Buci) from February until Palm Sunday, and the other in the Faubourg Saint-Laurent (where the Gare de l'Est stands today) from the end of July to the end of September. Still very lively in the early eighteenth century, these fairs were attended by both the common people and the elite, who found there a great variety of merchandise, from the everyday to the luxurious, as well as plays and fairground games. This was where the *opéra comique* had been invented, a theatrical genre that mingled pantomime, dialogue, singing, and dancing on temporary wooden stages.

However, the fairs collapsed in the second half of the century, their decline accelerated by a fire at the Saint-Germain fair in 1762. The Opéra Comique theater had just merged with the Comédie Italienne, migrating first to the Hôtel de Bourgogne, then to the Salle Favart near the Palais-Royal, where this genre would remain. In fact, all the festive activities associated with the great fairs gradually moved to permanent venues in the city. The boulevards were the principal beneficiaries of this evolution, but by settling down the fair theaters also found a home in a new sort of entertainment venue being created by private entrepreneurs.

These new spaces, situated for the most part at the western and northern edges of the city, were of two kinds: commercial pavilions and amusement parks. In the 1760s, the physicist and pyrotechnician Jean-Baptiste Torré, who was already mounting spectacles on the boulevards, opened a richly decorated pavilion with salons and a garden in the style of London's Vauxhall Gardens. His Fairground Festivals, nicknamed "Torré's Vauxhall" or the "Summer Vauxhall" by the public, offered fireworks shows, concerts, and parades, as well as housing boutiques. At almost the same time, a similar establishment (the "Winter Vauxhall") opened at the Saint-Germain Fair, offering its clientele balls, concerts, and lotteries. Other pavilions, like the Coliseum in the Champs-Élysées and the "Redoute chinoise" at the Saint-Laurent Fair (which we have already encountered), adopted the same formula with more or less success.

The amusement parks developed in parallel with the commercial pavilions. They combined the pleasure of gardens with the spectacle of the fairs. Several royal or princely gardens such as the Tuileries, the Luxembourg Gardens, the Palais-Royal, and the Royal Botanical Garden had been open to the public for a long time, and whole families came there for a stroll. Itinerant merchants sold beverages and sweets or offered to entertain visitors under the watchful eyes of guards. There was also the garden of the Arsenal to the east, as well as the Champ de Mars and the Champs-Élysées to the west. After 1770, Parisians enjoyed a new type of garden, called either "English," "Chinese" or "Anglo-Chinese," that was very different from traditional French gardens in both conception and function. These private parks, which charged an entrance fee, were created by princes and rich financiers in the western faubourgs and were designed for visual pleasure. The best known, situated on a hillside where the Gare Saint-Lazare now stands, belonged to the treasurer of the Navy Simon Boutin, who had christened it Tivoli in homage to the Roman gardens. Opened to the public in 1771, Tivoli (or "Boutin's Folly"), offered a series of picturesque gardens ornamented with fake ruins and pavilions spread across fifteen acres, with entertainment, carousels, swings, and a cabinet of mineralogy.

Three times larger, the Monceau Garden (or "Chartres's Folly"), designed by the painter and engineer Louis Carrogis, known as Carmontelle, for his master, the Duke de Chartres, was also constructed like a stage set, with pavilions and fabricated marvels. Finally, the Ruggieri Brothers, pyrotechnic artists to the king, opened a garden with an Oriental theme at the Château des Porcherons near Boutin's Folly in 1765. They mounted very successful fireworks shows there during the summer; in the 1780s they hosted pantomimes and balloon flights there as well. Although of a very different order, the new Palais-Royal was a direct descendent of these pleasure gardens and commercial pavilions that had flourished in Paris for twenty years. But one of the advantages it had over its predecessors was its location near the boulevards.

It was in fact on the boulevards (where the Summer Vauxhall had been built) that the fairground theater found its public again. Laid out at the end of the seventeenth century at the northern edge of the city where the ancient city walls of Charles V and Louis XIII had stood, by the end of the following century the boulevards formed a long promenade planted with trees and lined with gardens, mansions, boutiques, cafés, and theaters. A large alley that ran down the center was reserved for carriages. Those out for a stroll took advantage of two shady paths along the sides paved with sand. At night, streetlamps lit the scene. The *Almanach du Voyageur à Paris*, the popular guide for visitors, noted that "games, pleasures, refreshments, spectacles, and music seem to have found a permanent home there."

Street theater gripped the crowds that pressed around the trestle stage. On the Boulevard du Temple, the theater was flourishing. Troupes from the fairground that specialized in pantomime would perform for their regular customers: the great rope dancers from the Nicolet Theater (renamed the Gaîté in 1792), the actors from the Ambigu-Comique directed by Audinot, and the Variétés-Amusantes, as well as many other smaller spectacles of marvels and entertainment. People of quality went slumming on the boulevards, rubbing up against the popular classes there as they had previously at the great fairs and would later at the Palais-Royal.

At the Merchant Palace

After 1783, this whole pleasure industry poured into the Palais-Royal, which thereby achieved a synthesis of fair, promenade, and garden, but with a fundamental difference: the new center of attraction lay at the heart of Enlightenment Paris and under the direct patronage of a prince of the blood. Many observers (and historians after them) described for their readers the life of the new Palais-Royal; its fame spread quickly across Europe. Here I will mention only those things either directly or indirectly related to science that could be found there. Note that at the Palais-Royal, as at the fairs and the other entertainment venues, three kinds of activities came together in one place: commerce, spectacle, and conversation. But mercenary activity dominated and directed the ensemble—hence its nickname, the "Merchant Palace."

As we have seen, all kinds of shops could be found on the ground floor of its sixty pavilions and in the wooden galleries, but their common denominator was the superfluous, ranging from true luxury to cheap imitations. For example, there was Verrier's shop in the Valois gallery, a sort of consignment store selling "precious goods at a fixed price," and Pascal Nozeda's physics shop next door to the Café du Caveau, where optical instruments and barometers were sold, as well as many independent booksellers in the wooden galleries, who sold banned literature under the counter. Through its luxury trade, the Palais-Royal was an integral part of the elegant rue Saint-Honoré, celebrated for its stores that specialized in novelties, like À la Couronne d'Or (At the golden crown), owned by the mercer Dominique Daguerre. On the first floor of the Regency Café on the Place du Palais-Royal, Henry Sykes offered for sale all sorts of scientific instruments imported from England. There were also physics instruments for sale at the Grand Balcon, the store owned by Rond, and at the one belonging to the optician Molteno on the rue du Coq Saint-Honoré, as well as at the Widow Bianchi's shop near the Croix-du-Trahoir fountain on the corner of the rue de l'Arbre Sec.

Someone strolling around the Palais-Royal would also find theaters there, the two main ones being the Théâtre Beaujolais, a marionette theater transformed into a show that featured children singing offstage, known as the "Little Actors," and the Théâtre des Variétés-amusantes, which moved there from the boulevard in 1784. In addition to these theaters that seated several hundred people, there were also small-scale spectacles: enchanting ones like Séraphin's Chinese Shadows, two marionette shows, Castagna's authentic Italian fantoccini and the French Pygmies, and instructive ones such as Tessier's Children's Museum. The cabinets of curiosities enjoyed great success: at number 8, near the Théâtre Beaujolais, was Curtius's wax museum, where the future Madame Tussaud was taught the art of wax modeling as a young girl; on the second floor of number 44, François Pelletier's cabinet of mechanics, physics, and hydraulics was soon replaced by Michel Adanson's natural history cabinet; in the wooden galleries, small shows like the wax "Belle Zulima," the giant Butterbrodt, and mechanical billiards offered less refined entertainment.

People came to the Palais-Royal not only to look but also to drink, eat, gamble, and converse, as well as to find prostitutes. The arcades sheltered restaurants and several cafés, on the ground floor and in the basement: Foy's Café and the Café du Caveau, which were already there before the renovation of the garden, as well as new ones such as the Café Corazza and the Mechanical Café. Some patrons stopped by to sip a liqueur or a coffee, while the regulars read the newspapers there. Certain cafés had games or small performances. At the Caveau, "the rendezvous of politicians, wits, spreaders of news, schemers, moneychangers, speculators, and ambitious idlers," according to Mayeur de Saint-Paul, the old dark cellar had given way to a lovely gallery ornamented with busts of great musicians, mirrors, and two large landscape paintings. On a marble table crowned with the medallion of the Montgolfier brothers could be read in gilded script: "Two subscriptions were opened at this table: the first on July 28 [1783] to repeat the experiment at Annonay; the second on August 29, 1783, to pay homage with a medal to the discovery made by Messieurs de Montgolfier." Upstairs, above the cafés, were the salons and clubs: gambling clubs like the Salon des Arcades and the Chess-Lovers Club, social clubs like the Olympic Club (founded by the masonic lodge of the same name), and learned and literary societies like the Salon des Arts, whose reading room was right above the Café du Caveau. Finally, at the very top was the realm of prostitution (see map 6.1).[5]

In 1786, Philippe, now the Duke d'Orléans, opened a subscription to construct a great subterranean circus in the middle of the garden intended to become the lynchpin of the Merchant Palace. A long arena surrounded with overhead lighting would host balls, concerts, and equestrian spectacles. All around it an Ionic colonnade would enclose a gallery on two levels and a rooftop terrace with pavilions, kiosks, and trellises (see plate 6). The Circus, lit by Quinquet lamps, would also include boutiques, an exhibition hall for artists, two billiard halls, a grand café, and a restaurant. Opened in 1789, it was reminiscent of the Coliseum and, at the same time, inspired by Astley's English Amphitheater on the rue du Faubourg du Temple. It could not have opened at a worse moment and was a commercial fiasco. The Confederation of the Friends of Truth, which aimed to be the rendezvous of all the fraternal societies, held its meetings there in 1790; in 1792 the Lycée des Arts moved in. But the building was destroyed in a fire in 1798.

The collapse of the Circus was merely a setback. The immense and durable success of the Palais-Royal was striking, and it rapidly became a symbol and a myth. Tourists came from all over; publicists described it as a place of learning, pleasure, and vice. Mercier imagined the Palais-Royal hosting (as in China) the

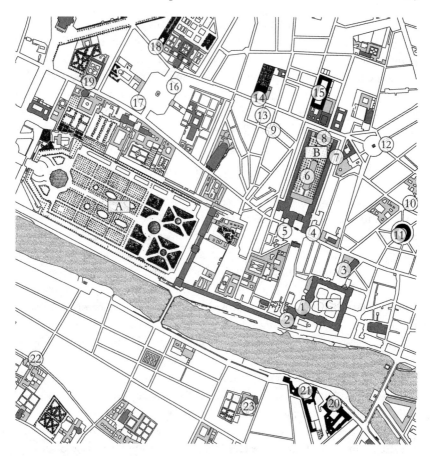

A. Tuileries Garden. B. Palais-Royal. C. Louvre.

1. Royal Academy des sciences. 2. Royal Society of medicine. 3. Chair of hydrodynamics. 4. Lycée.
5. Sykes, dealer of scientific instuments 6. Palais-Royal's circus. 7. Forster, natural history dealer. 8. Café du Caveau.
9. Helvétius, salon. 10. Société de l'harmonie universelle (Mesmer). 11. Wheat Hall. 12. Charles' cabinet of physics.
13. Holbach's salon. 14. Finance Ministry. 15. King's Library. 16. Gigot d'Orcy's cabinet of natural history.
17. Joubert's cabinet of natural history. 18. Townhouse of the lieutenant of police. 19. Madame Geoffrin's salon.
20. Mint. 21. Collège des Quatre-Nations. 22. Julie de Lespinasse's salon.
23. Townhouse of La Rochefoucauld d'Enville.

MAP 6.1: The Palais-Royal in Enlightenment Paris

spectacle of a city in miniature. The informer Restif, strolling in the evening under the arcades, followed the streetwalkers and extracted their secrets for the pleasure of his readers. Paris in its entirety was performing on its stage in a setting in which the pure and the impure, the high and the low, all social conditions and circles mingled together. "The art of making a ragout," Mercier noted, could be found "right next to the most advanced sciences. The brilliant rags of libertinage hang near the surgical instruments that it will make necessary."[6]

These unexpected encounters, real or imaginary, defined a space in which merchandise and information were constantly being exchanged. In this condensed public sphere under the protection of the House of Orléans, a rallying point for artists, politicians, speculators, entertainers, and luxury merchants, a veritable anti-court to Versailles, where public opinion was shaped, a modern Paris was born: enterprising, selfish, and playful. This sense of it as a microcosm explains the dreamlike power of the place that inspired writers from Restif to Balzac.

The Spectacle of Balloons

The Palais-Royal was made in the image of Paris, just a more intense and more concentrated version of the spectacular character of the city itself. The urban setting, with its ceremonies, prospects, and monuments, was designed to exalt authority and traditional hierarchies. Paris, on which the monarchy had mapped out its royal will, was no different. The layout of Place Louis XV after 1750 and its extensions (the Louis XVI Bridge to the south, the Champs-Élysées to the west, and the Madeleine to the north) provide proof, should any be necessary, that the monarchy had not renounced imposing its mark on the city. But in the spectacle that Paris offered its visitors and its inhabitants, what was most striking was not these monumental constructions but rather the multiple and fleeting signs that Parisians themselves were inscribing on the urban space through their own actions: posted bills, shop signs, store windows, lighting, odors, street cries, and traffic jams. Every contemporary observer noted the bubbling tumult of Parisian street life. None of this was new, of course, except perhaps the livelier sensations, the deeper imprint of urban life and social relations upon minds and spirits, a more sustained interest in the city's resources and their effects. This change in sensibilities must be correlated with the emergence of a public sphere. At the same time, at first spontaneously and then consciously, the urban collective became aware of itself as a living, suffering, and thinking reality, and thus gradually elevated itself to the level of an autonomous political entity.

The sciences accompanied this movement through their own performances. Always surprising, often droll, sometimes adulterated, they were part of the city's spectacle. And so the Palais-Royal launched the balloon craze. The subscription opened by Faujas at the Café du Caveau in July 1783 made possible the construction of a balloon in less than a month, ahead of Étienne Montgolfier, who had just arrived in Paris. The first hydrogen balloon rose in the Parisian sky on August 27. It was twelve feet in diameter, and its envelope was made of taffeta coated with rubber. To produce the "inflammable air" (that is, hydrogen), the Robert brothers poured vitriolic acid on iron filings, which was the usual procedure. This delicate and dangerous operation took place in the courtyard of Charles's laboratory on the Place des Victoires, a stone's throw from the Palais-Royal. Four days later, the filled balloon rose to a hundred feet, attracting a huge crowd.

This popular success convinced the organizers to move the experiment site from the Périer Brothers' property in Chaillot to the Champ de Mars. The balloon, escorted by a police detachment, was transported through the streets during the night of August 26–27. The next day an enormous crowd estimated at close to one hundred thousand people witnessed the first release. Subscribers and their guests were in the first rows of seats; other spectators massed along the quay and on the Chaillot hillside. At five o'clock in the afternoon the balloon took off in the rain to the sound of a canon. Within two minutes, it was lost in the clouds. The experiment had attracted the curious—and also the savants. The engineer Jean-Baptiste Meusnier, assisted by the astronomers Joseph d'Agelet, Guillaume Le Gentil, and Edme-Sébastien Jeaurat of the Academy of Sciences, had undertaken to observe the balloon's trajectory, with the intention of verifying the laws of movement of a heavy body in a resistant fluid, but the measurements failed due to the bad weather. As for the balloon, it provoked terror among peasants when it landed in the village of Gonesse, about eighteen miles from the Champ de Mars.[7]

Faujas had launched the experiment and the Robert brothers had carried it out, but the inventor of the balloon was Jacques Charles. A man of the world and a lover of music, this supposed illegitimate son of the Duke de Castries, was said to have been encouraged by Benjamin Franklin to open a public course in experimental physics in 1780. When the silk balloon took flight on August 27, his rival Étienne Montgolfier, financed by the government and supervised by the Academy, was still preparing his balloon made of paper in the Faubourg Saint-Antoine, at the Réveillon wallpaper factory. On September 19, the machine was finally finished and taken by wagon to Versailles. With a rooster, a duck, and a sheep on board, it took off from the courtyard of the ministers and rose to 1,500 feet before landing in the village of Vaucresson, less than six miles from where it took off.

The race between the two teams was now underway: Which of the two, the "Charlière" or the "Montgolfière," would carry the first human being into the sky? On August 30, Pilâtre de Rozier, one of Charles's rivals in the small world of Paris demonstrators, had offered himself to the Academy of Sciences as a guinea pig. On October 15, he rose in a *montgolfière* over the Faubourg Saint-Antoine, with the lieutenant of police and the archbishop of Paris as witnesses. But Étienne Montgolfier had very discreetly preceded him. Réveillon undertook other tethered flights at the Folie Titon in the following days, but the balloon had not yet been released. On November 21, the Marquis d'Arlandes and Pilâtre de Rozier took off from the garden of the royal château of La Muette, northwest of

Paris, on a trip that would take them over the south side of the city. The aeronauts landed without damage in a field on the Butte-aux-Cailles.

Montgolfier's team had been the first to take off, but Charles's team did not remain idle. Two days before the La Muette flight, the Robert brothers announced in the *Journal de Paris* that they had constructed a hydrogen balloon capable of carrying two men. The venture had been financed by a subscription, with Lieutenant of Police Lenoir himself at the top of the list. This balloon had a volume ten times greater than that of the Champ de Mars. A valve and ballast controlled its altitude, estimated with the aid of a barometer. The "cut loose" ascension took place in the Tuileries Gardens on December 1 before an immense crowd—half of Paris, it was said. Again, subscribers were seated in the first row. Charles and the younger of the Robert Brothers, Noël-Marie, had taken their places in the gondola. The flight lasted almost two hours. The Duke de Chartres and his friend the Duke de Fitz-James, following on their horses at breakneck speed, had barely reached the landing site in the prairie of Nesle when Charles took to the air again solo in order to make some physical and meteorological observations. He rose to almost ten thousand feet, but not without anxiety and suffering from earache. This painful exploit put him off, and so the physicist never again went up in a balloon.[8]

None of the savants, or those who claimed the title, were eager to go up in balloons. From the start, Étienne Montgolfier, the first of the aeronauts, had avoided publicity. Charles had let his assistants, the Robert Brothers, organize the Tuileries flight. Meusnier, who had developed the theory of aerostats, never set foot in a gondola. Flight was now the business of demi-savants, mechanics, entertainers, and adventurers. Balloon flight became part of the economy of spectacles and entertainment. The amusement parks made such flights a specialty. Alone among recognized physicists, Joseph-Louis Gay-Lussac and Jean-Baptiste Biot would undertake scientific balloon flights in Paris in 1804.[9]

The Academy of Sciences' commission on aerostats, which delivered its report on December 23, praised the respective merits of hot air balloons and hydrogen balloons. Étienne Montgolfier went back to Annonay in June 1784 with the title of corresponding member of the Academy of Sciences, a medal, and a prize to share with his brother. Charles, too, won laurels: in 1785 the king granted him an apartment in the Louvre, and, with the halo of his title as founder of aerostatic science, he continued his physics demonstrations with great success.

Science for Everyone

With balloons, everyone could participate in the spectacle. From this perspective, aerostatic experiments marked a new stage in the spectacle of science: more than the effect of hydrogen gas, they presented the public with the sight of a human being raised into the sky and, with this ascent, with the image of human perfectibility. The spectacle of the sciences and their triumph replaced the spectacle of nature. At the same time, the exhibition of marvels gave way to a parade of inventors and discoverers. By exalting their prowess, balloon flights drew science closer to the traditional forms of street entertainment. Their festive character gave rise to engagement, communion, and collective emotion. In this way, the business of science entered fully into the popular imagination.

However, the participation of spectators in ballooning was totally passive. Only a rare few could actually fly; the others had to be content with admiration from afar. Physics demonstrators like Charles offered the public much more, claiming to instruct people by amusing them. By this means they took part in the great movement of Enlightenment. Pedagogy was one of the great passions of the eighteenth century. It expressed a basic optimism, a confidence in human nature that represented a complete break with the Christian conception of the fall and redemption. From it followed the conviction that men and women, because they were naturally good, could improve themselves individually and collectively through education. And the vocation of science was precisely that: to enlighten them. In Paris as elsewhere, this project was realized in many educational enterprises, both serious ones and others that were just for fun, among them those of the physics demonstrators.

In the educational landscape of Paris at the end of the eighteenth century, public courses were a phenomenon of massive proportions. Their number and their diversity contributed to making Paris the capital of science (see map 6.2). "There is no other city in which the resources poured into all kinds of instruction have increased so much," noted Thiery in his visitors' guide to Paris. Alongside the free courses given at official establishments like the Jardin du Roi and the Collège de France or the subscription courses offered by private institutions like the *lycée*, many courses offered by individual lecturers for a fee were announced on posters and in the newspapers. Foreigners came to study in Paris more for its many public courses than for its university. Taken as a whole, they formed a complex network through which people could roam, going from one course to another according to their means, their desires, and their needs.[10]

This market in scientific education was itself inscribed in a space where knowledge circulated freely. Public courses varied greatly: at one end were those oriented toward professional training, which complemented official courses, while at the other extreme were courses that deliberately mixed spectacle and instruction. The cost of subscriptions ranged widely, but all courses had in common audiences that were quite mixed: students and the simply curious, professionals and amateurs, young and old, even men and women. Among the public courses that offered professional training, the most popular were in surgery, anatomy, and obstetrics, of which there were many in the Latin Quarter. The surgeon Desault

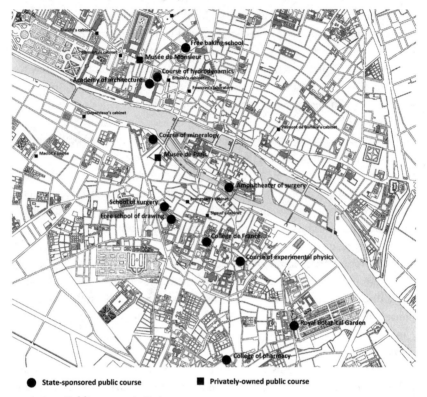

Map labels:
- Blondel's cabinet
- Free baking school
- Bienvenu's cabinet
- Musée de Monsieur
- Course of hydrodynamics
- Academy of architecture
- Brisson's cabinet
- Fourcroy's laboratory
- Deparcieux's cabinet
- Vernon de Bomare's cabinet
- Course of mineralogy
- Maclot's course
- Musée de Paris
- Amphitheater of surgery
- School of surgery
- Brongniart's cabinet
- Free school of drawing
- Sigaud's cabinet
- Collège de France
- Course of experimental physics
- Royal Botanical Garden
- College of pharmacy

● State-sponsored public course   ■ Privately-owned public course

MAP 6.2: Public courses in Paris

enjoyed an extraordinary success, attracting as many as three hundred students to his private lecture hall on the rue des Lavandières. Among this crowd, there were some people who were not repelled by dissections and came simply out of curiosity. Interest in chemistry also led to the appearance of many specialized public courses, both for amateurs and for artisans working in relevant fields. One of the most celebrated of these in the 1780s was one that Fourcroy gave in his laboratory located on the square in front of Notre-Dame, and later on the rue des Bourdonnais, where after 1785 he popularized the new chemistry of Lavoisier.[11]

Even more than chemistry, physics interested lovers of marvels. Every year Charles's magnificent cabinet hosted two courses announced in the *Journal de Paris*: one starting in early December, the other in early March. They attracted an elite and largely female audience, seduced by the eloquence of the demonstrator and the number and splendor of his instruments. Charles showed projections with his "megascope" and performed electrical experiments. The physicist had to interrupt his courses during the Revolution, which he survived without a problem. When he gave his collection to the nation in January 1792, his cabinet was moved from the Place des Victoires to the Louvre, and then moved again in 1807, to the Conservatory of Arts and Trades, where it can be found today.

The Physics Demonstrators

On the eve of the Revolution, Charles dominated the world of physics demonstrators in Paris, but he was not alone. Dozens of self-proclaimed "physicists" offered spectacles and public courses, competing ferociously to attract customers. Two things were required to achieve success: respectability and talent. The first was acquired from the official institutions that dispensed social and cultural credit. Everyone sought recognition from either the king, the Academy, the university, or some other princely or scholarly authority. Invention was often sufficient, as we saw in the case of Pilâtre de Rozier, who at first called himself chief apothecary of the Prince de Limbourg, or Joseph Pinetti, who claimed to be a pensioner at the Prussian court.

In truth, at this time the trade in titles was not limited to con men. In principle, respectability was based on scientific merit, but in reality it was primarily a matter of appearances: the public demanded brilliance, eloquence, an honorable demeanor, and fine equipment. As for talent, what counted was a talent for seduction and entertainment; above all, one had to cause a sensation. The physics demonstration, even when it was part of a public course, was actually more spectacle than scientific instruction. The difficulty for the demonstrator was to reveal marvels without deceiving his audience, because for the public there was an irreducible difference between physics, which put nature on the stage, and comedy, which mimicked and parodied it. Charlatanism, which passed off the latter as the former, represented the major sin; demonstrators themselves constantly flung that accusation at each other.

Among all these "physicists" there was in fact little variation in the content of their experiments: everybody offered mechanics, optics, and electricity, to which was sometimes added a little chemistry, with explosions of gas and illuminations of phosphorus. What distinguished them from one another was above all their status, their location, and the public they targeted. Traditionally, there were two kinds of demonstrators: official ones, who lived in the Latin Quarter and who worked in the schools, and itinerant demonstrators, who practiced their trade at the fairs. With the development of new neighborhoods for spectacles and entertainment, the fairground "physicists" tended to settle down, enlarging their clientele and gradually effacing the distinction, which had never been absolute, between instructive demonstrations and spectacular demonstrations.

This evolution was accompanied by a certain homogenization of the audience. Demonstrators had long sought to diversify their offerings by adapting to various types of clientele. Abbé Nollet, for example, was equally capable of addressing the court at Versailles, the audience at the College of Navarre, and students in the military schools. Similarly, his enemy Delor, a demonstrator at the university, gave courses to amateurs in his private cabinet (first at the Estrapade, then on the Quai d'Orléans) and demonstrations to students at the École Militaire. Under the name of "Comus," Delor's disciple Nicolas Le Dru managed to extend his fairground activities to the court, passing through high society along the way. But the audiences scarcely mingled, and the barrier between the colleges and the fairground remained insurmountable for a long time.

In the 1780s, the situation became much more fluid. The sciences were very much in fashion, and "physicists" were multiplying. The boulevard and the Palais-Royal confused everything and everyone. The Latin Quarter demonstrators like Brisson, the successor to Nollet at the Collège de Navarre, and Rouland, who held the post of demonstrator at the colleges connected to the university, lagged behind. While the Academy was concerned about the surge in fake science, the commercial competition among practitioners was sharpening, provoking a boom in publicity and in the quest for endorsements from the savants. It is in this context that we have to consider not only the passion for balloons but also the conflicts between the Academy and the supposed charlatans.

From this perspective, Jean-Paul Marat presents the most remarkable case. The man who would call himself "l'Ami du peuple" (the friend of the people) during the Revolution had a medical degree from St. Andrew's University in Scotland and had made a name for himself in Paris by caring for the posh clientele of the Faubourg Saint-Germain. He lived on the rue de Bourgogne in the home of his patron, the Marquise de l'Aubespine, whom he had saved, it was said, from a certain death. In 1778 he established a cabinet of physics there in which he undertook experiments on fire and light. His principal instrument was a solar microscope that he had purchased from Henry Sykes. With the Count de Maillebois as intermediary, Marat presented his work to the Academy, while asking his friend the Abbé Jean-Jacques Filassier to exhibit it at the Hôtel d'Aligre on the rue Saint-Honoré.

The *Journal de Paris* could not praise this "Maratist course" enough, but the Academy's commissioners appeared more circumspect. For a long time, they sought to sidestep the issue. But Marat was so persistent, going so far as to seek Franklin's support, that they had to decide. The verdict came from the pen of the geometer Jacques Antoine Joseph Cousin on May 10, 1780: the Academy refused to give its approval to experiments "contrary in general to the most well-known principles of optics," saying that it had not been able to verify them "with the necessary exactitude" and that in any case the experiments "did not prove what the author thought they did." A wrathful Marat disregarded the censure of his critics and published his memoir anyway: "If one must be judged, let it be by an educated and impartial public: it is to this tribunal that I appeal with confidence, this supreme tribunal whose decrees the scientific bodies themselves are forced to respect."[12]

Marat's reputation in society did not suffer much from his academic failure. He continued his experiments with the solar microscope, working now on electricity as well. He had his followers, like his friend the clockmaker Breguet, the mineralogist Romé de l'Isle, and the physicist Pilâtre de Rozier, and even Jacques-Pierre Brissot, a young lover of science who was full of ambition, attended the public courses, and would become his political enemy during the Revolution. Franklin and Volta deigned to attend Marat's demonstrations in 1782. And Philippe-Rose Roume de Saint-Laurent worked to have him made director of the royal physics cabinet in Madrid.[13]

Charlatans or Physicists?

Marat's adversaries were no longer the contemptible academicians, to whom he made sure not to present his research, but rather his fellow demonstrators. In 1781, the son of Nicolas Ledru had implied that he was a plagiarist, and two years later it was Jacques Charles, himself an expert in lanterns and projections, who drove the point home. In his lectures he took up the attack on Marat's theories, which he compared to those of Comus. A furious Marat rushed to the Place des Victoires to demand an explanation; the two men came to blows, and swords were drawn. Beyond Marat's volatility, this incident revealed the tensions underlying the world of Parisian demonstrators. Marat and Charles both claimed to be on the side of science, carefully distinguishing themselves from the "charlatans" of the boulevard. The case of Nicolas Ledru, who had provoked the incident, was more complex.

Ledru (whom we have already met) had begun his career under the name "Comus" as a physicist-entertainer on the Boulevard du Temple around the end of the 1750s, using many tricks to dazzle his audience. His specialty was magnets, with which he created surprising effects. One of his most famous tricks, according to Diderot, "consisted of establishing communication between two people without the perceptible assistance of any intermediary agent." He also had a clairvoyant mermaid, for which, according to Helvétius, he would have merited the title of genius, if his mechanism had not been "more curious than useful." "All of Paris rushed to the places where he performed," recalled the mathematician Jean-Étienne Montucla in 1778, adding that "the ignorant admired him, considering him practically to be a sorcerer, while the savants tried to figure out the secret behind his trick. We have to admit that it was impenetrable as long as nobody suspected that magnetism was the principal cause."[14]

Comus's reputation brought him to the court, where he amused the ailing dauphin, and put him into contact with the great nobles, who also sought to pierce his secret. A postal employee named Edme-Gilles Guyot, whom we have already encountered as the author of the popular *Recreations*, finally revealed the secret behind his trick in 1769, but that did not interfere with his success, since, as Grimm wrote, "*how* it was done was all that mattered; the content was unimportant, the form was everything."[15] With his son, Comus continued to mount his spectacle on the boulevard; announcements for them appeared regularly in the newspapers until 1783. But under the mask of the magician there was a physicist who was always careful to distinguish experiments from entertainments in his demonstrations. Montucla himself agreed that nobody else had "brought together in one prestigious location so much knowledge of physics."

During a stay in England in the 1760s, Ledru had met the scientific instrument makers Jesse Ramsden and Edward Nairn and had made improvements on an inclination compass used by navigators. Later, he started conducting more experiments on electricity, frequenting scientists like Rouelle and Darcet and publishing his results, with the help of his son, in the *Journal de physique*. Interested especially in electric therapy, in 1782 he opened a clinic on the rue des Rosiers in the Marais, where he applied electricity to some epileptics and others with nervous diseases. In the spring of the following year, he gave a preview of his theories and experiments on the influence of what he called the "electric fluid" on the "nervous fluid." After a preliminary examination by its commissioners, the Faculty of Medicine was eager to grant its approval to an enterprise that was actively supported by the government. Ledru and his son obtained the respective titles of physician to the king and physician to the Faculty of Medicine. Having permanently closed down the spectacle on the Boulevard du Temple, they solemnly and very officially installed their medico-electrical clinic in the Convent of the Célestins in November 1783, in the presence of Lieutenant of Police Lenoir, Franklin, and other invited guests.

The incident between Marat and Charles had erupted the previous March at the very moment when Ledru was the subject of the Faculty of Medicine's preliminary enquiry, which was evidently not a coincidence. People were surprised that an entertainer was obtaining the endorsement of administrative and medical authorities. Marat and Charles denied that medical electricity had any value and considered Ledru to be simply an imposter. What then could be more humiliating to Marat than to be compared to this charlatan by Charles?

The new commissioners charged by the Faculty of Medicine with examining the electro-medical cures thought pretty much the same thing. According to the *Mémoires secrets*, "they regard as mere charlatanism, sorcery, sleight of hand, everything done up to now by this supposed new Asclepius, still the Comus he was on the boulevards." Nevertheless, Ledru, supported by the court and other powerful aristocrats, easily survived this ordeal. It is true that Mesmer, another healer who cut an even greater figure, was at the time commanding the attention of those who were denouncing illusion. In the end, Ledru's medico-electrical clinic functioned without hindrance until 1810.[16]

Compared to Ledru, the other entertainers were rather bland. His main rival, François Pelletier, was far behind him. He was mostly a mechanic who was appreciated for his automatons. Having started out in the fairs, he opened a shop on the Boulevard du Temple before roaming across Europe, including performances at the courts of Vienna and Madrid. After watching him do his tricks before Empress Marie-Theresa in Vienna in 1769, the engineer Wolfgang de Kempelen created his famous chess player, a fake automaton that later toured Europe. Returning to France in 1778 with the pompous title of mechanical engineer to His Royal Highness Don Gabriel, heir to the Spanish throne, Pelletier, who called himself an inventor, moved into the Palais-Royal in 1784. But a year later he sold his mechanical cabinet and withdrew to the Boulevard du Temple; he was still in business on the rue Saint-Antoine in the 1800s. Some makers of scientific instruments also gave more or less spectacular demonstrations in their workshops in order to promote their wares. But most of the entertaining physicists operated in the theaters of the boulevards or the Palais-Royal. And for many of them, Paris was merely one phase in an itinerant career.

The Italian Joseph Pinetti, for example, whose real name was Giuseppe Merci, began his international career in Rome. Arriving in Paris in 1783 by way of Berlin, the elegant magician presented his spectacle at court (then in residence at Fontainebleau), and then appeared at the Theater of Menus-Plaisirs on the rue Bergère, where, it was said, he attracted a "top-flight audience." His show included fake automatons like the "little Turkish savant" that guessed the cards chosen by spectators. Perhaps his finest trick was a card nailed to the wall by a pistol shot (Fig. 6.2). His adversaries insinuated that he betrayed the trust of his admirers and that he even accepted invitations to consult as a clairvoyant. Encouraged by Lalande and Montucla, a young physicist named Henri Ducremps undertook to unmask the imposture in a book called *La magie blanche dévoilée* (White magic unveiled). In response Pinetti published his own little book containing some salon amusements, before he left Paris for London, where he soon boasted that he enjoyed the patronage of the whole royal household of France! Returning to

Professeur et Démonstrateur de Physique amusante, qui
après avoir réduit en cendres une Carte choisie au
hazard jette le Jeu en l'air pour la faire reparaître
en la clouant au mur d'un coup de Pistolet.

FIGURE 6.2: The spectacle of physicist Pinetti. Frontispiece of Henri Decremps, *La Magie blanche dévoilée*, Paris, Langlois, 1784, from the collection of François Bost.

Paris in February 1785, he gave a show at the Variétés-Amusantes theater, before resuming his itinerant life as an entertainer in Spain, in Rome, and again in Paris (where he performed at the Circus of the Palais-Royal), then in Berlin, and finally in Russia, where he seems to have died in 1800.

Notes

1. *Journal de Paris*, December 8, 9, 11, 14, 18, 19, 22, 24, and 26, 1783, and BACHAUMONT, 1777–1789, vol. 24, 68, 82–83, 89, 91–92, and 94–95.

2. *Journal de Paris*, August 28, 1783; BACHAUMONT, 1777–1789, vol. 23, 116–118 (August 24 and 25, 1783); and B. FAUJAS DE SAINT-FOND, *Description des expériences de la machine aérostatique de MM. de Montgolfier et de celles auxquelles cette découverte a donné lieu*, Paris, 1783, 7–9.

3. Description of the new garden by DULAURE, 1787, vol. 2, 249–255, and by MERCIER, 1782–1788, chap. 819: Palais-Royal, followed by chaps. 820 and 821: Suite du Palais-Royal.

4. Quotations taken from MERCIER, 1782–1788, chap. 819, and GUIGOUD-PIGALE, *Le Baquet magnétique*, London, 1784, 45–46.

5. Descriptions of the Palais-Royal in THIÉRY, 1787, vol. 1, 236–287; *L'Almanach du Palais Royal utile aux voyageurs . . .*, Paris, 1786; and F.-M. MAYEUR DE SAINT-PAUL, *Tableau du nouveau Palais-Royal*, 1788, quotation at 31. On the cafés, see D. CHRISTOPHE and G. LETOURMY, eds., *Paris et ses cafés*, Paris, Action artistique de la Ville de Paris, 2004.

6. MERCIER, 1782–1788, chap. 820.

7. See LILTI, 2015, 22–25; for a "literary" definition, see GOODMAN, 1994, 90–135. On men of letters in high society, see LILTI, 2015, 91–133, and particularly on Voltaire, 109–111.

8. FAUJAS DE SAINT-FOND, *Description des expériences*, 268–280, and *Première suite de la description des expériences de la machine aérostatique de MM. de Montgolfier . . .*, vol. 2, Paris, 1784, 11–55.

9. "Sur la 'folie des ballons,'" in LYNN, 2006, 123–147, and THÉBAUD-SORGER, 2009. On the activities of Meusnier, see G. DARBOUX, ed., "Mémoires et travaux de Meusnier sur l'aérostation," *Mémoires de l'Académie des sciences*, 2nd ser., 51 (1910), 1–128, and LAVOISIER, *Correspondance*, vol. 4 (*1784–1786*), Paris, 1986, 293–303.

10. On the public courses in Paris in the decade of the 1780s, see THIÉRY, *Le Voyageur à Paris*, 8th ed., 1790, vol. 1, 201–206, as well as the announcements in the *Journal de Paris* and other Parisian newspapers, and more generally B. BELHOSTE, "Un espace public d'enseignement aux marges de l'université: Les cours publics à Paris à la fin du XVIIIᵉ siècle et au début du XIXᵉ siècle," in T. AMALOU AND B. NOGUÈS, eds., *Les Universités et la ville, XVIᵉ–XVIIIᵉ siècles*, Rennes, URA, 2013, 217–236. On the effect of the consumer economy on cultural practice in eighteenth-century France, see GOODMAN, 2009.

11. On the surgery courses, see DELAUNAY, *La vie médicale aux XVIᵉ, XVIIᵉ et XVIIIᵉ siècles*, Paris, 1935, 331–333.

12. J.-P. MARAT, *Découverte de M. Marat sur la lumière . . .*, London, 1780, 6.

13. On Marat's science, see GILLISPIE, 1980, 290–231, and COQUARD, 1993, 122–156.

14. TORLAIS, 1953. Diderot mentions Comus in a letter of July 28, 1762, to Sophie Volland. The clairvoyant mermaid is mentioned by Helvétius in *De l'homme, de ses facultés intellectuelles et de son éducation*, 1773, section 3, chap. 2, 218. Montucla

praises Comus in volume 4 of the new edition of *Récréations mathématiques et physiques d'Ozanam*, 1778, 290 and 331.

15. GRIMM, 1829–1830, January 1770.

16. On the electro-medical clinic, see THIÉRY, 1787, vol. 1, 663–665, and vol. 2, 688–689, and TORLAIS, 1953, 17–24. The quotation from the *Mémoires secrets* is from BACHAUMONT, 1777–1789, vol. 24, 74 (December 6, 1783).

7

Inventions

BEAUMARCHAIS'S *THE MARRIAGE of Figaro* was one of the most successful plays of the Paris theater. First performed by the actors of the Comédie Française on April 27, 1784, after a long ban, this bright, cheerful play was enthusiastically welcomed. The audience was delighted to see privilege and arbitrary power mocked. The court itself was not the last to laugh: the aristocracy flocked to it, and the play was performed seventy-six times to full houses the first year. This triumph definitively launched the new Théâtre Français in the Faubourg Saint-Germain, which the queen had inaugurated in person two years earlier.

Designed by architects Marie-Joseph Peyre and Charles de Wailly, the new theater was just down the street from the Luxembourg Palace and was built thanks to the financial support of Monsieur, the king's brother. It combined the comfort of a space dedicated entirely to the public with the solemnity of a monument raised to the glory of the dramatic arts. The surrounding area was arranged for carriages; numbered arcades "so that masters and servants can find each other easily after the play" housed shops. A water pump, reservoirs, and an iron curtain that could insulate the stage if necessary provided protection against fire. With seating for almost two thousand spectators, the theater was designed for their pleasure and comfort. Circular in design and arranged in the Italian style, it included loges, a richly decorated stage apron, and a pit with benches (Fig. 7.1).[1]

Special attention had been paid to the lighting. Candles replaced the footlights in front of the stage, and a chandelier representing the sun, surrounded by the twelve signs of the zodiac, lit the hall. This custom-made chandelier was placed under a reflector that sent luminous rays both downward and toward the stage. Candles, no doubt placed beyond the view of spectators, illuminated the sun from which their light radiated, an arrangement directly inspired by

FIGURE 7.1: Sectional view of the Théâtre Français in the Faubourg Saint-Germain. Engraving by Antoine-Joseph Gaitte, n.d.

considerations on theater lighting that Lavoisier had presented to the Academy of Sciences at its public meeting in November 1781.

The chemist, who had already taken an interest in the use of lamps with reflectors to light public roads, had proposed replacing a ceiling chandelier that obstructed the view of some of the spectators with elliptical reflectors hidden in recesses in the ceiling, which could also accommodate fans to cool the building. Thanks to support from the Count d'Angiviller, Lavoisier conducted lighting experiments over the course of the following months in a mock-up of the theater installed in the picture gallery of the Louvre. Despite encouraging results, he did not pursue his research any further, relying on craftsmen for possible applications and ceding priority for the invention of the lamplight to Peyre and Wailly, who, for their part, claimed that they had thought of the idea long before.[2]

Lavoisier's research on reflector lamps illustrates the attention paid by eighteenth-century savants to matters of public interest. The Academy of Sciences itself did not neglect inventions useful to society. Since its foundation, it had as one of its missions the improvement of the arts as well as the responsibility for evaluating machines and new processes of production. Every inventor and schemer sought its approval, which would validate their ideas, reassure the public, and most importantly carry a lot of weight when it came to asking for support and financing. Finally, without such approval it would be hard to secure an exclusive privilege from the government, a sort of patent before its time. Thus, savants came to play a decisive role in the politics of industrial support. They were also called upon to arbitrate conflicts and rivalries between inventors, in which powerful public and private interests were involved. The affair of the Quinquet lamp furnishes one example that is . . . illuminating.

The Quinquet Lamp

The lighting for the new Théâtre Français, which had required so much effort, at first proved rather ineffective. "No matter how large the chandelier is that illuminates the theater, it cannot light it sufficiently; it is impossible to see anything in the boxes clearly; everything fades into everything else, and the women, who are made to adorn the spectacle, are reduced to the pleasure that they generally care least about: that of looking and listening," noted the misogynist Grimm in his *Correspondance Littéraire* when the theater opened. In addition, the architects were pursuing their experiments under the eyes of the public, so to speak. The exasperated actors demanded that these interminable trials be abandoned and went so far as to have those doing the work stopped. Fortunately, a new invention presented to the company of the Comédiens-Français by Antoine Quinquet and Amboise Lange gave them hope for a noticeable improvement: they proposed oil lamps with an air draft that would be ten times brighter than the candles they would replace.[3]

A test was conducted on March 5, 1784 (a month before the premiere of the *Marriage of Figaro*), and it seemed promising. On opening night, the theater was therefore illuminated by a chandelier bearing forty "Quinquet" lamps. The audience appreciated being able to read and to see each other. Unfortunately, the powerful chandelier also offered valuable assistance to those who were writing down Beaumarchais's text on the fly in order to publish pirate editions. The new lighting had other disadvantages: while the spectators no longer complained about the poor light, they had new fears of dripping oil and bursting pieces of glass that fell to the floor. It would take several years to solve these problems and for oil lamps to replace candles permanently in theaters.[4]

The adoption of their lamps at the Théâtre Français was a great coup for the two inventors, Quinquet and Lange, who had just presented their invention to the Academy of Sciences and, through an announcement in the *Journal de Paris*, invited the public to admire their lamps in Daguerre's shop on the rue Saint-Honoré.[5] Pilâtre de Rozier, on a roll with novelties, quickly installed Quinquets in his musée, while at the Palais-Royal the new lamps lighting the cafés spread a "soft light, a kind of half-light that rendered beauty more interesting and even set off ugliness to advantage." Individuals did not lag behind, and the lamps quickly proved a commercial success, as Jefferson, newly arrived in Paris, testified to his American friends.[6]

However, in November 1784, the two inventors had a public spat. Lange, or rather L'Ange (the Angel), as he liked to call himself, complained about his name being dropped: "Everywhere I hear talk of Quinquet lamps; why not call them Angelic lamps? It seems to me that this name sounds better, because at least nobody could challenge my share in the invention."[7] Soon, he claimed full credit for the invention, leaving out entirely the name of his associate Quinquet, who protested. A third man, however, had more reason to complain: A year earlier Ami Argand had presented to the Paris authorities an air-flow lamp of which Quinquet and Lange's seemed to be only a copy.

The three would-be inventors knew one another. They had been seen together at the end of the summer of 1783, gathered around the balloon that Étienne Montgolfier was constructing at the Réveillon factory. It was there, in the feverish atmosphere surrounding the preparations for the flight, that Quinquet and Lange had casually become aware of Argand's invention. Without denying it, the two partners swore that they acted in good faith. Argand had shown them his lamp without revealing its secret, and they had then reinvented and improved it. In this priority dispute in which everyone's interest was mainly commercial, no one was completely right or wrong. But before examining the affair, let us see who these supposed inventors really were.[8]

Quinquet and Argand

Antoine Quinquet, who would give his name to the lamp in France, was a Parisian pharmacist. He was born in Soissons, where his father was a grocer. In 1779, after spending about ten years in various dispensaries, primarily that of Antoine Baumé, one of the most famous apothecaries of Paris and a member of the Academy of Sciences, he had bought his own pharmacy on the rue du Marché aux Poirées at the corner where the gate of the old Wheat Hall stood. He had also spent some time in Geneva, where he had worked for Colladon, and maintained his connections with apothecary circles there. He had supported his compatriot Pierre-François Tingry, who had moved to Geneva in 1773 and made a brilliant career there as a pharmacist and naturalist. When the young Ami Argand, son of a Genevan clockmaker, left for Paris two years later to study the sciences, he carried with him letters for Lavoisier and Fourcroy and a recommendation to Quinquet signed by Tingry.

The year 1783 appears to have been a turning point in the life of our pharmacist. Until then all his time had been taken up by his business; now here he was, almost forty years old, enthralled by science. His specialty was electricity, atmospheric electricity specifically, which might explain his interest in the science of balloons. He contributed to the subscription for Charles's balloon launched by Faujas in July 1783. On September 4, fifteen days before the Versailles flight, he presented a paper at the public meeting of the Collège de Pharmacie in which he claimed to have produced hail and snow artificially by means of electricity. He thus demonstrated, and even expanded on, an idea of the young Argand—or at least he believed he had. At the same time, more concretely, he also appropriated the idea for the lamp.[9]

The invention had germinated in the mind of Ami Argand far from Paris, in the Languedoc. Since his arrival in the French capital in 1775, he had made himself known. He had taken public courses, presented a paper to the Academy of Sciences on the causes of hail attributed to electricity, published articles in the Abbé Rozier's *Journal de Physique*, and taught chemistry. In 1780, he left with his brother Jean for Montpellier with a view to establishing a distillery there for eau de vie using a process of his own invention. Then he conceived of a lamp to illuminate at night the workshops set up on the premises of Monsieur de Joubert, treasurer of Languedoc and a great lover of science. The originality resided in the burner, today called the "Argand beak," formed of two concentric pipes serving as a chimney and housing a wick made of a circular ribbon whose height could be adjusted by means of a pin. Such a simple arrangement increased the intensity of the light spectacularly. In 1782 the lamp was presented successfully to the Estates of Languedoc. On his return to Paris, Joubert showed an example to the finance minister. The intendants of commerce were also interested; they asked to see it and charged the chemist Pierre Macquer with examining it.

Why did Argand leave Montpellier for Paris in the summer of 1783? Perhaps it was to negotiate a privilege or a financial reward for his distillation process, or perhaps it was to launch the fabrication of his new lamp. On the way he stopped at Annonay, where his friends the Montgolfier brothers had just performed their first balloon experiment. Ami Argand had met Etienne during his time in Paris, and had probably met his brother Joseph when the inventor was living in Avignon and came up with the idea of a hot-air balloon They were both men full of new ideas and projects, and it seems that they had already exchanged ideas and theories about the effects of heat and its applications.

And thus it was in the company of the aptly named Ami that Étienne had taken the diligence to Paris on July 11 to present his invention to the Academy of Sciences at their request. As soon as he arrived in the capital with a model of his lamp fabricated by a Montpellier tinsmith in his luggage, Ami Argand set to work. At the same time he was busy with the balloon experiments, helping Charles and the Roberts with the production of hydrogen, and helping Étienne Montgolfier even more with the construction of the balloon at Réveillon. He had of course resumed contact with Quinquet, whom he brought to the Folie Titon to help with the preparations. It is very probable that Argand spoke to him too about his projects, about the distillery and also about his lamp. With fifteen years' experience as a pharmacist, first with Baumé and then on his own account, Quinquet could give him good advice.

In fact, the pharmacist presented him with a business contact: the grocer Ambroise Bonaventure Lange, living on the rue du Petit-Pont, a distiller by trade and, like him, a dabbler in electricity in his spare time. The main purpose of the meeting was no doubt the distillation process. But they also spoke about the lamp. According to Quinquet, who perhaps deliberately confused the two inventions, Argand had wanted to sell his secret "to the grocers' guild." What is certain is that the two Parisians were more interested in the lamp than the distillation process. Intrigued, Lange accompanied Quinquet to the Réveillon factory. Citing as witnesses Faujas, Étienne Montgolfier, and Réveillon, the inspector of manufactories Louis-Paul Abeille would later accuse the two men of having at the time employed "all the skill, all the perseverance imaginable to succeed in stealing from M. Argand a secret and a mechanism that was in fact for them more than just a curiosity."[10]

Argand's Failure

Argand, who evidently feared that his idea would be stolen, jealously held on to his secret. With great prudence, he negotiated with the government for some kind of award for his invention: an exclusive privilege either to produce it himself or to sell it to a manufacturer. The inventor also wanted to make improvements on his lamp. For a long time, he said, he had nurtured the idea of equipping it with a glass chimney, and for that purpose he had entered into a relationship with the physics-instrument-maker Antoine Assier Perricat, perhaps even using Quinquet as an intermediary.

While the lamp mechanism was very simple, its construction was very delicate. The metal parts had to be adjusted and brazed with great care to avoid oil leaks; most important, the chimney required a type of heat-resistant glass called "flint glass," which was difficult to procure in Paris. Argand thought he would have to go to London to find the workers he needed, and even partners—something he had planned to do since leaving Montpellier. In early November, even before Pilâtre's flight, he went to England, where he soon found entrepreneurs interested in his lamp: the mirror maker William Parker and James Watt's partner, the manufacturer Matthew Boulton. We will only follow his adventures across the Channel (where he would also end up failing to defend his rights) from afar and continue the story in Paris.

As we have seen, in Argand's absence Quinquet and Lange got started on getting the lamp to market. Although forced to admit the proven priority of the Genevan, they claimed to have invented the glass chimney and turned to the Academy of Sciences for the recognition they needed. Although not manufactured very well by Parisian tinsmiths, their first lamps began to appear in public

PLATE 1: Jacques Louis David (1748–1825), portrait of Antoine-Laurent Lavoisier (1743–1794), and his wife, Marie-Anne-Pierrette Paulze (1758–1836). New York, The Metropolitan Museum of Art © The Metropolitan Museum of Art, Dist. RMN / image of the MMA.

PLATE 2: Jacques Louis David (1748–1825), *The Tennis Court Oath at Versailles*, June 20, 1789. Versailles, châteaux de Versailles et de Trianon © RMN (Château de Versailles) / Gérard Blot.

PLATE 3: Louis Carrogis (1717–1806), known as Carmontelle, *Messieurs d'Alainville and de Montamy*. Chantilly, Musée Condé © RMN (Domaine de Chantilly) / René-Gabriel Ojéda.

PLATE 4: Pierre-Antoine Demachy, *The Mint seen from the tip of the Île de la Cité*, 1777. © Musée Carnavalet, Paris.

PLATE 5: Jean-Baptiste Hilair, *The Royal Botanical Garden: the new greenhouse*, 1794. © BNF.

PLATE 6: Antoine Meunier, *The Garden and Circus of the Palais Royal*. Paris, Musée Carnavalet © Musée Carnavalet / Roger Viollet.

PLATE 7: Jacques Louis David (1748–1825), *Portrait of Doctor Alphonse Leroy*. Montpellier, Musée Fabre © RMN / Hervé Lewandowski.

venues and wealthy homes. "Quinquet lamps" were the talk of the town. Ami Argand, stuck in London, where he was having trouble getting the manufacturing of his lamps off the ground, was still too far away to challenge his competitors in France on either the merit or the market for the invention. However, with the support of Étienne Montgolfier, his brother Jean had already intervened on his behalf with Finance Minister Calonne. At the end of 1784, the situation remained uncertain: Argand, who had filed a patent, began to make lamps in England with his associates; on their side, Quinquet and Lange, having the field to themselves in Paris, began to attack each other, with the latter claiming his share in an invention that public opinion seemed to attribute entirely to the former. The masks would fall the following year.

Argand returned to Paris in the spring of 1785, confident and committed to defending his rights. His lamps manufactured in England of silvered brass were of much higher quality than those made of lacquered tin by his competitors. He benefited from the support of influential friends, "savants" like Faujas and Montgolfier, and the Languedoc financiers with whom he was associated for distilling the eau de vie. Finally, the Bureau of Commerce came around to his side. During the following months, he made the rounds of the salons and ministries in Paris and Versailles to promote his lamps among the powerful. What he wanted was an exclusive privilege to manufacture them in France and, while waiting to begin production there, a license to import them from England. Calonne arranged a personal audience with Louis XVI for him on July 31. The king, very satisfied with the lamps, quickly granted Argand the exclusive privilege he sought, for fifteen years, although he refused to grant him the import license. The council's decree was issued at the end of August.

So Argand triumphed. Any copy or counterfeit would be banned in the kingdom. His factory would be built in the region of Gex near Geneva, where Voltaire had settled down in the 1760s. For Louis XVI, who also granted him a subsidy of twenty-four thousand livres, it was a matter of promoting an underdeveloped part of the kingdom; for Argand, the advantage was being close to his hometown, and above all near the border, since he intended to have the parts that in his view could not be produced in France smuggled in from England.[11]

At this point, Argand thought he had made his fortune. The new lamps were in high demand. He had perfected the design over the course of the two preceding years, and after a few disappointments, he and his English associates had mastered the fabrication process. Finally, and above all, his English patent and his French privilege now gave him a monopoly in the two major markets. Unfortunately for him, he did not take into account his rivals, who did not intend to let this godsend slip away. In London, his patent was contested in the courts. In Paris, Quinquet and Lange, having entered the market before him, had not laid down their weapons. Of course, the two associates were not getting along, but Lange, who had started out as a stooge, proved to be intelligent, stubborn, and a skilled manipulator. For some time he had been publicly claiming his share in the invention: the glass chimney. On January 29, 1785, he presented on his own a new model of his lamp to the Academy.

Argand, quite wrongly, was not worried: distinguished witnesses like the Marquis de Cubières and Faujas, had seen his lamp; the chemist Macquer had examined it (although unfortunately he had just died); as far back as October 1783, according to Abeille, Argand had also spoken of a "chimney lamp" to the engineer Meusnier, with whom he had exchanged information on the combustion of lamps. Lastly, even before his return to Paris from England, he had sent Faujas an exemplar of his lamp, to show to the Academy of Sciences along with a claim for priority. The Academy had acknowledged it on December 15, 1784, and those in attendance had been struck by how brightly it shone. So the inventor was stunned when he learned of the report presented by the Academy commissioners who had been charged with examining the paper prepared by his rivals on September 6, 1785.[12]

In this document, from which Quinquet's name was completely absent, Argand's role was only mentioned in passing in connection with the burner; the primary invention, it claimed, was the external channel formed by the glass chimney, and this was attributed exclusively to Lange, whose lamps, the commissioners concluded, "deserved the approval of the Academy all the more in that it was due to this glass chimney that the light achieved such brightness." Such an evaluation obviously challenged the exclusive privilege that Argand had just obtained from the government and thus reopened the struggle in both London and Paris over who would be recognized as the true inventor, and hence the main beneficiary.

Before turning to the final episodes of this saga, let us return to the judgment of the commissioners. How can we explain a report so favorable to Lange, when Argand's priority appears to be beyond doubt? In his private correspondence Argand claimed that there had been an actual plot against him. He suspected that the lamp Faujas had presented to the Academy in December on his behalf had been examined and copied by Lange with the complicity of the commissioners. It must have been this lamp that the distiller had submitted for examination on January 29 and was then evaluated in the report, and not those lamps presented a year earlier by the two so-called inventors. Argand accused the academician Vandermonde of being the main instigator of this fraud, but without providing any proof or explanation. In fact, Vandermonde had withdrawn from the commission and had not signed the report, which, Argand also noted, had been presented in the absence of Meusnier, who, he said, knew the truth. Yet it was indeed to Quinquet and Lange that the engineer would later attribute the paternity of the glass chimney, which he judged of little value for the lamp. Because he had not had a model of his lamp made before his departure for London, Argand had great difficulty proving that he had invented it.

In fact, because it considered the lamp to be a collective invention, the Academy of Sciences must have decided that it would be too generous to grant an exclusive privilege for its exploitation to anyone, and especially to a Genevan! From the Academy's perspective, there had been no invention, only an improvement made to traditional lamps that other inventors like Franklin and more recently Meusnier had already thought of. Moreover, the Academy was annoyed by the secrecy with which Argand had shrouded his lamp. Count de Milly, for example, while conceding to him the initial idea of the use of an air draft, compared him to Lange, whom he praised for having made no mystery of either his methods or the construction of his lamps. More generally, by attributing the invention of the glass chimney to the distiller, the Academy was helping out Parisian industry, since it was the capital's tinsmiths who were making his lamps and the Sèvres glassworks his flint-glass chimneys.

The Academy's verdict marked a turning point. Despite the active support of Calonne, who had inspector Abeille draft a memorandum in favor of his protégé, Argand's position deteriorated. The battle lasted several months, with witnesses and arguments circulating constantly between England and France. In London, where the Genevan was defending his patent in the courts, Jean-Hyacinthe Magellan, a correspondent of the Academy of Sciences who was close to the Parisian savants and was their principal supplier of scientific instruments from across the Channel, ultimately lost him his case by giving false testimony. Disgusted, Argand quickly withdrew from the British market, which he left to his former associates. In Paris, Lange, buoyed by the Academy's report, challenged the council's decree before the Parlement of Paris, which, because it was in conflict with the minister, refused to register the exclusive privileges he had granted. In the end, there was a compromise. In the autumn of 1786, in response to a direct request from Calonne, Argand agreed to share his privilege with Lange, a move that as if by magic removed the obstacles to registration.[13]

The Lamp Belongs to Everyone

The two new partners divided up the work: Lange took charge of marketing the lamps in Paris, while Argand oversaw their manufacture in the Pays de Gex. He left in the spring of 1787 to establish the royal manufactory at Versoix, where production began in September. Always busy, the inventor was at the same time, more concretely, pursuing his plans for a distillery (which he had never abandoned) in partnership with Joubert.[14] He only came back to Paris occasionally. A good salesman, Lange was now very successful in selling the "lamps with an air draft and a chimney." Yet he still had to face lively competition: lamps made in England were being smuggled in, and others were being produced in Paris by tinsmiths who loudly contested the privilege of the two partners. In any case, that privilege, along with all others, would be abolished by the Revolution. Lange, reasserting his freedom, submitted an application for a patent in his name alone in 1791 and continued in business. Argand kept for himself the factory in Versoix, which he sold to his cousin Bordier-Marcet in 1800. Ruined, in poor health, and almost insane, he died three years later in Geneva.[15]

Among the Parisian competitors of Lange and Argand was, in a small way, Quinquet himself. After being abandoned by Lange, the pharmacist had not remained idle. He found a new partner in the person of a physician named Caullet de Vaumorel, with whom he wrote a paper on air-flow lamps that was read at the Academy of Sciences in December 1784. By this time, however, the two men shared an even more powerful interest: medical electricity and animal magnetism. While Quinquet was marketing a knockoff of Nairne's electrical machine, as well as the bar magnets and cream of tartar recommended by Mesmer, Dr. Caullet published two books that were for sale in the pharmacy: a translation of a book by an English instrument maker and *Mesmer's Aphorisms*, which revealed to the public the secrets of the doctrine that the magnetizer was claiming to reserve for his initiates alone in exchange for hard cash. This was the occasion for a new polemic with Lange, since the distiller—what a surprise!—was also an amateur physicist, also sold electrical machines, and claimed that he had invented the model being sold by the two partners back in 1776.[16]

Finally, to market his lamps Quinquet brought in another partner, George Palmer (alias Giros de Gentilly), who lived in Paris on the rue Meslée. This Englishman, who was a dyer and chemist, had published a very interesting book in London that was quickly translated into French. In it he theorized that the surface of the retina was composed of three types of fibers, each of which was sensitive to one kind of light, and attributed color blindness to their dysfunction. Noting that green and blue were confused in the light of a candle or an ordinary lamp, Palmer proposed to correct the problem by the use of a blue-tinged glass that would cast a light similar to daylight, which would be especially useful to painters and artists working at night. He presented the idea to the Academy of Sciences on December 4, 1784, a week before Quinquet read his lamp paper. The pharmacist, who should at least be given credit for being able to recognize the good ideas of others, got together with the Englishman to bring to market lamps with blue-tinged chimneys. While the idea was ingenious and was imitated by others, the two partners seem not to have pursued it for long. In fact, a little later, we find Quinquet partnering with the Marquis d'Arlandes on the production of "economical lamps with glass chimneys," without much more success, however. But the pharmacist never totally abandoned the invention that bore his name in France (and still do today!), since in 1795 he proposed (on the same principle) a portable stove.[17]

Air-flow lamps had become such a success that their manufacture soon gave birth to the new trade of lamp maker. By 1810, there were around fifty of them in Paris; fifteen years later, twice that many were making and selling all sorts of lamps based on Argand's invention: the astral lamp, the Carcel lamp, the Gotten lamp, the Duverger lamp, the "ascienne" lamp, the "sinombre," and other patented lamps. While the pale gleam of gas may have begun to replace the soft light of the Quinquets in public areas, behind casement windows the latter continued to illuminate nights of study and mischievous pleasures. In the portrait of Dr. Alphonse Leroy that Jacques-Louis David painted at the end of the 1780s, he placed the marvelous invention next to his model (see plate 7). In the following century, it served for a long time as a muse to poets, who sang of this faithful companion and "too little oil to feed it" (Béranger), and of early mornings "when like a bloodshot eye, throbbing and quivering / The lamp makes a red stain on the day" (Baudelaire).[18]

Parisian Trades and Inventions

To his misfortune, Ami Argand had collided with the world of Parisian trades, which was quite hostile to royal privileges. His counterfeiter Quinquet belonged to the corporation of apothecaries, renamed pharmacists in 1777; his other competitor, Lange, was a merchant grocer and distiller; last but not least were the Parisian tinsmiths who made ordinary lamps, who had also vigorously contested his monopoly. Two conceptions of technical innovation thus came into conflict over the lamp.

For the artisans, innovation was a collective achievement from which the entire community should benefit—in effect, each inventor simply enriched a common treasury of expertise and experience. The corporative system, which was abolished by Turgot in 1776 but soon restored in both Paris and the provinces, rested on the existence of shared practices and traditions, transmitted and enriched from generation to generation. While the idea of improvement was not foreign to it, it assumed collective control and a sharing of technical innovation. By contrast, for inventors like Argand invention was the fruit of genius and personal labor, which deserved protection. The royal administration, which shared this point of view, had for a long time willingly granted inventors exclusive privileges, that is to say, temporary monopolies on the exploitation of their inventions. At issue was not just the protection of legitimate rights by royal power but also the encouragement of technical innovation. Although the granting of a privilege depended on royal goodwill and presupposed patronage in high places, it was rarely arbitrary, since it relied on savant expertise that was entrusted either to the Academy of Sciences or to savants attached to the Bureau of Commerce. As we have seen, this was the procedure followed for the invention of the lamp.

Generally speaking, the administrative and savant élite, who had been won over by the ideas of free-trade economists, judged the corporative system and its conventional and routine mindset harshly. Liberals like Turgot who had complete faith in the functioning of markets wanted to see the corporative system disappear completely. Others like Necker who wanted to preserve a significant role for the state in oversight and regulation sought to reduce the corporations to a simple tool for policing the labor market. At the same time, everyone was convinced that obstacles to free trade, industrial competition, and technical progress had to be removed. Despite the restoration of the corporations after the abortive attempt of 1776, in Paris things were effectively moving in this direction.

In many sectors, production was increasingly organized in relation to demand. Powerful contractors, in direct contact with local or international markets, imposed their own technical and commercial requirements on the workshops and then forced them to compete for their business. This pressure in turn gave rise to complex chains of subcontracting. The work was thus divided up according to a logic foreign to the traditional divisions among trades. As a result, Parisian manufacturing proved to be a highly integrated and very competitive system, guided much more by market forces than by the rules of corporatism. In the luxury industries, for example, it was the merchant-mercers, driven by a wealthy and demanding clientele, who oriented and organized production by mobilizing many interdependent trades. Commercial, artistic, and technical innovation was a common goal.

For trades regulated by guilds such as jewelry, clockmaking, optics, cabinet making, and perfumery, therefore, there was a constant demand for creativity. The corporations themselves, even as they defended the principle of collective improvement for the benefit of all, were not hostile to technical change; they just wanted to contain it within the regulatory framework of organized trades, which might bother certain craftsmen, such as those who made scientific instruments. In any case, their control over the arts and trades was by no means absolute. In fact, there were many privileged sites in Paris where trades could be exercised freely, or almost so, such as the Louvre, the Temple, the Arsenal, within the precincts of religious orders and hospitals, and the entire Faubourg Saint-Antoine. Moreover, we have seen that the government was happy to grant privileges that exempted their beneficiaries from corporative constraints if doing so was considered profitable to the kingdom. By this means the state encouraged both inventions and technology transfer.[19]

Argand had benefited from this policy for his lamp, as had many other inventors. For example, in 1784 the engineer Charles-Emmanuel Gaullard Desaudray set up a royal manufactory for gilding and English-style veneering in the Faubourg du Roule, with an exclusive privilege, a pension, and housing for foreign workers. The enterprise, which competed directly with the factory of the goldsmith Daumy, was not successful.[20] Another example (contemporaneous with Argand) was the mechanic François Jean Bralle, secretary to the Count d'Artois, who in partnership with the clockmaker Vincent secured an exclusive privilege for a royal clock manufactory in Paris. The factory would produce the mechanisms that Parisian clockmakers were at the time importing en masse from Geneva, other parts of Switzerland, and England, and would train good workers, and not only for clockmaking but also for making scientific instruments. Major clockmakers including Berthoud and Bréguet were consulted and approved the idea. The manufactory opened its doors on the rue du Buisson Saint-Louis in 1787, but, beset with financial difficulties, it soon went out of business.[21]

Watt versus Périer

While they were outsiders to the Parisian trades, Argand, Bralle, and Desaudray were well integrated into the much more mixed world of industrial speculation. Argand himself offers the best example of these inventive and cosmopolitan entrepreneurs who sought patronage and capital for their projects. He had secured both of them from a few great financiers when he was working on the distillation of eau de vie. When he returned to Paris with Montgolfier, another industrialist-inventor, he had gotten the crucial support for his lamp from the Bureau of Commerce and the finance minister himself. His time in London had allowed him to extend his business network across the Channel, where he had made valuable friendships with Boulton and Watt. His ultimate failure makes even more clear the difficulties such industrial ventures faced in Paris. What he had needed in order to succeed was the financial support of either the court nobility or Parisian bankers (the forced alliance with Lange had eventually taken their place), as well as enough contacts within the circles of savants (so stingy with their support) who were willing to stand up for him. Despite their contacts, neither Bralle nor Desaudray succeeded any better, but other Parisian entrepreneurs did have more success.

In fact, the high nobility and the financiers linked to it had a lively interest in industry. The Society of Emulation, founded in 1776 by Abbé Baudeau, a physiocrat close to the Duke d'Orléans, on the model of the Society of Arts in

London, thus proposed to stimulate inventions that would improve the arts and trades. Among its two hundred or so members, there were enlightened aristocrats like the Duke de Chaulnes and the Count de Puységur, as well as court bankers like Bourboulon de Bonneuil; there were also technicians like Desaudray and Dufourny and savants like Condorcet and Lavoisier. Even though the society closed down in 1782 after the disgrace of Baudeau, the following years saw a flowering of industrial enterprises that combined financial speculation with technical innovation, often with international connections.

The government itself supported these initiatives, especially after the arrival of Calonne at the finance ministry in 1783. In three years, more than 550,000 livres were granted to new industries in the form of loans and subsidies. The ministry was most interested in the transfer of technology from abroad and its implantation in French soil. Argand, Desaudray, and Bralle all took advantage of this policy, as did many others. Calonne also helped the Milnes (father and son), two mechanics from across the Channel, establish a textile factory equipped with the new English machines known as "spinning mules" at the Château de la Muette. Finally, on a suggestion from Argand, he sought to lure Watt and Boulton themselves to France.[22]

In 1778, the two entrepreneurs from Soho had negotiated a contract for steam engines designed to pump water from the Seine with the Compagnie des Eaux (Water Company) started by the Périer brothers. The agreement called for their manufacture by the Périers, but at first only covered the steam pump at Chaillot. Meanwhile, Watt and Boulton had at the same time secured an exclusive fifteen-year privilege for the dissemination of their invention in France. But the two Parisian mechanics, bypassing this privilege, undertook to construct and sell steam engines themselves. With the financial support of the Duke d'Orléans they set up a factory for this purpose at Chaillot, a stone's throw from the reservoirs of the Water Company. A year and a half later, Jacques Constantin Périer presented a paper on his engine to the Academy of Sciences. The engineer and physicist Charles-Augustin de Coulomb wrote a very favorable report, and the entrepreneur soon entered the Academy as a supernumerary assistant mechanic. A publicity campaign launched the factory, which was a great success. In addition to pipes, it produced machines of all sorts, including several dozen steam engines, for which, of course, no royalties at all were paid to Watt and Boulton.[23]

Realizing that they had been burned, Boulton and Watt asked Argand to represent their interests when he returned to Paris in the spring of 1785. By this time, they seemed to have already given up their exclusive privilege, which was being flouted with impunity by the Périers. Instead, they continued to seek the payment they were due for the Chaillot pump according to their agreement with the Périers. Not only did Argand succeed in this mission, but, taking advantage of his access to the government, he arranged for Watt and Boulton to come to France to replace the hydraulic system at Marly, which supplied Versailles with water, and eventually to negotiate other deals. For Calonne, this invitation was part of a larger commercial and industrial policy of rapprochement with England that was crowned by a trade treaty signed in September 1786. Watt and Boulton arrived in Paris two months after the treaty was signed, as much for personal reasons (their two sons were visiting France at the time) as for business. They received a spectacular welcome: they visited Marly and Chaillot, gave their opinion on the improvement of the pumps of Notre-Dame and the Pont-Neuf, and advised Calonne on the installation of the forges at La Charité-sur-Loire, in which the minister had an interest.

While nothing tangible came out of this visit, Watt came home enchanted. Fêted like a hero, "drunk from morning till night with Burgundy and undeserved praise," he wrote, he had finally gotten the recognition he had never received at home.[24] He especially appreciated the reception given him by the savants, who welcomed him like one of their own. He dined with Lavoisier and had long discussions with Périer, Monge, and Berthollet. He talked about mechanics and chemistry but was careful not to mention the crucial improvements he had just made to his steam engine: a double-acting piston, parallel motion, and a centrifugal governor. The effort was wasted, however: a young Spanish engineer living in Paris named Agustin de Betancourt, managed to figure out Watt's improvements in 1788 when he visited the Albion Mills in London, where Watt had just set up his new machines. As soon as he returned to Paris, Betancourt made a scale model, the description of which he presented to the Academy of Sciences, and which was the inspiration for Périer's successful execution at full scale of machines designed for the flourmill on the Île des Cygnes, in the Seine near the Champ de Mars, which began operating in the winter of 1791. The following year, the skillful counterfeiter was awarded a patent to import and improve the double-acting engine.[25]

Javel Water

During his stay in Paris, James Watt witnessed with great interest the laboratory trials that the Academy's chemist Berthollet was then conducting on a new process for bleaching fabric. As soon as he got home, he undertook his own trials in a Glasgow laundry. Berthollet's process relied on the use of an acid that the Swedish pharmacist Scheele had obtained in 1774 by adding magnesia (magnesium oxide) to marine acid (hydrochloric acid). According to Scheele, the action of the magnesia consisted of releasing the "phlogiston" from the marine acid, hence the name "dephlogisticated marine acid" that he had given his discovery. Berthollet then undertook a series of experiments on this new product.

In a paper read at the Academy in the public meeting of April 6, 1785, Berthollet maintained that the new acid was in fact produced by the marine acid combining with the "vital air" (oxygen) drawn from the magnesia. In short, he abandoned the generally accepted phlogiston theory for the new theory of oxidation. Berthollet was thus the first chemist to come over openly to Lavoisier's heterodox ideas. As a result, in the new nomenclature, dephlogisticated marine acid was renamed "oxygenated muriatic acid" ("muriatic acid" designating the old "marine air"). In 1810 Humphrey Davy would demonstrate that this acid was not an oxygen compound (as Berthollet had thought) but a new element, which he called "chlorine." The question, although fundamental from the perspective of chemistry, was fairly unimportant from a technological perspective: the theory of oxidation accounted well enough for the bleaching action of chlorine.[26]

Scheele himself had already noted (without dwelling on it) that chlorine bleached fabrics. But Berthollet was particularly interested in this property, no doubt because he saw in it a means to test color fastness. He had in fact, following the death of Macquer, just been appointed commissioner for chemistry at the Bureau of Commerce, charged with the quality control of dyeing processes on behalf of the government. He quickly saw the possibility of using chlorine to bleach cloth, that is, to remove from natural fibers their yellowish tinge. Bleaching was one of the longest and most delicate operations for the preparation of fabric. The most common method consisted mainly of washing the fabric with detergent and then exposing it to the sun and fresh air in a bleachfield, a process that was then repeated ten to twenty times. Fabric could also be soaked in baths of acid or sour milk. However, these labor-intensive and complicated operations took at least three or four weeks.

In his first small-scale trials, conducted toward the end of 1785, Berthollet utilized a chlorine solution in water, with mixed results: if the solution was too strong, it destroyed the fabric; too weak, and there was only a temporary effect and the yellowish color returned after a few washings. However, Berthollet soon figured out how to counteract the irritating vapors from chlorine: he only had to add lye (called at the time fixed alkali) to make the solution odorless without reducing its bleaching power. Lavoisier judged the process sufficiently promising to recommend it to the Finance Ministry's Agriculture Committee.

In the course of 1786, Berthollet undertook his first large-scale experiments. He first improved his process by taking inspiration from the traditional bleaching operations: he now alternated chlorine baths and alkaline washings. He soon noticed that chlorination (which he supposed to be a kind of oxidation, as we have seen) not only bleached the cloth but also made the yellowish tint soluble in an alkaline bath. A little later, he added baths in sour milk and diluted sulfuric acid to the production cycle. In the end, the operations in Berthollet's process were identical to those in the traditional method, with one difference (fundamental, of course): baths in chlorinated water replaced drying in bleachfields, which spectacularly reduced the total time of the bleaching operations.

Berthollet immediately renounced any rights he could claim to the exploitation of his process. He was content with making it known, seeking neither privilege nor reward. In any case, his position as a savant and a commissioner at the Bureau of Commerce seemed to him incompatible with an industrial career. He let Watt exploit his process across the Channel; in France, he supported especially his former assistant Bonjour, who had helped him with his first trials and had now opened a laundry in Valenciennes. Other industrialists—in Rouen, Lille, and elsewhere—employed his process with varying degrees of success. In Paris, Lavoisier supported the establishment of chlorine-bleaching workshops near Montparnasse. But the most interesting initiative came from the chemists Léonard Alban and Mathieu Vallet, who proposed a major improvement to Berthollet's process in 1787: the replacement of the chlorine solution with potassium hypochlorite, in other words, by what would come to be known in France as "Javel water."[27]

Vallet and Alban were the managers of a chemical factory that had been established in 1778 in Javel, as the plain of Grenelle on the south bank of the Seine outside the city gates was called. Founded by a company under the patronage of the Count d'Artois, this manufactory produced sulfuric acid and other chemical products like aqua fortis, vitriols, alum, and Glauber salt. In 1783 Charles and Robert had produced the hydrogen necessary for their balloon flights with sulfuric acid supplied by Javel. Alban and Vallet in turn undertook experiments on steerable balloons, or "dirigibles." They organized free flights for enthusiasts in their balloon, the *Comte d'Artois*, emblazoned with the prince's coat of arms.[28]

Famous now, the Javel manufacturers were on the lookout for all kinds of novelties. Berthollet's discovery naturally attracted their attention. The chemist himself had come to Javel twice to explain his process for bleaching cloth, most likely in 1786. Sometime later, Alban and Vallet announced that they too had discovered a fluid that was capable of bleaching cloth soaked in it for a few hours. They had produced this chlorinated wash by adding lye to the water before the chlorine rather than afterward, as Berthollet had done. By this simple stroke, they produced potassium hypochlorite, which had the advantage of being odorless, stable, and easy to produce while having a bleaching power superior to that of diluted chlorine. While the cost of lye made the product much more expensive, they later swore that a suspension of chalk in water was a satisfactory substitute.

Alban and Vallet launched their "Javel water" with strong advertising in the newspapers. An irritated Berthollet reminded them of his visits to the manufactory and cried plagiarism. Moreover, he doubted the efficacy of their process, judging (wrongly) that the alkaline solution used by the two Javel chemists weakened the bleaching power of the chlorine. His mistake was related to his discovery in the spring of 1787 of a new salt, potassium chlorate (or over-oxygenated muriatic of potash, in Lavoisier's nomenclature), which he obtained precisely by dissolving chlorine in a solution of potash lye. Berthollet noticed not only that this salt had no bleaching power but that it was very explosive. Even as he was denouncing the supposed error of the Javel chemists, he was teaming up with Lavoisier in the Arsenal to develop a new kind of gunpowder based on potassium chlorate. The tests ended tragically in the autumn of 1788 in an accident that resulted in two deaths at gunpowder factory in Essonnes, south of Paris.

In fact, as Balard showed much later, all that was necessary was to introduce enough lye into the solution to convert potassium chlorate into potassium hypochlorite, which was effective and not dangerous: so Alban and Vallet were right. Their "Javel water" was soon being sold in England, where one of the company's owners, the financier Bourboulon, accompanied by Matthieu Vallet, had taken refuge after the banker's resounding bankruptcy in March 1787. Together they opened a factory near Liverpool to produce their product. Vallet, who had settled in England for good, later played an important role as an expert chemist, contributing to the development in Lancashire of the new bleaching industry.[29]

The Gobelins Dyers

As chemist-commissioner for the Bureau of Commerce, Berthollet had been charged by the government with preparing a treatise on the art of dyeing, which it took him six years to write. Hellot and Macquer, who had occupied the same post before him, had already published partial studies, the former on wool dyeing, the latter on silk dyeing. Published in 1791, Berthollet's work, which was based on a broad investigation, aimed to be both more ambitious on a theoretical level and more complete in its description of practical processes than either of these. On this latter point, his book was still far from exhaustive, however. As Berthollet himself recognized, "the mystery created in most of the workshops is a great obstacle to those who work to shed light on the arts."[30] In particular, there was no mention of the processes used in the Gobelins dyeing workshop, one of the most important and most advanced in France and all of Europe. Despite the fact that Berthollet was not the dyeing inspector for this manufactory (contrary to what all his biographers assert), this silence is still surprising, since very innovative research on colors had been conducted at the Gobelins during the last decades of the Ancien Régime.

The Royal Manufactory of the Gobelins in the Faubourg Saint-Marcel mainly produced very expensive tapestries for the royal residences. It housed three workshops, two for high-warp looms and one for low-warp ones, which were run by contractors whom the government paid by the piece. Since 1749 Jacques Neilson had run the low-warp workshop. Neilson, who was himself both inventive and artistic, had made crucial improvements in the production process thanks to the assistance of Vaucanson. He was later made the director of a small school within the Gobelins site to train tapestry workers. While the contractors were responsible for production, the overall operation of the manufactory, which fell under the administration of the King's Household, was the responsibility of the painters of the Academy of Painting and Sculpture, who prepared and selected the cartoons and supervised their execution. Under their influence, the art of tapestry had evolved a great deal since the Gobelins was founded.

Originally working in the Flemish style, the producers were content with a few strong colors that did not fade. This "tapestry coloring" had the advantage of being stable, but it often deviated from the colors in the cartoons. The painters demanded greater fidelity, gradually imposing the idea that a tapestry was merely a woven painting. The colors used in the cartoons themselves came to be more and more varied and subtle, in particular for everything that concerned skin tones. Consequently, a conflict arose between painters like Oudry, who complained of the resistance and bad taste of the workers, and the contractors, who, citing "production issues," maintained that "to paint well and to have tapestries executed well were two absolutely different things." It was the painters who won out in the middle of the eighteenth century. However, the fashion for tapestries with nuanced and light tones in the style of François Boucher encountered a serious problem: the balance of colors tended to deteriorate over time, with each hue weakening and changing at a different rate. In just a few years a tapestry might lose not only its brilliance but also its original harmony. Any solution demanded improvement in dyeing.[31]

Since its foundation in the seventeenth century the manufactory had had a dyeing workshop started by a Flemish dyer, Josse van der Kerkove (Fig. 7.2). Both wool and silk were dyed there with fast dyes that used scale insects called kermes or cochineal for the scarlet reds (following the "tin" or "Dutch" method), indigo for the blues, and *bois d'Inde* (campeachy wood from the West Indies) for the blacks. But the workers, ignorant and conventional, were unable to meet the challenge posed by the multiplication of tones and shades. This is why in 1769 Jacques Neilson, as entrepreneurial as ever, offered to take over the management of the dyeing workshop with his son. Since he lacked the necessary knowledge, he went in search of a talented dyer. Not until 1773 did he find this rare bird in the person of Antoine Quémizet. Neilson recognized in this simple uneducated worker—"of dubious morals and a very difficult person"—an exceptional dyer: "I found in him," he noted enthusiastically, "an exact observer who did not miss a

FIGURE 7.2: Dyeing workshop at the Gobelins, *Encyclopédie,* Plates, vol. 10, *Teinture des Gobelins,* plate 8, *Service du tour.* Engraving by Robert Bénard after Radel.

single detail and who submitted all operations to the rules of chemistry, without which, as he said, no dyeing process was certain."[32] Originally from Rouen, where he had given a public course on dyeing in 1769, Quémizet had taken the courses taught by chemists in Paris.

At the Gobelins, the new dyer threw himself into the work. Jacques Neilson, who financed his research, was soon encouraging him to prepare a textbook on dyeing, which he presented to the Bureau of Commerce in May 1775 under the strange title of "Tinctoresque Principles." The chemist Macquer, impressed by this "kind of comprehensive treatise on the art of dyeing wool," noted that "the author, although a very experienced practitioner in his craft, has not neglected theory." He especially praised "the large chart of all the compound colors and shades," which was an "entirely new piece of work, all the more useful in that this whole essential part of dyeing was [previously] an obscure chaos consisting merely of trial and error with no other rule than instinct and routine."[33] The entire work, which included four volumes or registers of samples, presented an ordered sequence of all the prime and compound colors, with a description for each one of the process by which it was produced.

Despite Macquer's favorable opinion, a distrustful administration balked at financing this research, so Neilson and his dyer continued to bear all the expenses. The government did finally call for expert assessment, though, which took place in the winter of 1777–1778. It seems that in the interim Quémizet had made significant improvement in the fastness of colors by the development of new production processes. What is certain is that the results of the tests conducted by Macquer, Montucla, and Soufflot were very favorable. In his report, Macquer did not hesitate to declare the work conducted at the Gobelins to be "the finest, the most extensive, and the most necessary that has ever been done in the art of dyeing." And he added: "It is indeed so essential, in particular for the Gobelins manufactory, that it is astonishing and annoying that no one in the history of the establishment has thought of undertaking it before. We believe that it must be concluded from these facts and considerations that it is of the greatest importance that this work be continued and watched with all the care that it deserves."

Following this report, Quémizet received a reward of 2,400 livres, and work began to reorganize the dyeing workshop to accommodate the new processes for the production of dyes. In all other respects, however, the still wary administration ignored Macquer's advice. The research, which it considered to be an unnecessary extravagance, was to be discontinued, and a chemist would henceforth be attached to the Gobelins for the general inspection of dyeing, which meant that Neilson and Quémizet would now be under supervision. The court chemist Claude Melchior Cornette of the Academy of Sciences was appointed to this new post.

The situation at the dyeing workshop began to deteriorate. Quémizet, who was seriously in debt, was ready to leave Paris and France. In the end, he was obliged to sign an agreement whereby he promised to give all his attention to the dyeing workshop, no doubt in exchange for the 2,400 livres offered by the government. However, he was coming into work less and less. He was a man without attachments, jealous of his independence, a spendthrift, quarrelsome and violent. The drama came to a head on December 22, 1779, when Jacques Neilson's son Daniel, who was his partner in the management of the workshop, died suddenly. Was it an accident? Was it a crime? We do not know. Whatever the case, the next day, even as Daniel was being buried, Quémizet left Paris incognito. Two days later, he was found dead in Gisors, forty-five miles from Paris, probably a suicide.

Relations between Quémizet and the Neilsons had been poisonous for several months. Contravening a formal government order, Neilson had resumed research into colors, this time on the dyeing of silk. For this purpose, they had entrusted to Quémizet a great quantity of silks, cochineal, and other ingredients. Quémizet had taken all this material, as well as his manuscripts, into the Temple precinct, where he was conducting new experiments unbeknownst to his employers. He was in cahoots with a dyer named Pinel, whom Neilson characterized as his henchman. Still according to Neilson, on the eve of his death Quémizet was prepared to sell his manuscripts to the Finance Ministry; this would have harmed not only the entrepreneurs but also the Gobelins manufactory and its patron, the King's Household, which was in a constant struggle with the minister of finance. D'Angiviller personally asked Lieutenant of Police Lenoir to find the dyer's manuscripts, but the documents, it seems, could not be found.[34]

After this terrible crisis, disorder reigned in the Gobelins dyeing workshop. The large registers of color samples had not been removed from the manufactory, but Jacques Neilson, profoundly depressed by the death of his son, was no longer interested in the problem. All he wanted now was to be reimbursed for his expenses. Moreover, no one took Quémizet's place at the Gobelins. The dyeing workshop was falling apart, under the distant eye of the chemist Cornette. The situation started improving in 1786, with Neilson's retirement and the appointment of a second dyeing inspector in the person of the chemist Darcet. A new dyer, a certain L'Écureuil, then took over management of the workshop. At this time, the Gobelins management was still trying to pursue Quémizet's work. Darcet thus tried to restore the dyeing processes based on documents conserved in the manufactory, but in vain. The registers themselves disappeared during the Revolution or shortly thereafter. When he arrived at the Gobelins in 1824, Chevreul saw only a few moldy fragments of them. As for the "Tinctoresque Principles" so appreciated by Macquer, nobody seemed to remember them. Berthollet made no mention of them in his *Treatise*.

In the end, it was a dyer named Homassel who saved Quémizet and his work from complete oblivion. He had worked under Quémizet's direction in 1779, and remained at the Gobelins workshop for a few years after he had left. He was thus familiar with both the register of color samples and the new processes. In 1798, he published a *Cours théorique et practique sur l'art de la teinture* (Theoretical and practical course in the art of dyeing), inspired to a large degree by his experience at the manufactory, which enjoyed some success. In his preface, Homassel praised Quémizet, whose work he claimed to be completing, and settled his own accounts with the savants. They had written only "novels about dyeing," of which, as he made clear, the most implausible was Berthollet's treatise.

It is not certain that Quémizet, had he lived, would have appreciated the diatribe. In fact, the Gobelins' chief dyer was a strong critic of the routinization of the trades and the stupidity of both workers and masters. He preferred to put everything to the test, to proceed to experiments, and to look for chemical and physical reasons for the success of processes and tricks of the trade. At bottom, like the savants whose support he sought, Quémizet had wanted to use theory to shed light on practice. By taking as a model the technical treatise written by a commissioned academician, he did not hesitate, this modest journeyman dyer, to place himself on the level of great chemists at the Bureau of Commerce like Macquer and Berthollet. This did not mean that his relations with the savant world were free of tension—or possibly resentment. But Quémizet's activities do not fit into a schema that opposes the know-how of artisans to the science of savants.[35]

In this respect, Homassel's book marks a change of both era and tone. The relationship between elite craftsmen and savants deteriorated sharply in the final years of the Ancien Régime. Inventors complained of being looked down on, and some of them now directly challenged the authority of the academic tribunal for securing patronage and privileges. For their part, the savants condemned pedestrian attitudes in the routine of the trades and the ignorance of artisans. The idea that improvement in the arts and crafts would result from the application of science to practice was gaining ground within the Academy of Sciences. The replacement of Macquer by Berthollet at the Bureau of Commerce fits perfectly into this general trend. While Macquer was a chemist who was a practitioner of dyeing, Berthollet was a pure savant who was primarily interested in theory. In order to write his *Éléments de l'art de la teinture* (Elements of the art of dyeing), he had conducted an investigation of workshops, but he quickly ran up against artisanal secrecy. In the end, his aim was above all to establish principles for the art of dyeing based on Lavoisier's theory of oxidation and the laws of chemical affinity; for the practical side of things he was satisfied with simply putting together short descriptions of the standard processes.

Such a penchant for theory, however, did not imply indifference to useful applications or to inventions. After all, Berthollet himself had derived a revolutionary process for bleaching cloth from his study of chlorine. Around the same time, along with Monge and Vandermonde, he also proposed a new theory on the nature of steel and cast iron. This was the kind of research that Condorcet evoked in 1785 in his eulogy of Duhamel du Monceau, when he referred to "the revolution in thinking that has directed the sciences in particular toward public utility." He was convinced that the revolution would endure. "The idea of the common good will guide savants in their research," he predicted; "perhaps they will learn to prefer it to their own glory, and the most enlightened men will also learn how to distribute glory in a manner more useful to their interests." Stimulating industry and enlightening the trades clearly derived from this concern. As we shall see, however, there were other, even more noble opportunities for public service, like beautifying Paris and making it healthier.[36]

Notes

1. THIÉRY, 1787, vol. 2, 387–391. On the creation of *The Marriage of Figaro*, see M. NADEAU, "Théâtre et esprit public: les représentations du *Mariage de Figaro* à Paris (1784–1797)," *Dix-Huitième Siècle* 36 (2004), 490–510.

2. The great chandelier in the Odéon theater is described in GRIMM, 1829–1830, April 1782. Lavoisier read his paper at a public session of the Academy of Sciences on November 14, 1781 (*Œuvres de A.-L. Lavoisier*, vol. 3, 91–102). On the complaint from Peyre and Wailly, see BACHAUMONT, 1777–1789, vol. 18, 169 (November 29, 1781) and 171–172 (December 1, 1781).

3. On the installation of Quinquet lamps in 1784, see AN O¹ 847 microfilm 3; the test was done for the first time on March 5, 1784.

4. In May 1785, the dissatisfied Comédiens Français demanded the return to candle-light (AN O¹ 847 microfilm 3).

5. Quinquet and Lange announced their invention in the *Journal de Paris* of February 18, 1784, and presented their invention to the Academy of Sciences on February 21.

6. On the chandelier at Pilâtre's museum, see the *Journal de Paris*, October 21, 1784, and L.-V. Thiéry, *Almanach des voyageurs*, 1785, 304; on the Quinquet lamps at the Palais-Royal, see GRIMM, 1829–1830, vol. 13, June 1784, 554; on Jefferson and the new lamp in November 1784, see ibid., 45–46.

7. Letter from Lange, *Journal de Paris*, November 10, 1784.

8. According to L.-P. ABEILLE, *Découverte des lampes à courant d'air et à cylindre par M. Argand, citoyen de Genève*, Geneva, 1785, 13–14, Quinquet and Lange stole the idea from Argand on the Réveillon premises.

9. On the balloon craze, see LYNN, 2006, 123–147, and THEBAUD-SORGER, 2009. On the works of MEUSNIER, see G. DARBOUX, ed., "Mémoires et travaux de Meusnier sur l'Aérostation," *Mémoires de l'Académie des sciences*, 2nd ser., 51 (1910), 1–128, and LAVOISIER, *Correspondance*, vol. 4 (1784–1786), Paris, 1986, 293–303.

10. L.-P. ABEILLE, *Découverte des lampes à courant d'air et à cylindre par M. Argand, citoyen de Genève*, Geneva, 1785, 8–16.

11. Ibid., 51–56.

12. The report of the commissioners of the Academy on Lange's lamp is reproduced in *P.-V.* (1785), September 6, 1785, fol. 203r–204v.

13. The letters patent granted to Argand are reproduced in ABEILLE, *Découverte des lampes*, 51–55.

14. On Quinquet, see DORVEAUX, "Les Grands Pharmaciens, VII. Quinquet," *Bulletin de la Société de l'histoire de la pharmacie*, 1919, no. 21, 35–49; no. 23, 65–82.

15. On the contestations of their privilege by the Parisian tinsmiths, see HILAIRE-PÉREZ, 2000, 279–280.

16. *Journal de Paris*, June 25, 1784.

17. The marketing of the lamp by George Palmer was announced in the *Journal de Paris*, September 9, and December 16, 1785.

18. The quotations are taken from the poems "Ma Lampe" (BÉRANGER, *Chansons*) and "Le crépuscule du matin" (BAUDELAIRE, *Les Fleurs du Mal: Tableaux parisiens*).

19. SONENSCHER, 1989, in particular 130–173, and HILAIRE-PÉREZ, 2000, in particular 143–188.

20. L. HILAIRE-PÉREZ, "Des entreprises de quincaillerie aux institutions de la technologie: l'itinéraire de Charles Emmanuel Gaullard-Desaudray (1740–1832)," in J.-F. BELHOSTE, S. BENOÎT, S. CHASSAGNE, and P. MIOCHE, eds., *Autour de l'industrie, histoire et patrimoine: Mélanges offerts à Denis Woronoff*, Paris, Comité pour l'Histoire économique et financière de la France, 2004, 547–567.

21. SCHEURRER, 1996, and AN F^{12} 1325A.

22. HILAIRE-PÉREZ, 2000, 209–220.

23. On James Watt and the Périer brothers, see PAYEN, 1969, 99–135.

24. Citation in J. MUIRHEAD, *The Life of James Watt: With Selections from his Correspondence*, London, 1858, 399.

25. J. PAYEN, "Bétancourt et l'introduction en France de la machine à vapeur à double effet (1789)," *Revue d'histoire des sciences* 20 (1967), 187–198.

26. C. BERTHOLLET, "Mémoire sur l'acide marin déphlogistiqué," *HMAS*, 1785, 331–349. On the invention of bleaching by chlorine, see SMITH, 1979, in particular 113–190.

27. M. SADOUN-GOUPIL, "Science pure et science appliquée dans l'œuvre de Claude-Louis Berthollet," *Revue d'histoire des sciences* 27 (1974), 127–145.

28. SMITH, 1979, 18–20. Description of the Javel factory in THIÉRY, 1787, vol. 2, 642–644.

29. SMITH, 1979, 144–147.

30. C. BERTHOLLET, *Éléments de l'Art de la Teinture*, 2 vols, Paris, 1791. The quotation is taken from the introduction, vol. 1, xiv.

31. LACORDAIRE, 1853, 78–88, and REVERD, 1946. On Jacques Neilson, see A. CURMER, *Notice sur Jacques Neilson, entrepreneur et directeur des teintures de la manufacture royale des tapisseries des Gobelins au XVIIIe siècle*, Paris, 1878.

32. Letter from Neilson to d'Angiviller, May 22, 1775, AN O^1 2047.

33. Report of Macquer on the "Principes tittoresques par le sieur Quémizet, teinturier aux Gobelins" to the Council (of the Bureau of Commerce?), n.d., Bibliothèque du Musée national d'Histoire naturelle, Ms 283, I.

34. On Quémizet and his work, see BELHOSTE, 2015. See also CHEVREUL, 1854, 29–31, and LOWENGARD, 2006, chap. 16: "Neilson, Quemiset, Homassel, Dyers and Chemists at the Gobelins Manufacture."

35. Ch. HOMASSEL, *Cours théorique et pratique sur l'art de la teinture en laine, soie, fil, coton, fabrique d'indiennes en grand et petit teint*, Paris, year VII (1798). On Homassel, see CHEVREUL, 1854, 29–31 and AN O^1 2051.

36. CONDORCET, "Éloge de Duhamel de Montceau" (April 30, 1783), *HMAS*, 1782, 131–155 (*Œuvres de Condorcet*, vol. 2, 610–643).

8

Public Hygiene

DURING THE NIGHT of December 29–30, 1772, a brilliant light illuminated the Paris sky: the Hôtel-Dieu Hospital was burning. A fire that had been smoldering for a long time in a part of the basement where candles were made suddenly burst into flames around 1:30 a.m., engulfing the convent, the infirmary, and wards dating back to the Middle Ages. For hours, torrents of flames broke through the windows, and then suddenly the attic and the structural framework came crashing down. An enormous mass of fire shot up into the air: "The most magnificent and terrifying spectacle," according to one witness. The sick fled in their hospital gowns. Wandering around haggard at night in the freezing cold, most of them took refuge across the way in Notre Dame Cathedral. The fire lasted several more days. The fire pumps needed to continue operating until January 8.[1]

The catastrophe provoked intense emotion throughout Paris. There were rumors of five hundred killed, but in the end the death toll was fourteen. There were also nineteen people wounded, many of whom were firefighters. The scale of the destruction was also exaggerated: first estimates put the damage at two million livres, but the final total was only a third of that. While most people demanded that the hospital be rebuilt, old plans for moving it also resurfaced. Why concentrate so many sick people right in the heart of Paris? Would it not be better to disperse them among institutions that were smaller, healthier, and better laid out? The debate would occupy public opinion and the authorities for almost twenty years. In the end, the Hôtel-Dieu remained at the foot of Notre Dame where it had always been. But although on the surface nothing changed, the affair had exposed a remarkable change in sensibilities. Until then it had seemed normal and even desirable that poverty, sickness, and death should dwell amidst the healthy. Now this coexistence seemed scandalous. The Hôtel-Dieu fire had profoundly frightened everyone: although the public expressed pity for the sick, they were horrified by their ravaged bodies.

The issue of hospitals was only part of a larger problem, however. The threat ran deeper: in a Paris that was becoming more and more densely populated, waste removal had reached saturation. Underground areas were jammed with refuse, running water was polluted, and the air itself was full of unhealthy emissions. The public was alarmed and demanded that measures be taken. While philanthropists worried about the poor (out of compassion and also self-interest), the urban elites were moving westward, to newer and more airy parts of the city, leaving the insalubrious center to those with the fewest resources. The city had to be purified: the monarchy consulted physicians and scientists. The decade of the 1770s saw the beginning of public hygiene in Paris.

The City and Its Scientists

The capital of France, the metropole of the Enlightenment, held a unique position in the eyes of the state. It was the heart of the kingdom and also its showcase. The second largest city in Europe (though far behind London), since the middle of the century Paris had experienced rapid and disordered growth that shattered its old geographic and social frameworks. At the bottom of the social scale was a floating and poorly controlled population that had come from the provinces (and even from abroad) and flooded into the central *quartiers* and the faubourgs to the north and east, supplying the wealthy with unskilled labor that could be exploited at will. A little higher up, the complex world of the trades, generally incorporated into guilds, was the lifeblood of Parisian industry. They formed the socioeconomic base of the common people of Paris. At the top an educated and rebellious elite that increasingly challenged the authority of the king and the church was spreading its wealth in the new neighborhoods.

Even the old Parisian bourgeoisie of producers and shopkeepers found itself gradually dragged into the major shifts that were transforming the city into an economic and cultural metropolis. Pushed by a demanding and sophisticated clientele, they produced consumer goods of all kinds, both luxury and semi-luxury goods, clothing, home furnishings, and other material comforts. Their markets were both local and distant. Skillful entrepreneurs like Réveillon in the Faubourg Saint-Antoine, Alban and Vallet in Javel, and the Périer Brothers in Chaillot opened vast factories on the edges of the city. A notch above them, a stratum of financiers prospered by recycling the interest they made on loans to the monarchy and profits from overseas trade into the Royal Treasury and speculative investments.

Open to wider horizons, the urban fabric was changing rapidly. It was said that a third of Paris had been rebuilt in less than thirty years. The city had expanded right up to the Farmers-General Wall that since 1785 had established its fiscal and administrative boundaries. The center was becoming ever denser, a tendency that only Haussmann would be able to reverse in the latter half of the nineteenth century. At the same time, the city was expanding at its peripheries. The suburbs attached to Paris continued to become more urban, even if they would keep their semi-rural character until the middle of the next century. Vast speculative housing developments, especially to the northwest, on the Chausée d'Antin, at the Roule, and in the faubourgs of Saint-Honoré, Montmartre, and Poissonnière provided residential zones for the wealthiest inhabitants.

Within the walls of the city, the circulation of people, goods, and information was accelerating in a spectacular fashion. The old *quartiers* were opening up: mobility increased for purposes of work, shopping, and leisure activities, and the carriage traffic exploded, clogging the streets and bridges. Even the whirl of society was expanding, mingling social milieux, but without mixing them. News traveled faster and farther, by means of rumor and conversation, but also through posters, pamphlets, and the press. Public opinion—an abstraction whose existence nobody denied—gradually imposed its own law, including on those who ruled.

Government authorities went along with this vast movement that they could not prevent, doing their best just to guide and control it. Urban growth was viewed as both a danger and a blessing. Disorder, the degeneracy of morals, and public disobedience were feared, while civilization, the progress of Enlightenment, and civic spirit were praised. The end of the Ancien Régime saw the birth of a true urban politics in Paris, a concern for policing and urban planning on the part of both municipal and royal administrations. Provisioning became a permanent preoccupation. Road construction aimed to improve circulation and make the city healthier. Hygiene and public health were the focus of studies, debates, and official measures. In this attempt at enlightened administration the savants were asked to play a major role. In fact, they had never been consulted so much about urban problems as they were during the last ten years of the Ancien Régime. Ministers, the lieutenant of police, the Paris intendant, the provost of merchants, the city administration, even the Parlement of Paris—all of them, chaotically and in competition with one another, called upon the expertise of the savants.

A wide variety of institutions were drawn in. There was the Royal Academy of Sciences, of course. But on matters of hygiene, those in need of assistance could also address the Faculty of Medicine and the College of Pharmacy, as well as the Royal Society of Medicine. Founded in 1776 and registered by the Parlement in 1778, the latter had benefited from the start from the patronage of the king and his chief physician, Lassone, who presided over it. It also had close ties to the Finance Ministry, which shared responsibility for overseeing it. At the same time, its relationship with the Faculty of Medicine, which saw it as a competitor, was deplorable. The soul of the society was its permanent secretary, the anatomist Vicq

d'Azyr, a member of the Academy of Sciences who became known for leading an
investigation into a serious cattle epidemic on behalf of Minister Turgot.

The Royal Society of Medicine had more than one thing in common with the
Academy of Sciences: it met every Tuesday and Friday at the Louvre and published
regular volumes of its *Histoire et Mémoires*. Whereas the thirty ordinary members
were all doctors, the twelve associates were much more diverse: there were
statesman like Vergenne and Amelot, administrators like Lenoir, and scientists
like Daubenton and Lavoisier, in addition to non-resident associates from the
provinces and abroad. The society had as its mission the development of scientific
medicine, both experimental and practical. One of its warhorses was the struggle
against secret potions and dangerous remedies and against charlatanism: from
the moment of his arrival in Paris, Mesmer had found it to be a resolute adversary. Mobilizing its network of correspondents, the society promoted above all
studies on the ground, mainly of epidemics and endemic diseases. It also took a
close interest in matters of public hygiene, about which the authorities regularly
solicited their advice. The physicians Colombier, Thouret, and Hallé, as well as
the chemists Lavoisier and Fourcroy (who was also a physician), were among the
members most involved with these issues.[2]

The Hôtel-Dieu

The fire had reopened the general question of Parisian hospitals. At the end of the
1770s, there were more than forty hospitals of all sizes in the city, of which half
were hospices for the poor, while the other half also cared for the sick. The Hôtel-
Dieu was unique in that it cared only for the sick, but of both sexes, and was the
oldest and by far the most important. Situated in the center of the city near Notre
Dame Cathedral, it had partly burned down in 1737 and 1742 and then again, as
we saw, in 1772. Each time, there was discussion of moving it, to the Saint-Louis
Hospital or to Passy, or to the Île des Cygnes near the Champ-de-Mars. But faced
with the firm opposition of the administration of the Hôtel-Dieu and its nursing
sisters, the government had given up the idea.

Yet the largest hospital in Paris and in Europe was dilapidated and overcrowded.
In normal times about 2,500 sick people were crammed into two buildings along
the small arm of the Seine, one on the Île de la Cité, the other on the left bank,
connected to each other by bridges. But there were only a little over 1,200 beds
for them, some small, some large, distributed among twenty-two rooms on three
floors, so that most of the sick had to sleep head-to-toe, two or even three or four
to a bed. The living cohabited with the dead, operations were performed in the
midst of the sick, and women in labor had to share beds. In these conditions,
which were terrible for both comfort and hygiene, the mortality rate was awful.

After the 1772 fire, the situation at the Hôtel-Dieu had become truly scandalous in the view of enlightened opinion.[3]

Turgot had failed in his plan to move the hospital to the Île des Cygnes. When Necker became finance minister in 1777, he again attacked the problem. He established a commission to study the means of improving Parisian hospitals, and the following year he entrusted to his wife Suzanne the supervision of a small experimental hospital where each patient had his or her own bed: the charity hospital of Saint-Sulpice, located in the buildings of a former convent on the rue de Sèvres. The initiative of Necker and his wife was part of a powerful philanthropic movement that made possible the founding of several hospitals in Paris, in particular the Vaugirard hospice, the Cochin hospice in the Faubourg Saint-Jacques, and the Beaujon hospice in the Faubourg du Roule. But not wishing a confrontation with the hostile hospital administration—and perhaps also for financial reasons—Necker abandoned any idea of moving the old hospital.

However, he did have time (before his dismissal in 1781) to establish a department of hospitals in the Finance Ministry, which he entrusted to the intendant of finances Chaumont de la Millière, who was also in charge of the department of roads and bridges. He also named an "inspector general of civilian hospitals and houses of correction" in the person of Jean Colombier. This doctor of the Faculty of Medicine and member of the Royal Society of Medicine had already improved the military medical service and was involved in the creation of the Vaugirard hospice. He was assisted by François Doublet, a physician at Madame Necker's hospital, and by Michel Augustin Thouret, who would become his son-in-law and his successor in 1789. Like Colombier, they both belonged to the Royal Society of Medicine. Necker thus laid the foundation for a central administration of public assistance.[4]

Necker had not consulted the Academy of Sciences about the hospitals, but he did pass along Colombier's report on the establishment of new prisons in the Latin Quarter. Lavoisier, writing for the Academy, approved of the arrangements, but he did express some reservations, primarily concerning the hygienic conditions of the premises. Five years later, the Academy found itself again drawn into the question of the Hôtel-Dieu, this time not by the Finance Ministry but by the King's Household, which oversaw the Paris region. The initiative of Minister Breteuil, who took as a pretext the study of a new plan to move the hospital prepared at his request by the architects Coquéau and Poyet, was evidently directed against the Finance Ministry and its hospital department, which was hostile to the move and did not fail to criticize the project.

The Academy of Sciences immediately named a special commission to look into the matter chaired by the astronomer Bailly, a man trusted by Breteuil. If Bailly played a major role in writing the report, as advisor to Breteuil and as negotiator with the Archdiocese of Paris, the main expert on the commission was Jacques-René Tenon, founder and chief surgeon of the hospice at the College of Surgery. Despite the refusal of the Hôtel-Dieu management to collaborate with it, the commission undertook a through enquiry. It submitted its report in September 1786. In it the commission painted a harsh portrait of the Hôtel-Dieu, which it judged incapable of being reformed in its current state, but it rejected the plan to move it presented by Coquéau and Poyet, proposing instead the construction of pavilion-style hospitals to be located at the cardinal points of the capital. Breteuil immediately asked the commission to continue its work. A second report presented in June 1787 laid out a detailed plan for where the new hospitals should be built, followed by a third report in March 1788, which was based on a study conducted by Tenon and Coulomb in England and described the layout of the interiors. Tenon himself published under his own name a book that brought together the five individual reports that he had prepared for the Academy's commission.[5]

Breteuil's aim in entrusting this mission to the Academy of Sciences was two-fold: to put into the public sphere the problems of public assistance and welfare by calling attention to the deplorable state of the Hôtel-Dieu, and to take control of the administration of Paris hospitals by sidelining the team from the Finance Ministry. The Academy made a crucial contribution to the realization of the first goal by raising public awareness. The horrifying description of the Hôtel-Dieu in its first report moved people deeply. Louis XVI was so moved that he paid a visit to the hospital incognito. Tenon's work, whose main points Mercier incorporated into his *Tableau de Paris*, became a classic.[6]

Breteuil took advantage of public sentiment. He launched a subscription to finance the construction of the four new hospitals the commission had called for. Ten thousand copies of the prospectus, written by Bailly, were printed, in addition to being published in all the newspapers. It was quite successful: in two months, the subscription had raised more than two million livres, which were eventually swallowed up in the renovation of the old hospital or diverted to other purposes. For while the publicity campaign was a great success, the attempt to take control of Parisian hospitals ended in complete failure.

The affair of the Hôtel-Dieu had given the Academy of Sciences a starring role. It had not only advised the government; it had enlightened public opinion. Bailly himself achieved a level of popularity that would contribute to his election as a deputy of the Third Estate in 1789, and then to his nomination as mayor of Paris. But while the Academy was conducting its investigation, the hospital department of the Finance Ministry had the Hôtel-Dieu rebuilt in the same place. The hospital's administration, supported by the Finance Ministry and its inspector, Colombier, thus continued to oppose the construction of four hospitals ordered by the royal decree of June 22, 1787. To get around this obstacle, Breteuil wanted to transform the Academy of Sciences' commission into a permanent committee for the general inspection of all the civilian hospitals in the kingdom, but his resignation prevented him from doing so. His successor, Laurent de Villedeuil, gave the Academy's commission two new missions: first to investigate the feasibility of moving the Parisian abattoirs, and then to examine a plan to move cemeteries outside of cities. But Necker, now back in power, confirmed the authority of the Finance Ministry over the administration of hospitals and definitively rejected the plan for four new hospitals.

One might imagine that the conflict between the King's Household and the Academy of Sciences on one side and the Finance Ministry and its hospital department on the other derived from opposing ideas about public health and hospital organization. But that was not the case at all. The physician Jean Colombier, a member of the Royal Society of Medicine, was just as enlightened and innovative as the surgeon Tenon. He believed in the same principles of healthy conditions and hygiene, he was a strong advocate of single beds and airy wards, and he firmly supported the reform of hospitals with the aim of making them sites of care rather than of charity. In collaboration with Pierre Desault, a rival of Tenon's and the chief surgeon at the Hôtel-Dieu, he wrote regulations for the operation of the wards, which entrusted their management to the physicians and surgeons, and in 1788 he had a large auditorium built for instruction in clinical surgery.

If Colombier, like Desault and the Hôtel-Dieu administration, was hostile to the establishment of four hospitals on the periphery of the city, this was because he thought it preferable and less onerous to maintain one large hospital as close as possible to where the poor lived, which meant at the center of Paris. At the same time, he did not hesitate to question the power exercised at the Hôtel-Dieu by the Augustinian Sisters, provoking a sharp conflict that went all the way up to the Parlement on the eve of the Revolution. There was a certain irony in seeing the nuns, conservative in everything, prevail over the conclusions of the Academy of Sciences, and then turn around and demand the construction of four hospitals in order to better oppose Desault and Colombier.[7]

The Cemetery of the Innocents

Colombier died (of exhaustion, it was said) in the midst of the struggle against the Augustinian sisters. His son-in-law and successor, Michel Augustin Thouret, took up the cause even as the Revolution was beginning. By this time, Thouret was far from unknown. A doctor at the Paris Faculty of Medicine and a founding member of the Royal Society of Medicine, he got involved in the critique of the curative powers of magnets and the theory of animal magnetism popularized by

Mesmer. But it was above all his role in the evacuation of the Cemetery of the Innocents that would make his reputation. The presence of this cemetery in the *quartier* of Les Halles (the central market) in the heart of Paris had long been tolerated despite regular complaints from neighbors about its odors. In 1776, the apothecary Cadet de Vaux, whom we have already met, a man who was an all-around philanthropist and an advocate well ahead of his time of public hygiene, undertook (no doubt at the request of the lieutenant of police Lenoir) to measure the purity of the air—with very worrying results. "The air of the Cemetery of the Innocents," he concluded, "is the most insalubrious air one could breathe, equivalent to that inside the foulest hospitals."[8]

In order not to panic the population, Lenoir forbade the publication of the report, which was only presented to the Royal Society of Medicine. But moving the cemetery was still very much on the public agenda. Enlightened opinion was already worried about the dangers of airborne infection. In 1778, Vicq d'Azyr had recommended a ban on tombs in churches.[9] And then a serious incident in March 1780 led to the definitive closure of the Cemetery of the Innocents. Several burial vaults located on the rue de la Lingerie had been infected by emanations coming from a large common grave opened a few months earlier. Always on the case, Cadet de Vaux (having been promoted to inspector general for health affairs by Lenoir) claimed that a "cadaverous gas" was attacking people's nerves. The lieutenant of police sought both his advice and that of the Faculty of Medicine; based on their reports, the Parlement of Paris decided in November 1780 to end burials there and to close the cemetery. But for five more years the Innocents remained as it was. It was Thiroux de Crosne, Lenoir's successor, who finished off the cemetery by having all the tombs razed and tearing down the charnel houses and the church in order to establish a produce market on its site. The human remains were to be transferred to quarries. Thouret, assisted by the surgeon Marquais, was appointed to direct this delicate enterprise.

The government was rightly concerned about how Parisians would react. The Cemetery of the Innocents, an ancient and sacred precinct, was a site of worship and veneration whose destruction and transfer could prove even more offensive to popular sensibilities than that of the Hôtel-Dieu. The medical authorities had, thank God, the agreement and support of the religious authorities. The public was also worried about the quality of the air, because it would be necessary to open dozens of crypts and common graves. The operation lasted more than six months under the supervision of priests and ultimately ended without any incident. Day and night workers exhumed between 1,500 and 2,000 corpses with their coffins, and the charnel houses were emptied completely. Every evening, by torchlight, heavy funeral carts took away the debris to the Tombe-Issoire, beyond the barriers, where the inspector of quarries Guillaumont had had catacombs dug to accommodate them.

It was the terrifying grandeur of the operation that struck the public. Thouret, however, was most interested in drawing from it "interesting results for physics," for, as he wrote, "so much labor could not fail to offer results for science, and their utility in this regard might be the only positive thing to come out of these painful and macabre operations." To this end, he undertook a systematic study of the natural decay of cadavers, assembling a medical collection of mummies, bones, and viscera in various states of decomposition. His great discovery was the preservation of certain bodies in specific circumstances by the transformation of fat into adipocere, which his colleague the chemist Fourcroy quickly identified as a type of soap. For Thouret, this substance with a brilliant sheen that he compared to sperm oil was not a product of death but part of the "living animal economy." Believing that he had found it in the substance of the human brain, he theorized that this "principle" was nothing less than "the essential character of animal life, at the same time as it appears to form the primary mode of destruction, which after death decomposes all the parts."[10]

Returned to the land of the living, in 1790 Inspector Thouret participated in the work of the Constituent Assembly's committee on begging, with his patron La Millière (while Tenon was kept at a distance) and was largely responsible for initiating its projects. In accordance with contemporary philanthropic ideas, the committee recognized the rights of the poor and the duties of society to care for them and put forward the principle of a national welfare office. It recommended that charity be managed locally, methodically, and economically, as well as the development of an emergency response system, the separation of the sick from the poor, and the reorganization of the hospitals. While these prescriptions were not followed at the time, they would become the basis for the welfare policies implemented during and after the Revolution. Until his death in 1810, Thouret was closely associated with these ideas, playing a decisive role in their implementation and in the organization of public medicine in Paris.

Solids and Fluids

The distancing of the sick and the removal of the dead testified to a general anguish about organic decomposition. The hospitals and cemeteries were not the only institutions concerned. Enlightened opinion—of scientists, administrators, and journalists—was also alarmed by the stench emanating from night soil, sewers, and abattoirs. "Narrow streets that were difficult to cross, houses that were too tall and thus blocked the free circulation of air, butchers, fish markets, sewers, and cemeteries, all pollute the atmosphere, filling it with impure particles, and cause this stifled air to become heavy and malignant," noted Mercier in his *Tableau de Paris*.[11]

In the frightened imaginations of Parisians, everything was mixed up—animal carcasses covered with flies, putrefied cadavers in common graves, even the remains of dissections thrown by medical students into latrines—as if the whole city was rotting from within. "You have to have walked around these sites of infection," wrote Thouret in his report on Montfaucon, "to know what these residues or products are, what we might call the excrement of a great city, and to understand on a physical level the immeasurable increase of filth, stench, and corruption that results from packing people together in highly populated cities." At the same time, this terrible portrait completely overlooked the pollution of the soil and water by toxic products like arsenic, lead, and mercury nitrate.[12]

The main problem, according to the experts, was actually immobility. In enclosed spaces like theaters, schools, and (especially) prisons and hospitals, people were breathing air that was stale and smelled bad because it was not circulating. Stagnant liquids proved particularly dangerous: people were dismayed by the cesspool that had formed under the arcades of the Quai des Gèvres between the Pont Notre-Dame and the Pont-au-Change; they worried about the sluggishness of the Bièvre River, full of garbage and animal matter; they decried the blood of dead animals stagnating in the detritus around slaughterhouses. Even below ground Paris was saturated with organic remains. In 1780, excavations under the boulevards at the site of an ancient dump brought to light bones, horse dung, a green almond in its hull, and all sorts of putrid matter that had become indestructible. An even more surprising discovery was made on the rue Dauphine, where the wine merchant Paquet, breaking through a wall of his basement, exhumed an enormous bone, which he took at first for a tree trunk, and which the naturalist Lamanon identified as a whale bone. The same Lamanon also reported on strange fossil bones discovered by workers in the gypsum quarries of Montmartre at the edge of Paris.[13]

To cleanse the city, it was thought, one had to get rid of all this waste—dead bodies, sick bodies, waste, sludge, and excrement—and make all the fluids circulate, including the subtlest ones like electricity and magnetism. Was Paris not equipping itself at that very moment with lightning rods for atmospheric electricity? One writer jokingly proposed in the *Mercure de France* that animal magnetism be distributed by the Water Company: to receive the flow of animal fluids, there would be in each house "health cabinets, provided with picks and chains, etc., ready for any crisis, just as there are bath cabinets with pipes and faucets." Pharmacists, physicians, and chemists were at the forefront of this struggle to promote movement. Journalists relayed their investigations and condemnations to the public.[14]

In the grand circulation of fluids, that of water was essential. Paris drew potable water from the Seine in two ways: via ground water that was well supplied by the Seine and its small tributaries, just above ground at the foot of the hills on the Right Bank, and from an aqueduct to the south that brought water in from Arcueil. A hundred public fountains, thousands of wells, and a whole organization of water carriers ensured that the population was supplied with water. However, the quality of the water was poor: water from the wells was almost always polluted, water from the Bièvre was appalling, and water from the Seine (which Parmentier had praised) was spoiled by the effluents dumped into it as it ran through Paris.

The civil engineers of Roads and Bridges had studied a plan to divert the Yvette that was proposed by Deparcieux in 1762 and supported by the Finance Ministry. The calculations were based on future consumption of at least four times that of current usage. Consulted in 1775, the Academy of Sciences declared itself in favor of the project, but the authorities, afraid of the cost, estimated at twenty million livres, turned to other solutions. All of them consisted of drawing water directly from the Seine, either by means of a hydraulic pump, as the pumps of Notre Dame and the Samaritaine were already doing on a small scale, or else by steam pumps, as the Périer Brothers proposed in a plan presented to the Academy of Sciences in 1775.

The Water Company

Jacques-Constantin Périer and his younger brother Auguste-Charles were the Duke d'Orléans's mechanics. In other words, they had support. Their idea was to pump water at Chaillot, upstream from Paris, and to distribute it via a network of canals. Although the Academy of Sciences preferred the Yvette project, it advised granting them a fifteen-year privilege. The city administration also issued a favorable recommendation. Other competing projects were set aside, and the Périer brothers received their patent in February 1777.

They got to work immediately. They assembled a group of financial backers thanks to their connections and founded a limited partnership company, the Water Company (Compagnie des Eaux de Paris), of which they were the directors. At the same time, they prepared to set up their business in Chaillot close to the barrier, in the gardens of the Countess of Boufflers. Two pumps installed on the banks of the Seine were to pump water up a hill into four reservoirs that would feed the water network (Fig. 8.1). Recall that the elder Périer went to England to make a deal with James Watt and Matthew Boulton, who were beginning to market their machines. The steam pumps would be built in France from plans drawn up by the Englishmen, who would furnish the parts, since they could only

FIGURE 8.1: View of the fire pump at Chaillot. Engraving and drawing by Taré, 1784, in A. Pinard and H. Varnier, eds., *Commentaires de la faculté de médecine de Paris*, vol. 1, Paris, G. Steinheil, 1903, 323.

be manufactured in England. In return, Boulton and Watt would receive 24,000 livres, which would be doubled if the company's capital doubled (which it had by 1781), and the equivalent each year of a third of the value of the coal saved by their new steam engines. The *Journal de Paris* reported on the first trial run at Chaillot, which took place in August 1781 in the presence of the lieutenant of police and the provost of merchants.[15]

The Périers' venture was a great technical success, marking the introduction of steam engines in France, but the returns of the Water Company proved disappointing. It took a long time to set up the distribution network. By July 1782, water had reached the Faubourg Saint-Honoré and the Chausée d'Antin, then the rue Saint-Denis and the rue du Temple. Four years later the company installed two steam pumps at the Gros-Caillou on the Left Bank to supply the Faubourg Saint-Germain. Unfortunately, the number of subscribers remained low, much lower than the Périer brothers had counted on, and the monopoly they had secured in 1777 was fiercely contested.

The Yvette project had resurfaced in 1781, put forward by the civil engineer Defer, with the support of the Count of Provence, the king's brother, "Monsieur." Its cost had been cut to a seventh of the original estimate, and the Academy of Sciences had approved it. Partisans of this project did not hesitate to attack the

Périer Company. They questioned the quality of the water. The Périers had to call in the Royal Society of Medicine, which declared the company's water very healthy. The battle was also financial. The Périer Company needed capital, and the banker Clavière, who supported the Yvette plan, sent the price of shares plunging by betting against them. Mirabeau and then Brissot (both of whom were paid for their efforts) wrote pamphlets against the company; the playwright Beaumarchais was hired to refute them.

The authorities found themselves dragged into a conflict that pitted enormous private interests against each other. Great aristocrats and financiers—even the king himself—had invested in the company. Under Calonne, the Finance Ministry intervened to prop up the share prices. The stock stabilized, rising to its highest point in the summer of 1786, and the state was by far the principal shareholder. On the other side, the Yvette enterprise had powerful support, even though Monsieur had withdrawn in 1784. The intendant of Paris, Bertier de Sauvigny, was a proponent. Louis XVI's minister Breteuil himself pushed forward the purchase of the land to be traversed. The affair went as high as the King's Council, where the two ministers opposed each other. Finally, in 1788, when the price of its shares had again collapsed, the Water Company was bought up by the City of Paris and transformed into a royal administration for Paris water, from which the Périers were excluded. As for the Yvette project, it was authorized by the King's Council after yet another examination in 1787. A company was created and land purchased, but the Revolution put a stop to all its preparations.

At this time, the distribution of water in Paris remained an unresolved problem. Only a few hundred subscribers in the finer neighborhoods were benefiting from water being pumped into their homes from Chaillot and the Gros-Caillou. Everyone else had to resort, as before, to wells, fountains, and water carriers. The amount of water available for cleaning streets and sewers, for fire pumps, and for drinking troughs remained inadequate.

Waste Water

After provisioning, sewage disposal completed the water cycle. Apart from what evaporated or seeped into the ground, all water came back to the Seine, either directly or indirectly through drains and sewers: rain water and water from the fountains, as well as household and industrial wastewater. On the Right Bank, water from several drainage canals that served the northernmost *quartiers* emptied into the great sewer that ran from the Marais to the Seine below Chaillot, passing through the northern and western suburbs. The sewer had been entirely rebuilt, paved, and faced with stone at the end of the 1730s, then covered with a vault running almost its whole length. Some covered sewers flowing directly into the Seine were imperfectly draining the central *quartiers*; two other open sewers crossed the Faubourg Saint-Antoine. On the Left Bank, the network was much patchier. The Bièvre River itself was a veritable sewer, serving the Faubourg Saint-Marcel and its industries. Fetid odors emanated from these poorly cleaned cesspools of stagnant water. Moreover, except for a few residential neighborhoods to the west, the streets of Paris had no drains at all except for rainwater, and despite regulations that called for them to be cleaned at the residents' expense, they were infamous for their black sludge.

In principle at least, waste from latrines did not go into the sewers but into cesspits in each building. When these were full, specialized workers transported cartloads of their contents at night to a discharge site located three hundred yards outside the city, at the foot of the Butte de Montfaucon. While this system avoided contaminating the drains and ensured a recycling of the sludge, it caused a terrible stench when the pits were emptied, as well as near the discharge site. When the wind blew from the north, half of Paris was subjected to foul emanations. In addition, because they were not water-tight, the cesspits were seriously polluting the groundwater and the wells. Concerned, the administration asked scientists for their advice about what nobody dared call a sanitation operation. But since the problem *was* sanitation, physicians and pharmacists were consulted. The emanations from the cesspits, which were noxious and actually caused accidents, were their main concern, as we shall see below. Deliberate pollution of the water also stimulated investigations.

The first issue to be addressed was the Seine, into which ultimately all the waste liquids of the city were poured. Was the Seine not in fact the biggest sewer of all? It is true that the quality of its water had always been held in high esteem. Thus, it was really to demonstrate that the water of the Yvette was as good as that of the Seine in Paris, and not the other way around, that the academicians Hellot and Macquer undertook a comparative chemical analysis in 1762. The good quality of Seine water was confirmed by a commission of the Faculty of Medicine, then by the pharmacist Parmentier. When doubts arose after the installation of the Chaillot pumps near the outlet of the main sewer, Fourcroy and Hallé from the Royal Society of Medicine confirmed once again the healthiness of the water drawn from the river.[16] However, the administration was not completely reassured, because too much refuse clogged the sewers. Next, in 1790, the new Parisian municipal government became concerned about the sludge accumulating along the quays. Hallé proceeded to examine the banks above the Pont-Neuf; then he did a thorough study of the Bièvre River, which poured its black waters directly into the Seine. His report provided the authorities with a very complete description of the course of the river and its banks. Unsurprisingly, he stressed the presence of stagnant water, as well as of silt and refuse from factories, and he suggested measures for improvement and ongoing maintenance.[17]

For the physicians, the main risks came from the accumulation of organic waste, in particular that produced by the butchers. Blood seeped into the soil and entered the sewers. The administration had long wished to remove slaughterhouses from the center of the city. In 1788, Breteuil tried to mobilize public opinion by means of the *Journal de Paris*. His successor Laurent de Villedeuil submitted several papers to the Academy of Sciences that suggested relocating the slaughterhouses outside Paris. A commission on which Lavoisier played the principal role proposed as a site the Île des Cygnes near the Champ de Mars, but nothing was done before 1800. It was only under the Empire that five large abattoirs were built at the gates of the city.[18]

Attention was also focused on the waste disposal site at Montfaucon. A study done by the Royal Society of Medicine in 1788 at the request of the lieutenant of police, predicated on a conflict of interests between the owners of the property and the operator, examined the system of decanting waste and transforming fecal matter into *poudrette*: a fine powder that was used as fertilizer. The site had several open-air pools laid out along the slope of the hill. Thouret, who wrote the report, approved of the whole plan, while suggesting certain improvements. For example, rather than releasing the sluice into cesspools, which would pollute the ground water, he proposed letting it evaporate in the sun.

In fact, the most unsanitary thing at Montfaucon was not the sewerage but a shed where animals were cut up for meat and the pit into which the entrails were dumped that was located close to the pools. Horse entrails were putrefying in the sun to feed maggots used as bait by fishermen; guts from the slaughterhouses and butchers' shops of Paris spread an unbearable smell. The report also mentioned two factories in the Faubourg Saint-Martin where animal intestines were made into catgut and goldbeater's skin and whose polluted water spread into the surrounding streets. For a long time, Parisians had complained about the nuisance caused by these activities. Lenoir had tried to control them, but without success. In 1780, he had given the monopoly on meat cutting to Cadet de Vaux. The latter had established a "veterinary pit" in Javel, where the carcasses were treated with quicklime. But in the face of opposition from the meat cutters, the project was quickly abandoned and Lenoir had to be content with requiring that sheds for meat cutting be established on the outskirts of Montfaucon.[19]

Respiration and the Chemistry of Gases

The pollution of the air was even more disturbing than that of the water. The number of complaints about fetid odors increased in the 1780s: near Montfaucon, around the Cemetery of the Innocents, and at the Apport-Paris across from the Châtelet, as well as more generally about the effluvia from toilets. In this regard, the historian Alain Corbin has noted a "lowering of olfactory tolerance." What caught the attention of certain physicians, scientists, and administrators who dealt with this problem—and was then relayed to the public by journalists like Mercier and the editors of the *Journal de Paris*—was less the annoyance represented by foul smells than the underlying threat to the health of Parisians. To them these odors were signs: they indicated a dangerous stagnation of aerial fluids. Agitated air, like moving water, was purer and healthier; if it did not move, it became corrupted—hence the need to ventilate confined places properly. These old ideas, which neo-Hippocratic medicine had made fashionable again, seemed to be confirmed by the chemistry of gases, which had made remarkable progress since the 1760s.[20]

In 1774, Lavoisier had been the first to advance publicly the idea that atmospheric air was not a single element but a mixture of gases. He had forged this hypothesis, explained in the context of his work on the formation and reduction of metallic oxides, after having become aware of the recent experiments by Joseph Priestley on oxygen, a gas that the English chemist had just discovered. Having observed that a candle placed inside a bell jar burned brightly, Priestley had deduced that it was a result of atmospheric air being deprived of its phlogiston, that is to say, its principle of combustion—hence the name "dephlogisticated air" that Priestley gave it (the name "oxygen" itself did not appear in the nomenclature until 1787).

In fact, according to Stahl's theory, generally accepted by chemists at the time, phlogiston passes from the burning matter to the surrounding air. When the air is saturated with phlogiston, combustion becomes impossible; conversely, air entirely deprived of phlogiston leads to combustion. Atmospheric air, which in its ordinary state contains an average quantity of phlogiston, could thus, according to Priestley, pass from the state of entirely dephlogisticated air (our oxygen) that was very favorable to combustion to the state of phlogistic air (our nitrogen), in which nothing burns. Lavoisier, rejecting the phlogiston theory, interpreted Priestley's experiments completely differently: in his view, air that was supposedly "dephlogisticated" was a gas contained in atmospheric air, which combines with metals in calcination and also plays a fundamental role in the formation of acids.

In 1776, after having set it aside for a year, Lavoisier resumed the study of dephlogisticated air. In the meantime Priestley had shown that animals live longer in this air than in ordinary air. Lavoisier began by reproducing the experiment in the presence of several of his colleagues from the Academy. He continued his research after he moved into the Arsenal, where he set up a very well-equipped laboratory. He was then able to show that atmospheric air was composed of one-fifth "vital" or "healthy" air (the names he still gave to dephlogisticated air, or oxygen) and four-fifths unbreathable foul air, equivalent to Priestley's phlogistic air (i.e., nitrogen). He then turned his attention to the problem of respiration, at the intersection of chemistry and physiology.

People had long noted the presence of "fixed air" (that is, carbon dioxide) in the air exhaled in respiration. Starting in 1773, Lavoisier himself had undertaken some parallel experiments on the combustion of a candle and the respiration of a bird in a confined atmosphere. He soon discovered through a new series of experiments that in respiration not only was vital air absorbed, but the same amount of fixed air was exhaled. To reconstitute ordinary air, it would then suffice, by an operation that was the opposite of respiration, to eliminate the fixed air by means of caustic soda and to add an equivalent amount of vital air. As for what becomes of the vital air absorbed in respiration, Lavoisier hesitated between two theories: either the vital air passes into the blood, which would explain its red color, or else it is transformed into fixed air in the lungs. That autumn Lavoisier resumed his experiments with his friend and colleague Trudaine at his château of Montigny, putting nightingales under a bell jar, and he finished writing his first paper on respiration, which he never published.

The year 1777 was a particularly remarkable one for Lavoisier. His research on respiration, combustion, the constitution of acids, vaporization, and the substance of heat (or "caloric," a term created in 1781)—all closely connected— would give rise to numerous experiments and several important papers. He regularly collaborated with the physician Bucquet and the mathematician Laplace; both of them came to his laboratory in the Arsenal. It was at the end of that active and fertile year that he launched his first attack on the phlogiston theory at a public meeting of the Academy, a theory that he had been gradually moving away from, probably for a long time, but to which his colleagues still remained firmly attached. The event caused a sensation.[21]

Lavoisier's Bird

Historians of chemistry who have traced the history of these discoveries are mainly interested in the experiments conducted at the Arsenal laboratory. Indeed, it was there that Lavoisier achieved the precise measurements that were indispensable for establishing the basic principles of chemistry. However, he did not disdain either experiments on the ground or even experimental "demonstrations" that aimed to convince people more than to prove something. Thus, after having read a paper at a public session of the Academy in May 1777, in which he compared the results of his laboratory experiments on the calcination of mercury with those on respiration conducted in previous months, he gave a very different kind of presentation on the same subject a week later in the presence of Marie-Antoinette's brother the Holy Roman Emperor Joseph II (under the alias of the Count of Falkenstein), who was then visiting Paris. This time it was a matter of examining the alterations in atmospheric air caused by respiration, as well as how to measure and correct it. The presentation, which was meant to appeal to the public, was accompanied by a few "demonstrations."[22]

Lavoisier began by referring to Laplace's experiments to explain that atmospheric air is composed of different fluids capable of being vaporized "by the heat of our planet," but that only one of them (vital air, or oxygen) supports combustion and animal respiration. The others are "mephitic" (noxious), that is, unbreathable in a pure state. Reproducing an experiment by the Duke of Chaulnes, he poured fixed air, which is denser than ordinary air, into an open flask and demonstrated to the spectators its effect on a lit candle that was plunged into it (it was extinguished) and on a bird that was put inside it (it died in a few seconds, as if it was drowning). As we will see below, this last experiment gave rise to an unexpected incident. Lavoisier then moved on to the effect on a lighted candle first of air deprived of vital air (it still went out), then of vital air (it burned intensely), and finally of a mixture of the two in the ratio of three to one (it burned as in ordinary air). Once he had completed these demonstrations, Lavoisier returned to his initial observation and affirmed that the composition of air varied according to local conditions; in particular, in a closed space, respiration might significantly reduce the portion of vital air while releasing fixed air. Above all he believed that because these stagnant airs were of different densities, they could not mix perfectly, as demonstrated by experiments he conducted in various public places.

Lavoisier then explained that early one morning he went to the Hôtel-Dieu to gather air from the unhealthiest dormitory. He took two samples, one near the floor, the other as near as possible to the ceiling. The former proved to be only "slightly altered"; by contrast, the latter contained a higher proportion of "foul air" (nitrogen) and fixed air. Lavoisier had done a second test in the Tuileries theater, where the Comédie Française was then performing. One day when it was crowded, he collected air samples from the upper balcony, known as "paradise," as well as from the orchestra, into which he slipped at the end of the performance. His analysis would show once again that the upper air had a reduced level of oxygen; however, the air at the bottom was identical to ordinary air, which Lavoisier explained as a result of having been gathered too near the entrance.

In his presentation to the Academy, Lavoisier was categorical: the altered air in meeting rooms is always divided into three strata: the fixed air, denser than ordinary air, is concentrated at the bottom, and, as the tests demonstrated, noxious gas, because it is less dense, is concentrated at the top; only the intermediate stratum is little altered. From this research Lavoisier deduced that public halls should always be ventilated with openings at the top and bottom. Chemistry thus confirmed the old rule that air should circulate in order to be healthy. Lavoisier noted that in the Saint-Louis Hospital the wards had been built according to these principles, and he was worried about the reconstruction of the ward of the Hôtel-Dieu that had burned in 1772: "Perhaps the public will be surprised to know that the Academy was never consulted; no doubt it would be humiliating for the nation that even after the discovery of the theory of air founded on exact and sure experiments, we might in the eighteenth century commit errors in construction that were foreseen and avoided back in the sixteenth."

Lavoisier did not have enough time to describe to his listeners the tests that allowed him to evaluate the healthiness of air and the procedures for decontamination offered by modern chemistry. This was because time had been lost in an unexpected incident that occurred during his demonstrations. When the bird placed under the bell jar had collapsed as if dead, the academician Sage, who detested Lavoisier, demanded the animal. To the surprise of everyone, he revived it easily by rubbing its beak with ammonia. The little resuscitated bird started flying around the room and, once the windows were open, went off with a flap of its wings. The story, embarrassing for Lavoisier, made the rounds of Paris. A triumphant Sage claimed that the ammonia neutralized the effect of the noxious gases and that this was the most effective remedy for asphyxiation. A merchant had great success with flasks of ammonia that he sold as "Sage flasks."

Pierced to the quick, the physician chemist Bucquet, friend and collaborator of Lavoisier, undertook over the following months a major study of asphyxia and its remedies. He suffocated about two hundred birds and mammals with all sorts of gases. He counted the time it took for them to be asphyxiated, compared remedies, and dissected all the animals that died. He concluded that far from being a panacea, ammonia was even less effective as a remedy than vinegar or the vapor of sulfur, and above all that all these acids stimulated respiration without neutralizing the deleterious effect of the noxious gases as Sage claimed. Lavoisier encouraged Bucquet and even collected some samples of "marsh gas" from the ditch behind the Bastille for his experiments. In March 1778, he reported very favorably on his colleague's work to the Academy of Sciences.[23]

Cadet, Pilâtre, and Cesspits

In this context, Lavoisier became interested in the research of Cadet de Vaux on cesspools. As we have seen, in 1775 this well-connected pharmacist had been working on the problem of cemeteries. A little while later, Lieutenant of Police Lenoir, with whom he was close, asked him, along with Laborie and Parmentier, his colleagues at the College of Pharmacy, to look into the dangers of cesspits, particularly for those who emptied them. The despised corporation of masters of vile work—the dischargers of the cesspits, wells, and cesspools in Paris and its suburbs—indeed performed unpleasant work that also carried some risks. There had been accidents and work-related illnesses, such as the *mitte* (ophthalmia) that burned their eyes and the *plomb* (hydrogen sulfide intoxication) that suffocated them. Since 1755, the so-called Ventilator Company (Compagnie des ventilateurs) had claimed it could assure the emptying of cesspools safely and without inconvenience to neighbors thanks to a ventilation system that had been approved by the Academy of Sciences and awarded a privilege. The suppression of the corporation of dischargers in 1776 offered the city administration the opportunity to mandate the use of this system for emptying cesspits.

It was for this purpose that Lenoir had entrusted the investigation of noxious fumes emanating from cesspits to the three pharmacists. They took their mission very seriously. They stressed the danger of the method traditionally used by the dischargers and the superiority of the procedure that used bellows. However, they suggested the addition of a furnace to burn the vapors and improve air circulation, as well as the use of quicklime. They conducted conclusive experiments on a cesspit in the Latin Quarter that was particularly dangerous. The house had long belonged to an anatomy demonstrator who had filled the pit with debris from cadavers. Lavoisier, along with Milly and Fougeroux, examined the commissioners' report on behalf of the Academy of Sciences. By means of various experiments they were able to demonstrate that the danger came from the decomposition of "liver sulfur" (sulfurated potash) that formed in the cesspits, but they did not manage to identify hydrogen sulfide, today considered to be primarily responsible for the accidents and illnesses suffered by the workers. Lavoisier and his academic colleagues approved the conclusions of the pharmacists and confirmed the superiority of the combined system of ventilator and furnace.[24]

The government ordered the reports of both commissions to be printed at its own expense, and on April 10, 1779, it granted the Ventilator Company the exclusive privilege for fifteen years of emptying cesspits at the expense of the former master dischargers. In this case, as in that of water distribution two years earlier, the government granted a monopoly to a financial company in which the interests of the administration, the court, and scientific circles were closely intertwined. Cadet de Vaux himself took this opportunity to join the company. Over the following years, having become a health inspector, he pursued his sanitation campaign, regularly reporting accidents due to imprudence in the *Journal de Paris*. For example, he reported on the case of an upholsterer's assistant who died after trying to recover some coins that had fallen into a cesspit from a toilet.[25]

The Ventilator Company's monopoly did not remain uncontested for long. In 1781, the eye surgeon Janin from Lyon, who was well-connected at court, claimed to have invented a powerful "anti-mephitic." In fact, it was just vinegar. On the strength of a few experiments at some cesspits in Lyon and then Paris (in particular at the townhouse of the police chief Lenoir), the government rushed to publish Janin's observations, without even waiting for the opinion of the Royal Society of Medicine, which had undertaken experiments in Versailles that were inconclusive. On March 2, 1782, before the results were known, Cadet de Vaux's brother, the chemist Cadet de Gassicourt, presented some observations at the Academy of Science that were critical of Janin's procedure. Lavoisier, La Rochefoucauld, Macquer, and Le Roy were charged with examining the affair. For its part, the Royal Society of Medicine, dissatisfied with the first results of the Versailles experiments, named a new commission made up of Macquer, Fourcroy, Abbé Tessier, and Hallé. The two commissions combined their efforts, conducting parallel investigations.

The first experiment, conducted on the Quai Pelletier, was carried out without incident. Then on March 23 the commissioners went to the Hôtel de Grenade in the heart of the Latin Quarter for a second trial. This time, the pit was considered to be dangerous by the dischargers themselves. Students of surgery had lived in the building, and it was suspected that they had thrown waste from cadavers into the pit. Once Janin, very sure of himself, poured in the vinegar, the workers began to empty it. Tragedy struck while the work was still going on. When a bucket fell into the cesspool, a worker went down a ladder to retrieve it. He had barely descended past the top of the pit when he fell without a cry into the sluice. Two workers secured with ropes who went down to help him lost consciousness. A third managed to put a rope around the unfortunate victim, but he was already dead.[26]

The tragedy at the Hôtel de Grenade demonstrated the inanity of Janin's procedure. Yet he refused to concede defeat. He continued to argue with Hallé and Cadet de Gassicourt. For the savants, the investigation offered an excellent opportunity to study more precisely the cause of ophthalmia and hydrogen sulfide intoxication. Lavoisier studied the gas emitted by fecal matter and identified it as a mixture of fixed air (carbon dioxide) and inflammable air from marshlands (methane). He believed that it was possible to conclude from this identification that fixed air was the main cause of asphyxia in cesspits, which in his view justified the use of lime rather than vinegar.

However, Hallé's analysis of the Hôtel Grenade tragedy demonstrated nothing of the sort. The noxious gas produced by stirring up the contents of the pit dissipated rapidly by itself, which fixed air could not do, since it was heavier than atmospheric air. Moreover, the worker had collapsed as soon as his head had

gone below the opening, which demonstrated that the cesspit was full to the brim
with a noxious gas when the accident happened. Hallé also dismissed as possible
causes "putrid gas," or methane, since the noxious air was not flammable, and even
"hepatic gas," or hydrogen sulfide, since a dangerous pit generally did not have any
particular odor, and it was known to have a fetid odor. Hallé and Lavoisier at least
agreed, contrary to the commonly held view, on the absence of a link between the
fetid nature of cesspits and the danger they posed.[27]

Pilâtre de Rozier, the founder of the Musée de Monsieur, also contributed
to the campaign against cesspit accidents. In a letter published in the *Journal
de Paris* in January 1783, he paid homage to Cadet de Vaux for the invention of
the furnace but also announced a much simpler means of dealing with noxious
gases. But he gave no details about his procedure. The Royal Society of Medicine
soon set up a commission to investigate. First they went to Pilâtre's cabinet on
the rue Saint-Avoye. There the physicist breathed inflammable air (hydrogen),
blew it into a narrow glass tube, and then lit it with a candle (Fig. 8.2). Next he
swallowed fixed air (carbon dioxide), again without any danger (it passed into his

FIGURE 8.2: Pilâtre breathing hydrogen. Engraving by Joseph Collyer after a drawing
by John Russel, n.d.

stomach). He thereby showed that these two gases were not noxious, contrary to the opinion championed by several other physicists at the time. Pilâtre concluded that asphyxia can always be explained by a lack of breathable air and thus that to enter a cesspit or any other noxious place without risk requires only that breathable air be pushed into the lungs by means of a tube.

On April 8, the commission went to the Longchamp brewery on the rue Mouffetard to witness a demonstration. Pilâtre descended into a vat of beer two-thirds full of carbon dioxide, first while holding his breath, then with his respiratory apparatus. This consisted of a tube of gummed silk protected by an armature of iron wire wrapped around it that brought in air from outside and was attached to a mask fitted over his nose. It also included goggles and overalls. With his apparatus, Pilâtre was able to remain inside the vat for half an hour without suffering any discomfort. All he had to do was breathe in through his nose and out through his mouth. Fourcroy wrote a detailed report on behalf of the commission, in which he proposed that each guardhouse in Paris be equipped with a respiratory apparatus "so that they will have the means to rescue promptly anyone who has the misfortune of being asphyxiated by noxious fumes, by plunging immediately into the places where they have been struck." Lenoir decided that Pilâtre should be notified every time a dangerous cesspit was opened. For its part, the Academy of Sciences gave its approval to the physicist's invention.[28]

While it might be useful for rescuing those who were asphyxiated, Pilâtre's apparatus did not claim to replace the ventilation method of discharging waste. However, an "anti-mephitic" pump invented by Viot de Fontenay, based on the mechanic Thillaye's fire pump, did aim to compete directly with the ventilator. Experiments conducted at Versailles seemed to demonstrate the effectiveness of the procedure. In December 1785, Viot, who had the support of Baron de Breteuil, succeeded in having the Parlement nullify the Ventilator Company's exclusive privilege. The door was now open for the creation of the rival Anti-Mephitic Pump Company. Of course, the Ventilator Company went into action: once it had its privilege reinstated that August, it contested through its spokesman Cadet de Vaux any right by Viot to the invention of the pumps. Once again, the Academy of Sciences was consulted. A commission, of which Lavoisier was the linchpin, decided for the suppression of the monopoly, since "the administration should avoid getting tied up with privileges, especially if they are long-standing and liable to impede the progress of industry." The two companies decided to come to an agreement: after Viot died in 1787, his associate Thillaye brought the invention to the Ventilator Company, which was able to maintain its Paris monopoly until 1794.[29]

In the end, the campaign for hygiene in Paris resulted in few actual achievements before the Revolution, apart from the closing of the Cemetery of the Innocents. The affair of the Hôtel-Dieu showed the powerlessness of royal authority when it was deeply divided. Entrenched resistance, financial difficulties, and then Revolutionary events would defeat most of the projects. Those concerning hospitals, prisons, and abattoirs remained unresolved. At the end of the eighteenth century, Paris still did not have either a water distribution network or a sewage system. The intellectual ferment of the 1780s, however, did offer a field of experimentation for the new chemistry. And this movement would have consequences, since under the Consulate the group of hygienists, including Cadet, Hallé, Thouret, and Fourcroy, would relaunch it, and projects would flourish again. Only one member of the group was missing: Lavoisier, who had dominated all the others with his genius and industry.

Notes

1. FOSSEYEUX, 1912, 258–262, and *Relation de l'Incendie de l'Hôtel-Dieu de Paris, arrivé la nuit du 29 au 30 décembre 1772*, Paris, 1773.
2. On the Royal Society of Medicine, see THIÉRY, 1787, vol. 1, 354–356, as well as C. HANNAWAY, "The *Société Royale de Médecine* and Epidemics in the Ancien Régime," *Bulletin of the History of Medicine* 46 (1972), 257–273.
3. J. TENON, *Mémoires sur les hôpitaux de Paris*, Paris, 1788.
4. On Necker's policy and the creation of the Department of Hospitals in the Finance Ministry, see C. BLOCH, *L'assistance et l'État en France à la veille de la Révolution française*, Paris, 1908, 211–235.
5. The three reports of the Academy's commissioners are reproduced in LAVOISIER, 1862–1893, vol. 5, 603–668, 669–678, and 679–706.
6. MERCIER, 1782–1788, chap. 269: Hôtel-Dieu.
7. L. S. GREENBAUM, "Nurses and Doctors in Conflict: Piety and Medicine in the Paris Hôtel-Dieu on the Eve of the French Revolution," *Clio Medica* 13 (1979), 247–267.
8. A.-A. CADET DE VAUX, "Mémoire historique et physique sur le cimetière des Innocents," *Journal de physique* 22 (June 1783), 409–417, in particular 410.
9. F. VICQ D'AZYR, *Essai sur les lieux et les dangers des sépultures*, Paris, 1778 (free translation of an Italian work by S. Piattoli).
10. M. A. THOURET, "Rapport sur les exhumations du cimetière et de l'église des Saints Innocents," *HMSM*, vol. 8 (1786), Paris, 1790, 238–271.
11. MERCIER, 1782–1788, chap. 43: "L'air vicié ."
12. M. A. THOURET, "Rapport sur la voierie de Montfaucon et supplément," *HMSM*, 1786, 226.
13. On the cesspit on the Quai de Gèvres, see *Journal de Paris*, November 7, 1781; on the excavations at the demilune of the boulevards, see *Journal de Paris*, May 20, 1780; on Paquet's bone, see *Journal de Paris*, June 26, 1781, and R. DE PAUL DE LAMANON, "Mémoires sur un os d'une grosseur énorme qu'on a trouvé dans une

couche de glaise au milieu de Paris . . . ;" *Journal de physique* 17 (May 1781), 293–405; on the fossils of Montmartre, see R. DE PAUL DE LAMANON, "Description de divers fossiles trouvés dans les carrières de Montmartre près Paris et vues générales sur la formation des pierres gypseuses," *Journal de physique* 19 (March 1782), 173–194.

14. Quotation taken from a letter on magnetism published in the *Mercure de France*, May 18, 1784, 180–183. There are many articles from the *Journal de Paris* on public hygiene, generally under the byline of Cadet de Vaux, and several chapters in the *Tableau de Paris* by Mercier on water, the cesspools, butchers, veterinary pits, etc. On the regulation of urban and industrial pollution in Paris at the end of the Ancien Régime, see LE ROUX, 2011.

15. *Journal de Paris*, August 11, 1781.

16. F. GRABER, "La qualité de l'eau à Paris, 1760–1820," *Entreprises et histoire* 53 (2008), 119–153.

17. J.-N. HALLÉ, "Rapport sur l'état actuel du cours de la Bièvre," *HMSM*, 1789, lxx–xc.

18. "Rapport des mémoires et projets pour éloigner les tueries de l'intérieur de Paris," *HMAS*, 1787 and 1789, 19–42 (LAVOISIER, *Œuvres*, vol. 3, 579–602).

19. *Journal de Paris,* November 6, 1780.

20. CORBIN, 1986, especially 22–34.

21. A.-L. LAVOISIER, "Mémoire sur la combustion en général," *HMAS, 1777,* 592–600 (*Œuvres*, vol. 2, 225–233), dated September 7, 1777, and read in the public session of November 12, 1777, P.-V. (1777), fol. 542v. See the letter from the pharmacy student Deronzières in the *Gazette de santé*, 1777, no. 47, 195.

22. *Journal de Paris*, May 21, 1777, and A.-L. LAVOISIER, "Expériences et observations sur les fluides élastiques en général, et sur l'air de l'atmosphère en particulier," *Œuvres*, vol. 5, 271–281.

23. J.-B.-M. BUQUET, *Mémoire sur la manière dont les animaux sont affectés par différents fluides aériformes, méphitiques, et sur les moyens de remédier aux effets de ces fluides*, Paris, 1778.

24. L.-G. LABORIE, A.-A. CADET DE VAUX, and A. PARMENTIER, *Observations sur les fosses d'aisance et moyens de prévenir l'inconvénient de leur vidange*, Paris, 1778. On the Ventilator Company, see BOUCHARY, 1942, vol. 3, 93–116. On the work of the chemists, see M. VALENTIN, "Lavoisier et le problème du méphitisme des fosses d'aisance," in LAVOISIER, *Correspondance*, vol. 5, 287–290.

25. *Journal de Paris*, July 22, July 25, and September 23, 1781, and June 12, 1782.

26. J.-N. HALLÉ, *Recherches sur la nature et les effets du méphitisme des fosses d'aisance*, Paris, 1785, 34–61.

27. A.-L. LAVOISIER, "Mémoire sur la nature des fluides élastiques aériformes qui se dégagent de quelques matières animales en fermentation," *HMAS*, 1782, 560–575 (*Œuvres*, vol. 2, 601–615).

28. DELAUNAYE, *Description et usage du respirateur antiméphitique, imaginé par Pilâtre de Rosier, avec un Précis des expériences faites par ce physicien sur le méphitisme des*

fosses d'aisance, des cuves à bière, etc., Paris, 1786, and FOURCROY ET AL., "Rapport à la Société royale de médecine sur les moyens de rester sans aucun danger dans l'air méphitique," Archives de la Société royale de médecine, B, ms 14.

29. Report on a memorandum on cesspools written by Lavoisier, *P.-V.* (1787), fols. 69r–88r, March 10, 1787 (not reproduced in Lavoisier's *Œuvres*).

9

The Severe Science

ON SUNDAY, FEBRUARY 27, 1785, Lavoisier's laboratory at the Arsenal began a memorable experiment on the analysis and synthesis of water. Two years earlier the chemists had obtained one of the most surprising results in the new chemistry of gases: water, which had always been thought to be an element, was proven in fact to be a compound of inflammable air (hydrogen) and dephlogisticated air (oxygen). In 1781, Joseph Priestley, who had exploded a mixture of inflammable air and breathable (atmospheric) air in a closed vessel, observed, as others had before, droplets of water on the side of the vessel. Two years later, Cavendish was the first to demonstrate that this dew was indeed the product of the combustion of inflammable air, and he concluded that dephlogisticated air was water deprived of its phlogiston.

Lavoisier was aware of these results and sought to repeat the experiment in his Arsenal laboratory. On June 24, 1783, assisted by Laplace and in the presence of his colleagues Fourcroy, Dionis du Séjour, Vandermonde, and Legendre, as well as Meusnier, a corresponding member of the Academy, and the Englishman Blagden of the Royal Society, he too obtained a non-negligible quantity of pure water. Despite the imprecise nature of the experiment, he concluded that water is not a simple element but a compound, in the proportion of one to two, of dephlogisticated air and inflammable air. The next day, Lavoisier and Laplace announced the discovery to the Academy. A comparable, but more precise, experiment conducted by Monge a short time later at the engineering school in Mézières confirmed that water is indeed formed from air, pound for pound.

That water is a compound of two gases corroborated in spectacular fashion the theory of combustion that Lavoisier had been developing for several years. Nevertheless, many people, including chemists, continued to doubt it. The synthesis had to be confirmed by a subsequent analysis. In the autumn of 1783, Lavoisier began a series of experiments aimed at the decomposition of water. He obtained what he called at the time the "aqueous inflammable principle" by plunging red-hot iron or burning coals into water. He presented his first results at the public meeting that opened the academic year on November 12, 1783, a few days after the death of d'Alembert. A gloomy silence had followed Condorcet's impromptu eulogy. Then the audience listened patiently as four papers were read. Lavoisier's fell fairly flat, if we are to believe the *Mémoires secrets*, which commented drily: "Mr. Lavoisier, a friend of systems, claims that water is not an element, as has always been believed, that it decomposes and recomposes at will"; the following paper, "On the Fogs of Last Summer," "dealt with a more interesting subject," and the three eulogies with which the meeting closed (by Vaucanson, Bordenave, and Pinglé) "were much more enjoyable than these four papers." Lavoisier's severe science was definitely not made for social events.[1]

The Great Water Experiment and Balloons

While nobody, it seemed, was worrying in public yet about the nature of water, a new invention had just generated a lot of enthusiasm: the hot-air balloon, or "montgolfière," of the Montgolfier brothers, whose first flight had taken place in Annonay on June 4, 1783. As we saw in chapter 6, the physicist Charles had from the start proposed to inflate the balloon with inflammable air (hydrogen). To produce this hydrogen, Charles caused sulfuric acid to react with iron filings. His aerostat (a "charlière") ascended from the Champ de Mars on August 27 amidst an enormous crowd. The Academy of Sciences had already been captivated by this invention, naming a commission and summoning Étienne Montgolfier to Paris. Before the end of the year, the first humans had risen into the air: Pilâtre and the Marquis d'Arlandes in a *montgolfière* at La Muette on November 21, Charles and Robert in a *charlière* over the Tuileries on December 1. The Academy's commission delivered its report on December 13, rewarding the Montgolfiers with a prize of six hundred livres.

The very same day, on Lavoisier's initiative, a permanent commission on "aerostats" was formed to continue the research. Its members enlisted Jean-Baptiste Meusnier, who was well known to the Academy of Sciences, to join them. Soon after his graduation from the engineering school at Mézières in 1776, he presented a remarkable paper on geometry. Soon elected Vandermonde's correspondent in the Academy, he pursued his career as a military engineer in Cherbourg. At the same time he spent half the year in Paris to oversee the publication of the Academy's *Recueil des machines*. Aerostats had commanded much of his attention in 1783. He had followed the flight of the first balloon in Paris and studied the problem of its stability, in concert with Charles. He also worked with Lavoisier on the question of water. Present at the experiment on June 24, 1783, Meusnier had then invented, probably in September, a new apparatus, the gasometer, to measure the volume of gas in the experiments. Joining the aerostat commission made this collaboration official. A few weeks later he was elected assistant geometer at the Academy and received authorization from his superiors to stay in Paris all year round.

The aerostat commission had established several goals: the fabrication of the envelope, the production of large quantities of gas, and the control of altitude and direction. The work rested primarily on the shoulders of Meusnier. Over the course of 1784 the officer pursued research on the fabrication, locomotion, and behavior of aerostats and proposed the construction of giant dirigibles that would be capable of crossing the ocean. But the project remained on the drawing board. The question that most interested Lavoisier was obviously the production of hydrogen. Apart from the practical goal of producing the gas easily and in quantity, the chemist had in mind a theoretical goal: to confirm the composition of water by precise quantitative experiments and thus to deliver the final blow to the phlogiston theory. In March and April 1784, Lavoisier and Meusnier, assisted by Berthollet, performed experiments on the decomposition of water vapor by iron. They obtained large amounts of hydrogen by passing a stream of water through either a slightly inclined gun barrel that was heated red hot or a copper tube filled with iron flakes or shavings. But the experiments remained qualitative. Lavoisier now wanted to relieve all doubts about the nature of water by proceeding successively and quantitatively to its analysis and its synthesis. The experiment would be both a measurement and a demonstration.[2]

The gasometers delivered by the instrument maker Pierre-Bernard Mégnié at the end of December 1783 would first be used for experiments on the combustion of coal conducted in May and June 1784 by Lavoisier and Laplace. The preparations for the great water experiment began only in the last months of that year. Never before was an experiment in chemistry or physics conceived, prepared, and performed with so many precautions and so much attention to detail. In September Mégnié carefully dismantled and cleaned the gasometers. The fastidious operation of calibrating the apparatuses for different temperatures and pressures began at the end of December. Lavoisier and Meusnier measured the speed of the gas flow at the outlet and carefully weighed the two gun barrels that were to be used for the decomposition of the water vapor. Mégnié was responsible for reassembling the equipment. At the end of February 1785, the experiment could finally begin.

At Lavoisier's request, the Academy appointed a special commission of thirteen members, including all the chemists plus Bochart, Brisson, Bailly, Laplace, Monge, and the Duke de La Rochefoucauld. On Monday, February 21, after the Academy's meeting, Lavoisier invited the commission to the Arsenal to dine and examine the equipment. Other Academicians such as Lassone joined the party.

The experiment began the following Sunday and lasted three days. In addition to the commission, Lavoisier invited some colleagues to serve as witnesses, including honorary ones like Malesherbes and the Duke de Chaulnes, but also Laurent de Villedeuil, future minister of the King's Household—in total "more than thirty savants, physicists, geometers, and naturalists, both members and non-members of the Academy of Sciences, which had appointed a large commission for this purpose."[3]

The experiment had two stages: first analysis and then synthesis. For the analysis, the apparatus was identical to that used the year before for the production of hydrogen in quantity. Water poured into a funnel had to pass drop by drop into one or another of the iron gun barrels, numbered 1 and 2, placed over the brazier of a furnace. Copper pipes carried water into gun barrels filled with strips of iron and made watertight with a clay casing. The hydrogen produced by contact with the red-hot iron went out through the pipes, along with the remaining water vapor. A serpentine plunged into a coolant assured condensation of the vapor, and a bell jar over a trough made of mercury collected the hydrogen (Fig. 9.1).

A vacuum having been established in the apparatus, the decomposition operation, using barrel no. 1, began just before noon on February 27 and lasted until 6:30. It produced more than ten bell jars of hydrogen, which was immediately used to fill gasometer no. 2. Gasometer no. 1 had already been filled with oxygen. Now the synthesis operation could begin. The combustion would take place in a glass balloon with three openings: one to create a vacuum with the aid of a pneumatic pump, the second to let in the oxygen, and the last for the injection of the hydrogen. The balloon was filled with oxygen and then attached to the two gasometers. At 6:55, an electric spark produced by a Ramsden machine set fire to the jet of hydrogen. The combustion continued until ten o'clock at night.

The next morning, gasometer no. 1 was refilled with oxygen, and the combustion of hydrogen was carried out. During this time, a second decomposition operation was started, using gun barrel no. 2. The operation ended at forty seconds

FIGURE 9.1: *On the Decomposition and Recomposition of Water,* Engraving reproduced in A.-L. Lavoisier, *Œuvre,* vol. 5, illustrative plate.

after 4:43 and produced more than seven bell jars of hydrogen, which burned until 10:35. At 11:15, Monge sealed the balloon and the valves, and everyone left. On the third day, the residual gas contained in the balloon was transferred to a bell jar, and the examination of the water produced by the combustion began. The experiment properly speaking was over. On March 7, the commission analyzed the residual gas, composed principally of hydrogen, oxygen, and carbonic gas, and carefully weighed the gun barrels, the condenser, and the balloon. On March 12, they signed their report on the experiment.

This description gives only a pale idea of the extraordinary care with which Lavoisier, assisted by Meusnier, conducted the great experiment. Every precaution had been taken to reduce leaks from the apparatus. The purity of the water and oxygen had been verified, the flow of the fluids precisely controlled, the volume and weight of gas meticulously measured and calculated. The members of the commission oversaw every detail, took note of the measurements, checked the calculations, and initialed the experiment notebooks. But despite all this care and all these checks, the grand demonstration was not a success. The commissioners reported hydrogen leaks. On the second day, an error in experimental procedure caused an untimely splatter of vapor. Because not all of the hydrogen collected in the decomposition operation had been utilized in the reverse operation, the weight of synthetized water was much lower than that of the water used to start the experiment.

Finally, the proportion of the weight of hydrogen lost through the synthesis operation (about 15 percent) agreed with that given by the decomposition operation—this was the principal result of the experiment—but since the margin of error in the decomposition was about 6 percent, we must admit that on a quantitative level the demonstration was inconclusive. Perhaps this is why the commission did not publish its report. Moreover, all of its members were far from being convinced, and certain chemists like Sage and Baumé did not hesitate to say so. As for Lavoisier and Meusnier, they had to be content with publishing, a year later, a brief account of this experiment that had cost so much in money and effort in the *Journal polytype des sciences et des arts*, a brand new periodical launched by the bookseller François Hoffmann, which, as its title suggests, used a new process of stereotype printing.[4]

Chemistry at the Arsenal

Some thirty witnesses had attended the great water experiment. Such a crowd was not unusual at the Arsenal, which had become a meeting place for savants in Paris. When he was named director of the gunpowder administration in 1775, Lavoisier and his wife had moved out of their house on the rue Neuve-des-Bons-Enfants, where his laboratory was located, and into the Hôtel de la Régie, its headquarters at the Petit Arsenal. A huge apartment combined with a great fortune enabled him to receive as many visitors there as he wanted. His wife, Marie-Anne Paulze, who was also his collaborator, hosted a salon there. Coffee and tea were served; guests dined and listened to music, played games of chance, and laughed. Some evenings, a magic lantern projected glass slides painted by Madame for the company's enjoyment. It was perhaps here that a romance began between Marie-Anne and Pierre Samuel Dupont de Nemours in 1781.

Due to their fortune and lifestyle, the Lavoisiers circulated easily in high society, whose customs and conventions they shared. Their world, however, was not that of the Parisian aristocracy. While Madame was more light-hearted, Lavoisier had the seriousness of a financier and a high official. Above all, he thought like a savant, putting worldly sociability at the service of his scientific activities. Their guests were usually colleagues or foreign savants visiting Paris, such as Priestley, Blagden, Watt, Franklin, Van Marum, and many others. Discussions tended to be about the latest scientific news. Lavoisier often took his guests into his library, which contained 2,500 volumes, and especially into his laboratory.[5]

A mythical site in the history of chemistry, the Arsenal laboratory remains poorly understood. When Arthur Young was invited by Lavoisier to an "*English breakfast* of tea and coffee," on October 16, 1787, he marveled "to find this philosopher splendidly lodged, with every appearance of being a man of considerable fortune." He visited the laboratory and admired the construction of "the noble

FIGURE 9.2: *The Laboratory of Lavoisier,* drawing by Madame Lavoisier, n.d., private collection.

machine" used for the great water experiment (noting, however, that he did not really understand it very well).[6] According to Madame Lavoisier's drawings, the laboratory was set up in an attic under a mansard roof. On one side, it overlooked a courtyard and on the other a garden. We do not know the precise location, but it was probably in a building between the courtyard of the saltpeter works and the garden that ran long the ditch (Fig. 9.2). In any case, its comforts were minimal. The experimental equipment was placed on the tiled floor, and the items made of glass and ceramic were stored on shelves along the walls. The furniture amounted to a few tables and chairs; a stove provided heat. During the great water experiment, the temperature at the end of the day was below 10°C.

Lavoisier went to his laboratory at six o'clock each morning, before his workday began for the General Farm and the Gunpowder Administration at nine o'clock. He returned at seven in the evening and stayed until ten. He rarely worked alone there. No doubt he employed a few laboratory servants, about whom we know nothing, but he also had many collaborators: his wife, who often kept the records, a few young savants, and sometimes his colleagues. The impressive size of the laboratory and the quality of the instruments attracted the curiosity of amateurs and the interest of chemists. Lavoisier first worked with the physician and chemist Jean-Baptiste Bucquet of the Academy, who died young from cancer. In 1778, Lavoisier began a very fruitful collaboration with another young colleague from the Academy, Pierre-Simon Laplace, until then known for his work on celestial mechanics and probability. After Laplace came Meusnier, whom we have already met.[7]

At this point Lavoisier began to gain disciples. He won Monge over to his theory of oxidation in 1783, but the turning point came three years later with the great water experiment. In April, the chemist Berthollet, convinced by the demonstration, rallied to his side, arguing that dephlogisticated marine acid was a composite of oxygen and marine acid. A little later, Lavoisier made an oral presentation to the Academy of his *Réflexions sur le phlogistique*, in which, as we shall see, he made a frontal assault on Stahl's theory, which was currently accepted by most chemists. Fourcroy followed his example in his public courses, and finally, at the start of 1787, Guyton-Morveau joined the anti-phlogiston clan. Lavoisier's ideas gained ground, as attested by the immediate success of his new chemical nomenclature, which he published in 1787. However, the adversaries of his pneumatic theory did not lay down their arms. Within the Academy, Sage, Baumé, and Darcet continued to resist. And outside, the *Journal de Physique* opened its columns to defenders of phlogistics, to Sage, Marivetz, Lamarck, De Luc, Priestley, and others who were very hostile to the new chemistry. Its editor, Delamétherie, joined the fight himself, regularly attacking Lavoisier, whom he detested, but without naming him.

By then the Arsenal had become the center and meeting place for the true believers in the "chemical revolution." Every Saturday they gathered around Lavoisier. "It was a happy day for him," his wife later wrote: "some enlightened friends, some young people proud of the honor of having been allowed to participate in his experiments, met in the morning in the laboratory, where they had breakfast, talked, and created the theory that has immortalized its author; that was where one had to see and hear this man who was so fair-minded, with such a pure talent, such great genius; it was by his conversation that one could judge the loftiness of his moral principles."[8] Three young chemists, Jean-Henri Hassenfratz, Pierre-Auguste Adet, and Armand Seguin, helped with experiments and participated with their elders in the meetings about chemical nomenclature. Madame Lavoisier translated Richard Kirwan's essay on phlogiston from the English and made the drawings. In 1788, Lavoisier wrote his *Traité élémentaire de la chimie*. The following year, the Arsenal team launched the *Annales de Chimie*, a new journal that Adet had conceived in 1787. At first, they just intended to translate into French the *Chemische Annalen* published by Lorenz Crell in Germany, but very soon they decided to include original papers that could not be published in either the *Journal de Physique*, hostile as it was to the new chemistry, or the *Journal polytype*, which had vanished after a year.

Located near their apartment, the Arsenal laboratory was part of the Lavoisiers' private space. But it was also completely integrated into the professional environment of the director of the Gunpowder Administration. The experiments took place within the precincts of the Arsenal, sometimes even in the garden, as in the case of Lavoisier and Laplace's work on the expansion of metals in 1781 and 1782. The thousands of retorts, crucibles, balloons, and jars that he owned were stored in a building that belonged to the Administration. Did the men who worked at the Arsenal visit the laboratory? We know only that Philippe Gengembre, an employee of the Gunpowder Administration, came to work with Lavoisier, and perhaps even conducted the demonstration experiments that accompanied the classes Lavoisier taught. Lavoisier doubtless also used Arsenal personnel for some kinds of work, such as carrying the equipment or major carpentry and masonry projects. Beyond the use of such services, we suspect that there were close links between Lavoisier's activities at the Gunpowder Administration and his chemical research, for example, when he studied the detonation of saltpeter with charcoal and sulfur and the accompanying release of heat. But it must be acknowledged that his preoccupations considerably outstripped the practical domain, as much for gunpowder as for balloons, and that the experiments in his laboratory, like the great water experiment, had for him an essentially theoretical purpose.

Thus, it was close to the Bastille, far from the seat of the Academy and high society, that the new chemistry began its ascent. This distance was partially the product of circumstances: Lavoisier had begun his work a few streets away from the Louvre, and when he came to the Arsenal, it was to head the Gunpowder Administration. Yet nothing obliged him to make his home there. His activities as a tax farmer, the meetings of the Academy, and society life would have justified his remaining close to the Palais-Royal. In any case, the Lavoisiers had kept a pied-à-terre on the rue Croix-des-Petits-Champs. But having unlimited access to a huge laboratory any time he wanted must have been a major factor in his decision to move out to the Arsenal. That the chemical revolution would have as its theater a manufacturing site was not totally by chance, and it was symbolic that it took place at a distance from the Paris *quartiers* usually frequented by savants and amateurs of the sciences. The work of Lavoisier and his friends marked a break whose importance reached far beyond the domain of chemistry alone. With this revolution, research broke away from the worldly and public practices that had dominated the physical sciences and natural history in the eighteenth century. This separation, whose origins lay in a more demanding and more rigorous conception of doing science, was not only intellectual but social and spatial. The activities of professional and part-time savants were becoming clearly distinguished from those of amateurs, which meant a new relationship with the public. They also required the development of more specialized and more isolated work spaces, of which the Arsenal laboratory was a kind of prototype. Although the revolution had only begun, it was at least underway.[9]

Laplace's Celestial Mechanics

Despite only partial success and little public impact, the great water experiment was a significant event for the Arsenal group. It confirmed Lavoisier's pneumatic theory. It also offered the most remarkable example of a new experimental science, distinguished by the precision and sophistication of instruments and measurements. To conduct it well, Lavoisier had chosen as his assistant not a chemist or a physicist but an engineer who was also a geometer. The choice was not a matter of chance. Lavoisier had already collaborated with Laplace, first on the evaporation of liquids and then, in 1781 and 1782, on the specific heat of bodies. Together at the Arsenal they had performed celebrated experiments in calorimetry. The mathematician's contribution had been crucial to this common labor: it was he who had invented the experimental apparatus, carried out the calculations, and proposed a mechanistic interpretation of heat.

If with Meusnier Lavoisier borrowed what he needed from engineering, with Laplace he had introduced the culture of astronomy into the physical sciences. More than any other science planetary astronomy was at that time characterized by the precision of its observations and the mathematization of theory. With the law of universal gravitation, Newton had created the model for a science that was both quantitative and predictive, and since the beginning of his career Laplace had built on that foundation. Despite great success, "physical astronomy" (as celestial mechanics was called at the time) was in fact far from being able to account for the movement of all the celestial bodies. Newton himself had failed in his theory of the moon, and quite simply left out of his calculations any planetary perturbations. The result was a significant discrepancy between the theory's predictions and actual observations, of the moon but also of comets and the two largest planets in the solar system, Jupiter and Saturn. In the case of these two planets, observations taken over the last two thousand years had revealed an acceleration in Jupiter's mean motion and a deceleration in Saturn's, which Newton himself could not explain. Mathematical analysis was necessary to attack these problems seriously.

The Swiss mathematician Leonhard Euler, followed by Alexis Clairaut and d'Alembert in France, had obtained the first results in the 1740s. First, Clairaut succeeded in explaining by gravitation alone the apsidal movement of the moon, long known to observers. Ten years later, to bring an end to a famous controversy, Clairaut succeeded in accounting for the irregular movement of Halley's Comet due to the perturbations of Jupiter. Despite these achievements, most of the discrepancies remained unexplained, so much so that many people (starting with Euler) imagined a cause other than gravitation to account for these effects. Laplace himself, in order to explain the secular acceleration of the moon, hypothesized in 1773 that gravitational action was propagated at a speed that was very great but finite, an interesting assumption but inconsequential and unverifiable. In the same paper, he refuted at length Lagrange's theory on the "great inequality" of Jupiter and Saturn and suggested as an alternative explication the perturbations of comets.

Laplace did not revisit this great inequality that defied Newtonian theory for almost ten years. This was the period when, goaded by Condorcet, he developed the theory of probability and its application to the study of population and then, due to the contact with Lavoisier, turned to questions of physics. In astronomy, he was interested in the determination of cometary orbits. Thus in 1783 he showed that the celestial body discovered by Herschel two years earlier was not a comet but in fact a new planet (Uranus). For his part, Lagrange pursued his research on the theory of perturbations, extending, simplifying, and clarifying Euler's results and introducing in 1776 the fundamental concept of perturbation function. But the great inequality of Jupiter and Saturn continued to resist all his efforts.

The dramatic turn of events occurred on November 23, 1785, nine months after the great water experiment, when Laplace announced to the Academy of Sciences that he had solved the problem. While his method was borrowed from Lagrange, the discovery was entirely his. Taking up a suggestion from Lagrange, he showed that the great inequality resulted essentially from a phenomenon of orbital resonance between Jupiter and Saturn. More precisely, the quasi-commensurability of their mean motion produced very large perturbation terms in the calculation of their longitudes. What Laplace found was that the great inequality, which depends on the respective positions of the two planets, was therefore not in fact secular (that is, recurring every hundred years) but periodic, with a much longer periodicity of 877 years. While he was at it, Laplace explained the inequalities of Jupiter's moons, provided the solution to another enigma (the famous secular acceleration of the moon) and, for his grand finale, demonstrated (at least for first-order approximations) the stability of the solar system. On this occasion, Laplace only indicated the rough outlines of his theory, but six months later he presented a detailed analysis to the Academy with his mathematical formulas. The results were impressive: comparing an observation of Saturn made by the Babylonians in 228 BCE with the position derived from his theory, Laplace found a difference of less than one minute![10]

That same day, May 10, 1786, Laplace made the acquaintance of Jean-Baptiste Delambre, who would become his principal collaborator in astronomy. This young protégé of Lalande's had just observed the transit of Mercury across the sun thanks to a break in the clouds. Lalande informed the Academy of the observation during the meeting. Delambre, who was present, came up to Laplace to congratulate him on his great discovery and to offer to recalculate all the observations required by the new theory of Jupiter and Saturn. As tutor to the son of the powerful financier Jean-Claude Geoffroy d'Assy, Delambre had access to a private observatory at the home of his employers in the Marais and was not averse to doing calculations. These required a truly herculean effort: the observed oppositions of Jupiter and Saturn had to be reduced and corrected; those observations judged to be questionable had to be eliminated; and the orbital elements had to be derived through calculation and tables of them drawn up. Delambre completed the work in nine months. His *Tables de Jupiter et de Saturne*, published in 1789, opened a new era in the history of astronomical calculation. Astronomers had always distinguished themselves by their concern for exactitude and precision, but this time the construction of the tables rested on theory alone, and observational data entered only into the determination of the constants of integration. The last person to enter the Royal Academy of Sciences before it was shut down, Delambre was elected associate geometer on February 15, 1792.[11]

Laplace's work represented remarkable progress in planetary astronomy. Not only did it open the way to the construction of astronomical tables that were more reliable and precise, but it confirmed the extraordinary power of theory. It represented a retrospective triumph for Newton, but also for Laplace himself.

As he wrote with pride to the Italian astronomer Oriani about the moons of Jupiter: "It is a truly admirable thing to see how the law of universal gravitation explains all the variations of their orbits, which observations have allowed astronomers to glimpse without their being able to determine the laws governing them." Better still: casting a philosopher's gaze on the stability of the solar system, he discerned there "one of the phenomena most worthy of attention, in that it shows us in the heavens the same principles that nature has admirably followed on Earth to preserve the life of individuals and perpetuate the species, thereby maintaining the order of the Universe."[12]

The Empire of Calculus

With this last remark, was Laplace alluding to his research into the ratio of male to female births? Some people had interpreted the almost exact equality in the number of male and female births as the effect of providence. However, the statistics compiled by Jean Morand for births in Paris and its suburbs between 1709 and 1770 instead indicated a slightly higher number of boys than girls. With the same procedures he had used in celestial mechanics to demonstrate that in all probability the elliptical inclination of cometary orbits could not be the effect of any particular cause, Laplace now showed that that there was every chance that in the future the number of boys would always slightly exceed that of girls. On this occasion he introduced new mathematical techniques into the calculus of probability. Finally, in a paper read on November 30, 1785 (a week after the one on the inequalities of planets), he was able to confirm, again by the calculus of probability, that birth statistics provided a reliable means to estimate the population of the kingdom, as long as the birth multiplier was calculated on the basis of a large enough sample. (Laplace recommended a sample of at least a million people.)[13]

By evoking the order of the universe and the principles of nature in relation to preserving the life of individuals on Earth and perpetuating the species, it is plausible that Laplace was not thinking only of demographics. Physicists and naturalists were just as much concerned by the laws of nature. There was Lavoisier's work on heat, water, and respiration, for example, on which Laplace had himself collaborated. Perhaps he also had in mind the work of Charles Coulomb on electricity and magnetism. Like Meusnier, Coulomb was a military engineer, trained at the engineering school in Mézières. He made his name by winning the Academy's prize in magnetism for a magnetic compass, the starting point for his later work in physics and mechanics. In 1781, he was nominated to an engineering position in the capital that allowed him complete freedom to conduct his own experiments. That is when he entered the Academy of Sciences as an assistant in the class of mechanics.

In Paris, Coulomb undertook a long research program in physics by means of a balance derived from his first magnetic compass. Cassini had installed an instrument of this type at the Observatory of Paris to measure the variations in magnetic declension. It proved so delicate that he had to call upon Coulomb's help when the engineer arrived in the capital. This collaboration with Cassini was probably how Coulomb began to work on the torsion of strings and wires, which he presented to the Academy in September 1784, the same day as its report condemning Mesmerism was issued. On this occasion the engineer announced a set of experiments on electricity and magnetism, the details and results of which he presented to the Academy in a series of papers between 1785 and 1790.

These papers marked a rupture in research on electricity, a domain that had generated so much work during the eighteenth century. The torsion balance offered a way of measuring electrostatic forces for the first time. It thus shifted electrical science from qualitative observation to quantitative evaluation. This was a far cry from the spectacular demonstrations of a Comus or a Marat. The instrument used was certainly too sensitive to be very precise, and the experiments, performed under conditions not made explicit (probably in Coulomb's house), were difficult to reproduce. However, nobody—in Paris at least—challenged the results. Moreover, the change was not just of an experimental order; it was also and perhaps above all theoretical. By demonstrating interactions at a distance comparable to the force of gravitation, Coulomb brought electricity (as well as magnetism) into the field of Newtonian mechanics; in other words, he made possible its mathematization. Laplace quickly praised him for having discovered the law that is now called Coulomb's law.[14]

The Abbé Haüy made Coulomb's work known among amateurs in his adaptation of Aepinus's book on electricity and magnetism, published in 1787. Everything had seemed to separate this habitué of the Royal Botanical Garden from proponents of the new physics. A professor of Latin at the Collège de Navarre, then at the Collège du Cardinal-Lemoine, Abbé Haüy went into the sciences as an amateur himself, dividing his leisure time between music and botanizing. Interested first in botany, he turned rather late to mineralogy, on the advice of Daubenton. In 1781, he submitted his first papers to the Academy, which were devoted to the crystalline forms of garnets and calcites. Encouraged by Daubenton and Bézout, he continued his research, which opened the doors of the Academy to him in 1783. In this election, Haüy benefited from the active support of both physicists and mathematicians. Laplace was his principal advocate.

In his *Essai d'une théorie sur la structure des cristaux*, published the following year, Haüy stated the laws by which the structure of a crystal can be reduced to a stack of "constituent molecules," similar to a polyhedral nucleus, or "primary form," which in certain cases could be obtained by simple mechanical division.

His methods appeared quite simple compared with those of his physicist and mathematician colleagues: as his instrument he used the same applied goniometer as Romé de l'Isle and Arnould Carangeot before him; for his experiment he simply split crystals; and for his calculations he used a very elementary geometry. But the elegance of his theory, in which everything was precisely deduced from a few simple laws, made it a model for the new physics, whose enthusiasts Haüy naturally joined.

Over the course of the following years, the *abbé* pursued his research on the theory of crystals assiduously. He was able to account for a growing number of crystalline structures even as he reduced the number of possible primary forms to three types of "integrant molecules," which were themselves reducible to a single form, the parallelepiped, but at the cost of more complex modes of generation. He also undertook to establish a classification of crystals based on their structure, that is to say, on the shape of their integrant molecules rather than their chemical composition. Living a modest and retired life at the Collège du Cardinal Lemoine, he gave lessons in mineralogy that his colleagues would deign to attend.[15]

With Haüy, the idea of applying calculation to the study of beings was cautiously extended to natural history. Buffon's collaborator Daubenton encouraged him along this path and had himself envisaged comparative anatomy as the rigorous and systematic study of the morphology and structure of animals. Vicq d'Azyr, his disciple and nephew through marriage and permanent secretary of the Royal Society of Medicine, developed this idea in public courses in comparative anatomy at the Royal Botanical Garden in the 1780s, and then at the Veterinary School of Alfort. The instruction he gave was preceded and accompanied by precise research on the comparative anatomy of the wings and ears of birds, the arms and legs of mammals, the vocal organs, and especially the brain.

Until that time comparative anatomy had been seen as a secondary field, somewhere between human anatomy and natural history. Vicq's ambition was to make it the basis of a general reform of the life sciences, which would encompass human and animal medicine and the classification of species, as well as the laws of physiology. Vicq aimed to be methodical in his approach, making comparisons between both the various organs of a single species and the same organs of different species. Two major themes emerged from his analyses: the functional unity of anatomic structure, which determines the specific forms of every organ and its parts; and the morphological unity, which shows the repetition of the same plan for different organs. Even as he distinguished living bodies, whose essentially heterogeneous parts have forms determined mainly by their functions, from inert bodies like crystals, whose homogeneous parts have forms determined mainly by the laws of attraction, Vicq wished to give comparative anatomy the same status as Haüy's crystallography and Lavoisier's chemistry. Although his program did not go far beyond a statement of intentions, he did have a great influence on his successors Etienne Geoffroy Saint-Hilaire and especially Georges Cuvier.[16]

Vicq wanted to replace the superficiality of a natural history that appealed to amateurs with the severity of an anatomy that suited savants. Without naming him, he was targeting Buffon himself, more preoccupied with style than science, who had not included anatomy in his 1770 *Histoire des oiseaux* (History of birds), to the great regret of Daubenton. Buffon had dominated natural history for forty years, exercising supreme power over the Royal Botanical Garden. At the Academy of Sciences, by contrast, he had become increasingly isolated. D'Alembert and his heirs detested him and greeted his death without regret. Condorcet had to deliver the eulogy, which was a real chore for him: "Here I am, having to deal with another charlatan, the great Buffon. The more I study him, the more I find him empty and puffed up," he confided nastily to Madame Suard.[17]

The Physicists' Crusade

In the course of the 1780s, an informal group of savants was gradually taking shape around Lavoisier. The main figures were the chemists Berthollet and Fourcroy, the mathematicians Monge and Laplace, the physicists Coulomb and Haüy, the mechanical engineers Vandermonde and Meusnier, and the anatomist Vicq d'Azyr. Also attached to the group were the permanent secretary Condorcet, the naturalist Daubenton (who had distanced himself from Buffon), and the mathematician Lagrange, who had come to Paris from Berlin in 1787. On the eve of the Revolution, this group dominated both the Academy of Sciences and the Royal Society of Medicine. Even savants more at home in the salons like Bailly, Lalande, and Cassini had rallied to its banner. Without necessarily sharing the same opinions or sympathies, these men came together to defend a particular vision of scientific activity and the role of savants. All of them rejected the systematizing spirit and stressed the importance of experimentation and calculation. Newton was their hero. They believed that his rigorous method should be applied not only to the sciences of "inert bodies" but also to the life sciences and even to the study of human beings in society. Just as the neoclassicism championed by David has been called a severe style in painting, Lavoisier's group with its new style of experimentation was promoting a science that was all measurement and calculation—a "severe science," as it were.

For these savants, the search for truth represented a sacred vocation. It was also a battle. Learned societies were useful first as "a barrier constantly opposed to charlatanism of all kinds, which is why so many people complain about it," wrote Condorcet in commenting on Marat's experiments. The Academy of Sciences had long tolerated the "circle-squarers," the crazy inventors, and the builders of systems. Writing to the *Journal de Paris* in 1782 to protest against the space given in its columns to flying boats and divining rods, the astronomer Lalande stated that as a general rule, "if the savants remain silent, it is only out of contempt."[18] In this way, the Academy enveloped in the same disapproving silence the work of the neo-Cartesian Baron de Marivetz, a rich enthusiast of scientific projects and systems, and the work of the anti-Newtonian Marat a little earlier. Marivetz had teamed up with the draftsman Goussier, who had worked on the *Encyclopédie*, to realize a magnum opus that managed to ruin him: *Physique du monde* was an encyclopedic treatise for the general public in which Marivetz expounded a general theory of vortices. The first volume, which he submitted to the Academy, appeared in 1783. Contrary to Buffon, Marivetz asserted that the Earth was warming. Bailly, who was a Buffonian, spared him but had his doubts, Lalande stayed quiet, and the Academy ignored him entirely. Marivetz wrapped himself in his wounded pride but got his friends Romé de L'Isle and Jean-Louis Carra to denounce the savants' bias against him.[19]

The Mesmer affair, in 1784, marked a turning point. This time, the savant societies were not content with silent rejection. They publicly condemned a fraud, as they believed it to be, that found followers at the highest level of society. Mesmer asserted the existence of a universal magnetic fluid that naturally ebbed and flowed and whose effects were felt by every living organism. This fluid could be transferred, propagated, and increased. Magnetic treatment aimed to reestablish its natural movement, which was prevented or blocked by illness (Fig. 9.3). From the moment of his arrival in Paris in 1778, the Viennese doctor had run up against the silent hostility of the savants. This did not in any way hamper the success of his cures. The method also gained partisans among physicians; Charles Deslon, the regent doctor in Paris, fought ardently for recognition of animal magnetism by the medical faculty. Even the government, under pressure from the queen's entourage, offered the healer a pension of twenty thousand livres if he would divulge his method, an offer that Mesmer insolently rejected.

Until 1783, animal magnetism was simply a secret remedy which divided physicians and attracted the sick but was of little more interest to the public than medical electricity, which was in full swing at the time. The controversy became serious after the creation of the Society of Universal Harmony, a society inspired by freemasonry that really took off at the end of 1783. It had been created by Mesmer and his friends to spread the secret and practice of animal magnetism. Among the subscribers were physicians and important figures connected to the court, the army, diplomacy, and finance. These zealots had undertaken to spread the treatment to the provinces and even the colonies. Deslon, who had broken with Mesmer, opened a clinic in the rue Montmartre that welcomed many

FIGURE 9.3: Mesmer's baquet at the Hôtel de Coigny. Engraving after a drawing by Claude Desrais.

patients from high society as well as physicians interested in observing the extraordinary effects of magnetism.[20]

For reasons that remain obscure, the government was worried about these initiatives. Mesmer had suddenly lost the patronage he enjoyed at court, and Minister Breteuil ordered an investigation into his doctrine and his treatment. Four doctors from the faculty of medicine, Pascal Borie (later replaced by Michel-Joseph Majault), Jean-Charles-Henri Sallin, Jean Darcet, and Joseph-Ignace Guillotin, were appointed in March 1783. At their request, five members of the Academy of Sciences were added shortly afterward: Bailly, Jean-Baptiste Le Roy, Benjamin Franklin, Gabriel de Bory, and Lavoisier. The savants declared that they wanted to limit themselves to the question of the physical existence of animal magnetism and left it up to the physicians to determine its possible therapeutic effects. For its part, the Royal Society of Medicine had undertaken its own enquiry into animal magnetism, charging Thouret with pulling together everything relevant to it from ancient and modern literature. The society then got Breteuil to name from within its ranks a second commission, composed of Pierre Poissonnier, Antoine-Laurent de

Jussieu, Claude-Antoine Caille, Pierre-Jean-Claude Mauduyt, and Charles-Louis-François Andry.[21]

The Condemnation of Mesmerism

Mesmer haughtily refused to cooperate with the commissioners. Deslon, however, willingly agreed to communicate what he knew and to participate in their experiments on the magnetic fluid and the effectiveness of cures. The two commissions met together to organize their work. The investigation really got underway on May 9 at Deslon's home with the reading of a paper on the theory of animal magnetism and a demonstration of hypnotic "passes" (hand movements) and manipulations in front of all the commissioners. The lieutenant of police Lenoir, who was present, asked if a magnetizer could possibly abuse a woman, to which Deslon responded in the affirmative. The commissioners then witnessed a public treatment and saw the convulsions that overtook the patient. For Deslon, these spectacular effects were convincing proof of the existence of animal magnetism. For their part the commissioners required physical proof that Deslon could not give them. They reckoned that other causes might explain the patient's sensations and convulsions. And they concluded that beyond the public treatments it was necessary to conduct experiments that were private and better controlled.

The members of the first commission then decided to experiment on themselves. Except for Franklin, everyone went to the clinic on the rue Montmartre twice a week to be magnetized by Deslon, where a private room had been reserved for them. The result was negative: none of them felt any effects that could be explained by animal magnetism. The commissioners then undertook experiments on patients in private. Whereas the second commission (the one from the Royal Society of Medicine) was satisfied simply to examine a few cases, the Academy's commission conducted a series of carefully prepared experiments.

Everything suggests that Lavoisier played a major role in these experiments. "We have proceeded as we do in chemistry, where, after having broken substances down and discovered the elements of which they are composed, we ensure the accuracy of the analysis by reconstituting the substances from these elements," wrote Bailly in presenting the commission's results to the Academy. Lavoisier was more precise in his manuscripts: after having noted that animal magnetism should not be accepted "unless the effects it produces cannot be ascribed to any other cause," he proposed to find out "if the imagination alone, without magnetism, might be able to produce similar effects" and, as a consequence, to "conduct a series of experiments on magnetism in which the imagination is not involved and on the imagination in which magnetism is not involved"—in short, to "distinguish in magnetism between that which can be attributed to physical causes and what can be attributed to moral causes."[22]

Several of these experiments took place at Franklin's home in Passy. In one of them a young girl who was blindfolded spontaneously fell into convulsions, believing wrongly that she had been magnetized by Deslon. Similarly, a boy lost consciousness when approaching a tree that he had been told falsely had been magnetized. Similar experiments took place at Lavoisier's home at the Arsenal. Twelve porcelain cups were filled with water and presented one at a time to a woman who Deslon declared was very sensitive to magnetism. The woman fell into convulsions after the fourth cup, while only the last cup had been magnetized. When she recovered and asked for something to drink, she was given water from the magnetized cup, and this time she drank it without feeling anything! "The cup and the magnetism had therefore failed to have any effect, since the convulsions occurred without magnetism and was calmed rather than increased at the approach of the magnetized cup," Lavoisier stressed in his notes. On another occasion, the commissioners performed an even more decisive experiment. The door connecting two rooms had been replaced by a folding screen covered with paper and a chair placed in the doorway. On the pretext of needing some sewing done, they called in a seamstress who had been magnetized in three minutes in Passy. They arranged for her sit on the chair. One of the commissioners stationed in the other room magnetized her through the screen without her knowledge but according to the proper protocols. The woman continued the conversation gaily for half an hour without feeling anything. The magnetizer then came into the room where she was seated and openly offered to magnetize her. He proceeded openly, but this time without respecting the rules. Nevertheless, the woman went into convulsions after several minutes. Lavoisier concluded from this experiment that "the imagination alone produces all the effects attributed to magnetism, and magnetism without imagination produces no effects at all."[23]

The two commissions drew from their investigation the same lesson: magnetic fluid did not exist, and the effects attributed to it were nothing but the fruit of the imagination. The report of the first commission, of which Bailly was the official author, was the most complete. Lavoisier seems to have written the argumentation as well as the conclusions, but it was read to the Academy of Sciences by Bailly on September 4, 1784. The public report was complemented by a secret report addressed to the king, in which the moral danger of magnetic therapy was condemned.[24] For its part, the Royal Society of Medicine published a long historical and critical study prepared by Thouret, as well as the report of its commissioners. The physician and botanist Jussieu, who was also a member of the Academy of Sciences, disagreed with its conclusions and refused to sign it. A little while later he submitted his own report in which, without denying the effect of the imagination, he maintained that this effect was not sufficient to explain some of the cases. These cases, he believed, demonstrated the existence of a general agent in the air that would correspond to the vital force in organisms: animal heat in living bodies and electrical fluid outside them.[25]

The official reports condemned animal magnetism unequivocally. The Paris Faculty of Medicine quickly proceeded to purge its ranks. It excluded Deslon permanently and required all the regent doctors who had rallied to the doctrine of Mesmerism to make a retraction in due form.[26] For its part, the government had Bailly's report printed and distributed widely, and the *Journal de Paris* reproduced extracts from it in its columns. The academic condemnation aroused a reaction among the partisans of animal magnetism. Deslon protested. As for Mesmer, he calmly declared that the condemnation had missed its mark entirely, since his method had nothing to do with the one professed by his former disciple. The polemic between the Mesmerists and the anti-Mesmerists remained at the center of public attention for several months (Fig. 9.4). At the start of 1785, the government prohibited the *Journal de Paris* from mentioning Mesmerist propaganda.[27] Mesmer left Paris, then France, and Deslon died suddenly the following year. The partisans of animal magnetism were still active, but they were divided. The doctrine had ceased to interest the public.

Among physicians and savants, Mesmerism did have its followers. Several members of the Royal Society of Medicine believed in the effectiveness of magnetic therapy, that is, in the existence of a magnetic fluid. They had to remain silent. By contrast, at the Academy of Sciences, where Jussieu was very much alone, the condemnation of Mesmerism was virtually unanimous. Berthollet, who had been the only one among the savants to attend Mesmer's classes, had publicly denounced both his doctrine and his practice "as perfectly chimerical" even before the commissioners did. By pointing to the effect of the imagination, the company was encompassing all charlatans in its verdict. But for Lavoisier, who had largely inspired the report, there is no doubt that it was directed not only at those people at the margins of science—the healers, the entertainers, the visionaries—but also at certain colleagues who defended false doctrines within the Academy itself.

Phlogistics Ridiculed

At a meeting of the Academy only a year after the report on animal magnetism was issued, Lavoisier presented an orderly refutation of the phlogiston theory, to which the great majority of chemists, in the Academy and elsewhere, still adhered. According to this theory, elaborated by the German chemist Georg Ernst Stahl at the beginning of the eighteenth century, there exists in nature an inflammable earth element, phlogiston, which is released by the combustion and calcination

FIGURE 9.4: Mesmer magnetizing. Engraved plate in Jean-Jacques Paulet, *L'Anti-magnétisme ou origine, progrès, décadence, renouvellement et réfutation du magnétisme animal*, London, 1784.

of metals. As we have seen, in 1777 Lavoisier had proposed abandoning this principle. He explained the phenomena of combustion and calcination by the fixation in bodies of oxygen in the air. This new "pneumatic theory" offered the advantage of accounting precisely for changes in weight. It could also apply to respiration.

Despite his arguments, Lavoisier had few followers. His colleagues remained attached to phlogiston, whose existence seemed to be confirmed by the discovery of hydrogen. This very light gas, emanating from metals attacked by an acid placed under a bell jar, seemed to represent the inflammable principle in an almost pure form. In order to refute this idea, which was defended by Priestley, among others, Lavoisier had to prove that in the acid test the hydrogen came not from the metal but from the water contained in the pneumatic trough upon which the bell jar was placed, and for that he had to show that water itself was composed of hydrogen. This is in fact what he demonstrated in his great experiment of 1785.

The attack on phlogiston followed soon after. On June 28 and July 13, 1785, Lavoisier presented some thoughts on the question to the Academy. He exposed the contradictions in Stahl's theory, laid out all the arguments and proofs, and refuted the vain efforts of his colleagues to rescue it:

> The chemists have made phlogiston a vague principle that is not rigorously defined at all, and which consequently can be adapted to fit any explanation one might wish. . . . It is a veritable Proteus that changes form from moment to moment. It is time to bring chemistry back to a more rigorous kind of reasoning, to strip away from the facts with which this science every day enriches itself what reasoning and prejudices have added to them; to distinguish what is factual and comes from observation from what is deduced from systems and hypotheses; finally, to make sure to define the state of chemical knowledge today, so that those who follow us may be able to depart from this point and proceed with certainty to advance scientific knowledge.

He concluded that "Stahl's phlogiston is an imaginary being whose existence in metals, sulfur, phosphorus—in all combustible bodies—has been assumed without any basis," and that its doctrine is "a scaffolding more cumbersome than useful for the ongoing construction of the edifice of chemical science."[28]

This was the same tone that had been used the year before to condemn animal magnetism, a body of doctrine founded on a "purely hypothetical principle that even enlightened physicians have accepted." "Sound logic," Lavoisier had written, "does not allow us to accept new principles to explain the facts, when they may be explained by other principles that are already known. Thus, we shall not accept animal magnetism unless it displays effects that cannot be attributed to any other cause." And as we read in Lavoisier's conclusion, which Bailly had incorporated into his report on animal magnetism, the commissioners acknowledged that "this animal magnetic fluid cannot be perceived by our senses, that is, it had no effect, either on us or on the patients who were subjected to it"; that they had "demonstrated by conclusive experiments that the imagination without magnetism produces convulsions, and that magnetism without the imagination produces nothing," and hence that "this non-existent fluid is consequently useless."[29]

Here we see animal magnetism and phlogiston, two imaginary and useless entities, being placed on the same level. Even unspoken, the insulting comparison could not escape the chemists. At the Academy, the attack was received with a great uproar. Van Marum, who attended the meeting of July 13, 1785, described the incessant interruptions and the violent objections in his journal. Everybody was talking at the same time. On his side Lavoisier had only Berthollet, who had just rallied to his cause, and the mathematicians Laplace, Cousin, and Vandermonde. The old guard of chemists, Baumé and Sage, supported outside the Academy by Delamétherie, the editor of the *Journal de Physique*, were indignant. But at the Arsenal they openly mocked the opposition. A sarcastic Sage recounted much later an anecdote he had heard from Volta about a comic magic lantern show in the Lavoisiers' salon, possibly in 1782, the year that the Italian physicist was visiting Paris.

"You are going to see in these pictures the means I have dreamed up to throw ridicule on the Stahlians," the host announced to his guests. "Ah! Do you see in the first picture the character who represents phlogiston? On his head is a crown of thorns, and on his rear end the flame of genius; his attitude is that of a supplicant; his hands are joined. He is followed by the members of his sect, who are all in mourning. In this other picture, you see the arch-phlogisticist, his head on an anvil so that it can be reforged; but it is so steely that the hammer bounces off it. Ah! Do you see in the third picture the ditch in which phlogiston is going to be buried alive? He is followed by his partisans, who hold lacrymal vases in their hands [to catch their tears]." Sage continued with a final image, no doubt of his own invention: "Ah! You see this fourth picture? If shows the apotheosis of pneumatic chemistry, and a choir of parrots. All the Oxyphiles start to shout: 'Your invention, Monsieur Ludius (this is a word used by Cicero to refer to a buffoon, a joker, a juggler, Sage noted) will not fail to prosper'!" The severe science could be funny too.[30]

Savants and Instrument Makers

In the new experimental physics, as in astronomy, measurement and calculation went hand in hand. This required reliable and precise instruments. The savants could call on Parisian makers, whether clockmakers, opticians, jewelers, tinsmiths, mechanics, or makers of inlaid tables and game boards. A few talented craftsmen supplied the physics cabinets with fine objects and produced the small

instruments used for salon entertainment. However, for instruments of the very
highest quality, in both astronomy and physics, London had a quasi-monopoly
in Europe. Intermediaries such as the Portuguese physicist Magellan and the
merchant Henry Sykes handled the export of products to Paris from across the
Channel. For a long time, only a few rare Parisian elite craftsmen such as the
clockmaker Ferdinand Berthoud and the engineers of mathematical instruments
Claude Langlois and his successor Jacques Canivet, who were recognized by the
Academy of Sciences, were able to compete with the British in this highly specialized market. The end of the 1770s saw the appearance of a new generation of instrument makers in Paris, thanks to the support of the authorities. In effect, wrote
the *Journal de Paris*, "it is more important to the progress of the physical sciences
than is generally understood that the savants who use the instruments live in the
same place as the craftsmen who make them."

In 1777, the Academy of Sciences launched a prize competition for the
production of a quadrant. The winner would receive 2,400 livres and the
title of engineer to the Academy, vacant since the death of Canivet three years
earlier. Pierre Mégnié was awarded half the prize money in 1779, but not the
title. Originally from Dijon, Mégnié lived on the rue de l'Arbre Sec and made
astronomical instruments. He had been trained by the best Parisian masters
such as Hulot, the author of a famous *Art du tourneur mécanicien*. His quadrant used a micrometer of his own invention, which was also adopted for use
in other precision instruments. His brother Pierre-Bernard (called Mégnié
the Younger), who worked with him, specialized in physics instruments
like balance scales and barometers. This is the person Lavoisier called upon
to equip his laboratory and make his experimental apparatuses: the calo-
rimeter in 1783 and the gasometers for the great water experiment in 1784.
Unfortunately, the Mégnié brothers experienced serious financial difficulties
in 1786, and the elder went to Spain, where the government had asked him
to set up an observatory in Madrid. Lavoisier then turned to Nicolas Fortin,
a maker renowned for his precision balance, to serve as the main supplier of
his instruments.[31]

While Lavoisier was wealthy enough to buy very sophisticated apparatuses
and measuring instruments, out of necessity most physicists had to be satisfied
with experimental equipment that was very easy to fabricate. Haüy used a simple
practical goniometer, and even Coulomb's electrostatic balance, so delicate to
handle, was not at all complicated in his construction. Things were quite different in astronomy, where advanced instruments for the observation and measurement of time and positions were absolutely necessary. This is why the best
Parisian observatories were equipped with British equipment. In order for a scientific instrument industry to develop in Paris, it was absolutely necessary that

astronomers place their orders with local makers. Lalande, who was Mégnié's patron, was convinced of this.

Cassini carried out the idea when he relaunched astronomical observations at the Observatory. At the end of 1784, he proposed to Baron de Breteuil, the minister of the King's Household, to install three large instruments there, a mural quadrant, an equatorial, and a full astronomical circle, and to entrust their construction to Parisian makers. As he explained in a letter to the minister, it was a matter of "encouraging French elite craftsmen, of giving them a goal to strive for, an opportunity to exercise their talents, and finally to give them the means to rise to the level of British craftsmen, perhaps even to dislodge them, or at least to share with them a branch of commerce on which a rival nation seemed until then to have a monopoly."

Cassini selected three craftsmen: Pierre Mégnié for the equatorial, a certain Charité for the great quadrant, and Étienne Lenoir, a protégé of Borda's, for the astronomical circle. Of the three, Lenoir was certainly the most remarkable. He had supplied mathematical instruments for the Lapérouse expedition, and, most significantly, he had just undertaken for Cassini the production of an astronomical repeating circle, a sort of advanced goniometer that Borda had much improved in the 1770s. Very quickly, however, Cassini experienced considerable difficulties with his artisans. The problem was not so much their competence as their lack of means. They had neither the financial wherewithal nor the equipment to undertake such vast enterprises.

Of course, Cassini had foreseen having to supply capital and raw materials, and above all having to set up a workshop in the Observatory itself for the construction of the quadrant. But Charité had made exorbitant demands, and so he had to be replaced by Mégnié, who himself went bankrupt and left suddenly for Spain in October 1786. The case of Lenoir reveals other problems that were just as serious. He had barely begun to work when the corporation of smelters accused him of exercising their trade without having been received as a master, and they had his materials and tools seized. It took the personal intervention of the lieutenant of police and Breteuil to stop the prosecution.[32]

The affair revealed how much the Parisian corporative system, with its divisions into different trades and its system of subcontracting, was unsuited to the making of scientific instruments. To succeed, Cassini noted, the elite craftsman who dedicated himself to this work "had to bring together in himself alone the talents of ten different craftsmen. He had to use wood, metal, and glass, and to smelt, forge, turn, file, polish, varnish, etc. As soon as communities are formed for each of the trades, the engineer of mathematical instruments finds himself not belonging to any one of them in particular, but to almost all of them at the same time." Étienne Lenoir had already felt the wrath of the smelters in 1780; they had also gone after Billeau in 1782. The same year, the guild of upholsterers, mirror makers, and makers of eyeglasses had seized the microscopes and telescopes in Baradelle's workshop because he was a smelter. The only way to avoid these conflicts, thought Cassini, was to create a guild for makers of scientific instruments. With the help of his friend Bailly, he succeeded in winning Breteuil over to his views.

On February 7, 1787, a corps of twenty-four engineers in instruments for optics, mathematics, and physics, all to be nominated by the Academy of Sciences, was established by letters patent. The engineers would be free to make and sell all types of instruments. In September an "*artistes*' committee" made up of seven academicians (including Borda, Rochon and Cassini) selected six elite craftsmen (Lenoir, Carochez, Fortin, Charité, Baradelle, and Billeau) to be the first engineers. Subsequent elections (in 1788 and 1789) had succeeded in expanding the initial group, so that at its dissolution in 1791, the body included eighteen members. Organizing these skilled instrument makers into an engineering corps closely associated with the Academy of Sciences marked the take-off of a Parisian industry of scientific instruments under the aegis of the government that in time would have been able to compete successfully with the British.[33]

Other initiatives were moving in the same direction: in 1787, Rochon asked the optician Carochez to make an achromatic telescope like Dollond's for the cabinet at La Muette, as well as a reflecting telescope with a platinum mirror capable of competing with Herschel's. Cassini himself went to England to negotiate the making of a joint survey by the Paris and Greenwich observatories to measure their relative locations. He carried in his luggage the famous astronomical repeating circle made by Étienne Lenoir, an instrument that was far superior to the great theodolite constructed by Ramsden for the British trigonometric survey. Finally, as we have seen, Bralle and Vincent opened a manufacture for the production of English-style clocks in Paris protected by a royal privilege, where they planned to train workers for the fabrication of scientific instruments.[34]

Winning Over Public Opinion, Educating the Public

To ensure the triumph of their severe science, Lavoisier and his friends certainly relied on the support of the authorities. In fact, the imposition of a new scientific style was achieved thanks to tight control over savant institutions. The condemnation of Mesmer was conducted like an affair of state. The rise of an industry devoted to high-quality scientific instruments was part of a larger state policy to promote industry that had begun with the signing of the Anglo-French Commercial Treaty in 1786. But we must not conclude that the proponents of the new physics were indifferent to public opinion. Just like their adversaries, they understood that they had to win its support. Consequently, nothing could be more false than to contrast a severe science barricaded in its palaces with a public science, mistress of the salons, musées, and newspapers.

The savant élite used all the means at its disposal for its propaganda, starting with the tribune of the Academy. The presentation of the commission's conclusions on animal magnetism at the Academy on September 4, 1784, provides a very good example. Prince Henry of Prussia, the younger brother of Frederick II, had come to attend the meeting. In a speech aimed directly at him, Condorcet delivered a panegyric to the sciences and their patrons. He denounced "those who have an interest in fearing the progress of enlightenment" and who have tried "to make those who spread it odious." Several pensioned members then spoke: Tenon on "the human egg," Bailly on animal magnetism, Lavoisier and Meusnier on experiments relating to the combustion of alcohol), and Leroy on static electricity. At the end of the meeting, other academicians presented papers, among them Abbé Rochon on the measurement of angles and Coulomb on the torsion of strings and wires.

The meeting may have been closed to the public, but the *Journal de Paris* published Condorcet's speech and gave an enthusiastic account of the experiments of Lavoisier and Leroy. However, since it had already praised the report on animal magnetism and provided its readers with long extracts from it, the newspaper thought it useless to linger over Bailly's contribution and totally overlooked the papers by Rochon and Coulomb. Three years later, the Academy published its own account of the meeting in its *Mémoires*. Condorcet's speech was reproduced in full; the account then mentioned (in order) the papers by Lavoisier, Tenon, Rochon, Coulomb, and Bailly. Leroy's old-fashioned experiments, although much appreciated in 1784, had disappeared, replaced by those by Rochon and Coulomb, which were completely different. The solemn condemnation of Mesmerism thus closed a magnificent series of quantitative experiments, in the severe style, which gave the meeting a significance that it doubtless had not had for those who had witnessed it, but which corresponded perfectly to the idea of themselves that the Academy's physicists wished to leave for posterity.[35]

The great water experiment was also part of the enterprise of persuasion. But as we have seen, Lavoisier was satisfied with a brief article in an ephemeral publication, the *Journal polytype des sciences et des arts*. While the relative failure of the experiment might explain this discretion, Lavoisier's aim was also to help a little-known journal in which he hoped to publish his work much more rapidly and freely than in the Academy's *Mémoires*. Under attack by the ever-hostile *Journal de Physique* and its director, Lavoisier finally created a new journal, the *Annales de Chimie*. After having negotiated a privilege in the name of the whole class of chemists of the Academy, he received authorization at the end of 1788, and the first volume appeared in May 1789, with the Academy's approval and imprimatur.[36]

But to win the battle for public opinion, it would be necessary to reach a wider audience. On the initiative of Baron de Breteuil, the condemnation of Mesmer had given rise to an effective campaign in the course of which the proponents of animal magnetism had been ridiculed. Lavoisier himself—a serious man if there ever was one—had discreetly supported the mechanic Lhomont's plan to construct a great flying balloon in the shape of a grape picker, bearing a magnetic tub on his head and holding in his hand a banner on which would be written "Goodbye, tub, the grape harvest is over." By means of this experiment they would address "balloons and Mesmerism at the same time," and "Mesmerism would be made to look ridiculous" even as the benefits of pneumatic chemistry would be demonstrated.[37] The flight took place in the Tuileries before a huge crowd on March 13, 1785, but without the banner.

This kind of propaganda, which appealed to the imagination, in fact contradicted all the ideas of the savants, whose aim was to appeal to reason and not the passions as their adversaries did. But to be convincing, was it sufficient to show the utility of the new chemistry? Lavoisier played a decisive role in the hygienic enterprise in Paris. Berthollet applied the principle of oxidation to bleaching (or at least he thought he did). More generally, in all his speeches Condorcet praised the utility of the academic sciences. Yet the argument was not without its weak points, since the charlatans themselves loved to stress the effectiveness of their procedures. Thus, the partisans of Mesmer defended the theory of magnetic fluids based on the success of their cures. In the end, the only thing to do was to enlighten the public, which is to say, to teach it.

The savants of the Academy had until then shown little interest in matters of education. Some of them had jobs as professors, but almost always in establishments dedicated to advanced sciences. In a celebrated article in the *Encyclopédie*, d'Alembert himself had sharply criticized the colleges, but without offering any concrete solutions to remedy their defects. The details were left to the care of the professionals in the universities and the teaching orders. Moreover, neither Condorcet nor Lavoisier had any teaching experience. At the Lycée, where he held the chair in mathematics, the permanent secretary preferred to have others stand in for him, being content with giving ceremonial lectures. Among those who rallied to the severe science, only Fourcroy and Monge showed a real taste for teaching. By contrast, Sage, Baumé, and Darcet, the phlogisticists of the Academy, had cut their teeth lecturing in large auditoriums, while in physics the only people who taught were the demonstrators in public courses and the professors of philosophy at the colleges.

In the years preceding the Revolution, the new academic élite seems to have become aware of the role of teaching in the dissemination of good science. Paris saw many savants (or those who claimed to be) proposing to teach an avid and credulous public. As we have already seen, Condorcet, Monge, and Fourcroy accepted appointments as professors at the Lycée. In his inaugural lecture in February 1786, Condorcet drew a parallel between the chairs in practical mathematics created for artisans in Germany "who keep them from the errors into which their own imagination or the prestige of charlatans might make them fall," and his own course "for the fashionable elite" aimed to give them, he said, "a few sound ideas that will protect them from the ridiculousness of pretentiously pronouncing scientific words that they do not understand, or being duped by those marvelous systems, whose unintelligible principles (all the more fertile for being vague) claim to explain everything, from the formation of the planets to the cause of fever, and would also happily explain the order of another universe, if it had pleased nature to offer the spectacle of it to their inventors." The allusion to Mesmerism was obvious. But Condorcet did not go any farther in teaching.[38]

As for Lavoisier, he had long wanted to instruct the public in the principles of the new chemistry. Rather than give a course like his disciple Fourcroy, he decided to write a textbook and publish it himself. He had nurtured this plan since the start of the 1780s, when he was working with Bucquet on a *Traité élémentaire de la chimie*. The book was not finished and published until the beginning of 1789. In it Lavoisier brought together the main results of the chemical revolution that he had himself launched and which he now presented using the terms of the new nomenclature adopted two years earlier.

Shortly before the book appeared, the painter Jacques-Louis David finished a double portrait of the chemist and his wife (see plate 1), which is now in the Metropolitan Museum of Art in New York. This timing was not entirely coincidental. It is very likely that the painting was part of the strategy to promote the new chemistry—and the *Traité élémentaire* that expounded it. What do we see there? Lavoisier, in a black suit, seated with pen in hand, seems to be putting the finishing touches on his treatise. On the table and at his feet lie a few of the chemistry instruments described in the book: an aerometer, a glass balloon, a gasometer, and Fontana's eudiometer. Lavoisier turns his head, affectionately raising his eyes to his wife, who stands by his side with her hand artlessly placed on his shoulder. Behind her, on the left side of the painting, we can make out a large portfolio containing the drawings she made for the plates for the book. Her gaze is directed at the spectator, but also at the painter David, who was her drawing teacher.

There is little doubt that Madame Lavoisier had the idea for this double portrait, for which she paid the painter seven thousand francs. In fact, she is the main figure in the painting. The composition shows her as the collaborator of her husband, on whom she appears to lean lightly, and at the same time his intercessor, with the artist as much as with the spectator to whom she directs her gaze artlessly. The work's effect comes from a sort of equilibrium between grace and severity: the tender gestures, the gazes, a hand that is placed just so—all in a noble mise en scène with a theatrical decor of sumptuous fabrics and shining copper. We admire the ease with which intimacy and representation come together. Even so, David's task was not just to show the happiness and success of a couple; he also had to stage the new chemistry. From this perspective, the painting is susceptible to a very different interpretation: the severe science of Lavoisier is done with experiments and writing, with instruments, paper, and a pen. But this kind of science, austere as it may be, is meant to be spread and taught. It is toward Madame Lavoisier, a woman of taste, that the savant turns to incarnate this worldly and public version of the new chemistry, as if the meaning of the painting is that true science can be as interesting to the feminine audience of the salons as the vain effects of charlatans were.

The double portrait was a private commission. The Lavoisiers reached an agreement with David to have the painting exhibited in the Salon of 1789. This was also a skillful way of publicizing the *Traité élémentaire de chimie*. But events would decide otherwise. After the storming of the Bastille, the authorities judged that it would be inopportune to exhibit a portrait of the director of the Gunpowder Administration at the Louvre, and David's picture remained unknown to the public. An entirely new era was beginning, full of hopes and fears. From now on, for the Parisian savants the goal was no longer merely to spread good science and conquer public opinion. They had to contribute to the enterprise of national regeneration, to form new citizens, and to throw themselves into public education.[39]

Notes

1. BACHAUMONT, vol. 23, 251–263, November 12, 1783. See A. L. LAVOISIER, "Mémoire dans lequel on a pour objet de prouver que l'eau n'est pas une substance simple, un élément proprement dit, mais qu'elle est susceptible de décomposition et de recomposition," *HMAS*, 1781, 468–494 (*Œuvres*, vol. 2, 334–359).

2. A.-L. LAVOISIER and J.-B. MEUSNIER, "Mémoire où l'on prouve par la décomposition de l'eau que ce fluide n'est point une substance simple et qu'il y a plusieurs moyens d'obtenir en grand l'air inflammable qui y entre comme principal constituant," *HMAS*, 1781, 269–282 (*Œuvres*, vol. 2, 360–373).

3. A.-L. LAVOISIER and J.-B. MEUSNIER, "Développement des dernières expériences sur la décomposition et la recomposition de l'eau," *Journal polytype des sciences et des arts*, February 26, 1786 (*Œuvres*, vol. 5, 320–334), 321.

4. A.-L. LAVOISIER and J.-B. MEUSNIER, "Développement des dernières expériences sur la décomposition et la recomposition de l'eau," *Journal polytype des sciences et des arts*, February 26, 1786 (*Œuvres*, vol. 5, 320–334), and corresponding laboratory

registers (archives of the Académie des sciences, fonds Lavoisier, 10ᵉ registre). On the great water experiment, M. DAUMAS and D. DUVEEN, "Lavoisier's Relatively Unknown Large-Scale Decomposition and Synthesis of Water, February 27 and 28, 1785," *Chymia* 5 (1959), 113–129. See also "Les grandes expériences d'analyse et de synthèse de l'eau (27 février–1er mars 1785)," in LAVOISIER, *Correspondance*, vol. 4, 305–309.

5. On the social life of the Lavoisiers, see GRIMAUX, 1899, 44–49. On music at the Lavoisiers, see letter to Franklin in LAVOISIER, *Correspondance*, vol. 3, 744.

6. A. YOUNG, *Travels during the Years 1787, 1788 and 1789*, Dublin, 1792, visit to the Lavoisiers on October 16, 1787, 78–79.

7. On Lavoisier's use of time, see GRIMAUX, 1899, 44.

8. Biographical notice written by Madame Lavoisier, quoted by GRIMAUX, 1999, 44–45

9. On relations between the Arsenal laboratory and the Gunpowder Office, see BRET, 2002, 230–232.

10. C. WILSON, "The Great Inequality of Jupiter and Saturn: From Kepler to Laplace," *Archive for the History of Exact Sciences* 33 (1985), 15–290, and GILLISPIE, 1997, 124–145.

11. GILLISPIE, 1997, 142, and HAHN, 2004, 81.

12. Letter from Laplace to Oriani, March 5, 1788, reproduced in TAGLIAFERRI and PETUCCI, "Alcune lettere di S. Laplace a B. Oriani," *Quaderni di storia della phisica* 1 (1997), 5–34.

13. BRIAN, 1994, 256–286, and GILLISPIE, 1997, 93–95.

14. C. S. GILLMOR, *Coulomb and the Evolution of Physics and Engineering in Eighteenth-Century France*, Princeton, NJ, Princeton University Press, 1971, and Laplace's report, co-authored with Cousin and Legendre, on the work of Haüy, *Exposition de la théorie de l'électricité et du magnétisme selon les principes d'Aepinus*, in *P.-V.* (1787), fol. 293r–293v, July 21, 1787.

15. R.-J. HAÜY, *Essai d'une théorie sur la structure des crystaux, appliqué à plusieurs substances crystallisées*, Paris, 1784.

16. S. SCHMITT, "From Physiology to Classification: Comparative Anatomy and Vicq d'Azyr's Plan of Reform of Life Sciences and Medicine," *Science in Context* 22 (2009), 145–193.

17. Letter from Condorcet to Madame Suard, end of August or start of September 1788, in BADINTER AND BADINTER, 1990, 242.

18. Letter from Lalande (May 18, 1782), *Journal de Paris*, May 23, 1782.

19. Letter from Marivetz, *Journal de Paris*, no. 69, March 9. 1784; BACHAUMONT, 1777–1789, vol. 22, 17–18 (January 7, 1784).

20. On Mesmer in Paris, see GILLISPIE, 1980, 261–278 and B. BELHOSTE, "Mesmer et la diffusion du magnétisme animal à Paris (1778–1803)," in BELHOSTE and ÉDELMAN, eds., *Mesmer et mesmérismes: Le magnétisme animal en contexte*, Paris, Omniscience, 2015, 21–61.

21. On Mesmerism in a medical context, see G. SUTTON, "Electric Medicine and Mesmerism," *Isis* 72 (1981), 375–392.

22. J. S. BAILLY, *Exposé des expériences qui ont été faites pour l'examen du magnétisme animal*, Paris, Imprimerie royale, 1784, 9, and A.-L. LAVOISIER, "Sur le magnétisme animal," in *Œuvres de Lavoisier*, vol. 3, 499–527, in particular 508–510 (various handwritten notes found in his papers).

23. A.-L. LAVOISIER, "Sur le magnétisme animal," *Œuvres de Lavoisier*, vol. 3, 522.

24. The secret report written by Bailly was published for the first time in N. FRANÇOIS DE NEUFCHÂTEAU, *Le Conservateur.*, vol. 1, Paris, 1799 (year VIII), 146–155.

25. *Rapport de l'un des commissaires chargés par le Roi de l'examen du magnétisme animal*, Paris, Veuve Hérissant, 1784 (written by Jussieu)

26. DELAUNAY, 1905, 345–348.

27. *Journal de Paris*, March 2, 1785. On the government's interest in animal magnetism, see correspondence of Breteuil from 1784, AN O¹ 495; on the polemic aroused by the reports on animal magnetism, see PATTIE, 1994, 159–198.

28. A.-L. LAVOISIER, "Réflexions sur le phlogistique pour servir de suite à la théorie de la combustion et de la calcination publiée en 1777," *HMAS*, 1783, 505–538, 523, and 538 (*Œuvres*, vol. 2, 623–65, 640 and 655).

29. A.-L. LAVOISIER, "Sur le magnétisme animal," *Œuvres de Lavoisier*, 508 ("remarks by Lavoisier") and 526–527 ("résumé du mémoire"), as well as the *Rapport des commissaires chargés par le Roi de l'examen du magnétisme animal, Œuvres de Lavoisier*, 75–76.

30. VAN MARUM, 1970, vol. 2, 34 (English translation 222); B. S. SAGE, *Opuscules de physique*, Paris 1813, 129.

31. On the Mégnié brothers, DAUMAS, 1953, 361–363; on the training of Pierre Mégnié by Hulot, see *Journal de Paris*, July 12, 1781.

32. WOLF, 1902, 273–286, in particular 274, letter from J.-D. Cassini to Breteuil, January 18, 1785.

33. J.-D. CASSINI, *Mémoires pour servir à l'histoire des sciences et à celle de l'Observatoire de Paris*, Paris, 1810, supporting items, no. 11: "Mémoire pour les ingénieurs en instruments de mathématiques," 217–222, and no. 12: "Lettres patentes établissant un corps d'ingénieurs," 222–225. See DAUMAS, 1972, 269–274, and TURNER, 1989, 10–13.

34. On Rochon and Carrochez, see G. BIGOURDAN, "Un institut d'optique à Paris au XVIIIᵉ siècle," *Comptes rendus du Congrès des sociétés savantes de Paris et des départements tenu à Paris en 1921*, Paris, Section des sciences, 1921, 19–74.

35. *P.-V.* (1784), fol. 230v–233v, September 4, 1784, and *HMAS*, 1784. The importance of this session was pointed out to me by Peter Heering.

36. BRET, "Les origines et l'organisation éditoriale des *Annales de chimie* (1787–1791)," in A.-L. LAVOISIER, *Correspondance*, vol. 6 (1789–1791), 415–426.

37. Letter from Lavoisier to Bailly of October 15, 1784 (no. 529), in LAVOISIER, *Correspondance*, vol. 4, 43–44.

38. CONDORCET, "Discours sur les sciences mathématiques" (given at the Lycée on February 15, 1786), *Œuvres de Condorcet*, vol. 1, 453–481.

39. On David's portrait of the Lavoisiers, see M. VIDAL, "David among the Moderns: Art, Science and the Lavoisiers," *Journal of the History of Ideas* 56 (1995), 595–623, and M. BERETTA, *Imaging a Career in Science: The Iconography of Antoine Laurent Lavoisier*, Canton, OH, Science History Publications, 2001, 25–42.

10

Revolution!

TOWARD THE END of the winter of 1790, Jacques-Louis David began work on an enormous painting in the severe style that represented the Tennis Court Oath, one of the founding acts of the Revolution. A preliminary drawing, exhibited in June 1791, showed the overall composition: in the center, where gazes and arms converge, the astronomer Jean-Sylvain Bailly, member of three academies and president of the National Assembly, stands on a table; he raises his right hand and swears the oath in the name of all the representatives: never to separate until the constitution of the kingdom is established on a firm foundation (see plate 2). Like many of his fellow artists, David was enthusiastic about the Revolution. Since 1789, he had the assumed leadership of the opposition to the absolute power of the officers at the Royal Academy of Painting and Sculpture. He had presented a proposal for his great painting to the Jacobins, and a subscription to pay for it was soon launched. The work would hang in meeting room of the National Assembly. Bailly, too, was a revolutionary. He had participated in the writing of the *cahiers de doléances* of the Third Estate in Paris, and then was elected one of its deputies to the Estates General. On June 3, 1789, the Third Estate elected him its leader. When the Third Estate proclaimed itself the National Assembly on June 17, Bailly became its president. Three days later, on behalf of them all, he swore the celebrated oath.

Bailly's metamorphosis was that of the Republic of Letters as a whole. Under the Ancien Régime, he had been the very model of the courtier savant, the docile instrument of Minister Breteuil at the Academy of Sciences. But when his former patron returned to government service after the fall of Finance Minister Necker on July 11, 1789, he broke with him. On July 15, the assembly of Paris electors named him mayor of the city to replace the provost of merchants Flesselles, who had been assassinated the evening before. Bailly's responsibilities as mayor of Paris were mostly administrative, and in this role he proved both officious and self-effacing. A moderate patriot and ally of Lafayette, he joined the Feuillants after the royal family's flight to Varennes. He declared martial law and then on July 17, 1791, gave the order to shoot at the demonstrators on the Champ de Mars. This disastrous decision was his most important political act. Having become very unpopular in Paris, Bailly resigned as mayor three months later, abandoned public life, and left for the provinces. But his past caught up with him in September 1793: arrested in Melun, he was brought back to Paris, tried, and condemned to death for having been responsible for the fusillade on July 17, and then guillotined on the Champ de Mars, in tawdry conditions, on November 11, 1793.

By this date, David's painting was still not finished. The painter had been elected as a representative to the Convention, where he sat among the radical Montagnards. He was the one who had led the assault on the academies and succeeded in having them closed down during the summer of 1793. The case of Bailly was embarrassing: What to do about this condemned savant, like so many other fallen heroes of June 20? But David had not given up on his monumental painting, and he continued to gather the documents he needed to complete the individual portraits. He only abandoned the idea after Thermidor, when it became impossible to represent Robespierre, one of the major figures in his tableau, whom he had followed faithfully until his fall. Apart from a preparatory drawing, all that remained of the grand project was a vast white canvas from which stood out, like phantoms, a few pieces of heads and bodies.[1]

The Savants in Revolution

The chaotic destiny of David's painting paralleled the upheavals of the Revolution up to the fall of Robespierre. Barely four years passed between the Tennis Court oath and the closing of the Academy of Sciences—but everything had changed. The old system had vanished in the violence and clamor of arms, but the new one still had to be constructed. Like all men of letters, the savants had been carried away, either enthusiastic or helpless, in a moving current that kept speeding up and eventually overtook them. Paris had played the main part in the events of the Revolution. The fall of the Bastille on July 14, 1789, had closed the first act, with the triumph of the Third Estate and the entry of the Parisian people onto the political stage. The old powers had been swept away. Political clubs had succeeded the voluntary societies and musées. The same crowd that had amassed to gaze in wonder at balloons now gathered for the Festival of the Federation in 1790 and then for the return of the royal family from Varennes, where its attempted flight across the border ended the following year.

The political fever shot up and was fed by deep social and economic crises. Enormous transfers of wealth were taking place. Privileges and offices had been abolished in a single stroke of the pen. The royal court had disappeared, and the aristocracy suffered severe financial losses. The time of great industrial speculations had passed, and the colonial market collapsed. Amid all these upheavals, Parisian industry (particularly luxury goods) suffered badly. The clockmaker Breguet, friend of Marat and supplier to the old elite, complained of unpaid bills, the Gobelins manufacture had no more royal commissions, Ami Argand gradually sank into ruin, and the trade in scientific goods disappeared. The drop in demand seriously affected worksites and workshops, causing serial bankruptcies, skyrocketing unemployment, and labor unrest. The crisis was accompanied by upheaval in the organization of trades. In 1791, the National Assembly liberated labor by abolishing the guilds and passed a patent law that changed the rules concerning invention. Facing difficulties, the elite craftsmen turned to the public authorities for support. Their demands marked the emergence of the Parisian sans-culottes.

For men of letters, the situation was full of contrasts: on the one hand there were opportunities, and on the other great insecurity. The freedom to think, to publish, and to assemble was now recognized. The beginning of the Revolution saw an explosion of print and a proliferation of newspapers, while the old system of censorship was dismantled. Elected offices also offered new careers to men of the word and the pen. A crowd of intellectuals, both minor and great, stepped into the breach. At the same time, the Republic of Letters was hit hard by the disappearance of court society, the crisis in church and state, and the weakening of the nobility. Many emoluments disappeared with the patrons who disbursed them, while the nationalization of church property threw onto the pavement a throng of clerics, now without positions or livelihoods. Liberated from their former masters, many were radicalized. Writers who were once the beneficiaries of patronage, such as Brissot, Marat, Carra, and many others, were suddenly transformed into ardent revolutionaries.

These upheavals struck the sciences almost as much as the literary world. Parisian savants had welcomed with interest, often even with passion, the events that marked the end of the Ancien Régime. Almost all of them adhered to the principles of 1789, which were those of the Enlightenment, and liked the idea of reform. They took part in the new institutions that were being put in place in the capital. The constituent Bailly, now mayor of Paris, was not the only one to exercise a public function. In September 1789, Jussieu took charge of the hospitals in the new municipal administration. Condorcet himself oversaw the writing of a constitution for the municipality of Paris. Others participated in meetings of the municipal assembly, which at the time was dominated by moderates.

But there were few members of the Academy of Sciences in the Constituent Assembly. Apart from Bailly, there was only the astronomer Dionis du Séjour and La Rochefoucauld d'Enville, both elected from the nobility. As for

Condorcet and Lavoisier, they had both failed to be elected deputies. However, the savants did participate in national debates.[2] Some contributed to the work of the Society of 1789, a club of moderate patriots founded in early 1790, which met in a superb apartment in the Palais-Royal and attracted liberal aristocrats, financiers, and men of letters. The goal of the society was the advancement of what Condorcet dubbed the "social art:" the implementation of the principles of social science through laws and administration. Although there were only a handful of them, savants had a played a dominant role. Condorcet was the moving force of the society. Members included the academicians Lavoisier, Périer, Monge, Vandermonde, and Lacépède, as well as the honorary member La Rochefoucauld d'Enville. The society functioned like an academy, publishing a journal, maintaining a correspondence, and fostering useful inventions. It saw its purpose as providing inspiration for the politics of the assembly and the government, which corresponded perfectly with the advisory role that the Academy of Sciences had always played.[3]

The savants offered their services to the National Assembly's various committees. Thouret of the Royal Society of Medicine was an assistant to the Constituent Assembly's Committee on Begging, where he very effectively assisted its president, the Duke de La Rochefoucauld-Liancourt. Vicq d'Azyr submitted an important plan for the reform of medical education to the Committee on Public Health in 1790. Lavoisier provided an estimate of the national wealth for the Committee on Taxation in 1791. Finally, Condorcet and other academicians gave Talleyrand advice and information for the Constitutional Committee, both for his report on public education and for his plan to reform weights and measures, whose execution the Constituent Assembly entrusted to the Academy of Sciences.

Rifts truly began to form among the savants after the flight to Varennes, when the king's authority was called into question. The Society of 1789 did not survive these divisions. La Rochefoucauld d'Enville (who assumed the presidency of the Paris Department in February 1791), as well as Bailly, clearly leaned toward the moderate Feuillants. Conversely, Condorcet, Monge, Fourcroy, Meusnier, and Vandermonde were Jacobins. Some no longer hid their republican beliefs. Even Lavoisier distanced himself from the king. In September 1791, Condorcet was elected deputy from Paris to the Legislative Assembly, where he joined three other academicians (Broussonet, Tenon, and Lacepède). Several provincial representatives also belonged to the Republic of the Sciences: the chemist Guyton-Morveau (a close friend of Lavoisier); the mathematicians Louis Arbogast and Gilbert Romme, who taught in the colleges; the engineers Prieur de la Côte d'Or and Carnot, who had studied with Monge; and the naturalist and former assistant to Cagliostro Ramond de Carbonnières, a close friend of Broussonnet.

Except for Broussonnet, Ténon, and Ramond, all of them were very active in the assembly's Committee of Public Instruction.

The fall of the monarchy on August 10, 1792, marked a new stage in the Revolution. That very evening, on a proposal from Condorcet, Gaspard Monge entered the Provisional Executive Council as minister of the Navy. For the first time, a savant had become a member of the government. Condorcet and Lavoisier had already been appointed treasury administrators by the king the year before. Condorcet had even come close to being appointed minister of finance at the end of 1790, and Lavoisier had declined the post of minister of taxation (*contributions publiques*) in June 1792. What would have been unthinkable under the Ancien Régime was now almost normal.

Despite the formation of a government in the aftermath of August 10, power in Paris remained in the hands of the insurrectional Commune. Fear reigned in the capital: the sans-culottes were afraid of the arrival of the Prussians and of a counterrevolutionary coup. The massacres of suspects in the prisons began on September 2. Thanks to the cool head of the young Geoffroy Saint-Hilaire (a protégé of Daubenton), Abbé Haüy just barely managed to escape death at Cardinal-Lemoine. In Saint-Denis, just north of Paris, the astronomer Delambre, who was measuring the meridian, feared for his life. On September 10, Lavoisier, who had left the Arsenal for good, left Paris, where he felt under threat. La Rochefoucauld d'Enville, who had also fled Paris, was arrested four days later in the provincial town of Gisors, and killed in front of his mother and the mineralogist Déodat Gratet de Dolomieu, as he tried to escort them to safety. At the Academy of Sciences, where Fourcroy was demanding a purge, the honorary members had disappeared.[4]

The Rise of Voluntary Societies

The Academy had only one more year to live. Since 1789, it found itself confronted with a difficult choice: to transform itself or die. The political framework of absolutism in which the institution had functioned continuously since its foundation in 1666 had ceased to exist. For more than a century the Academy had benefited from a near monopoly over the evaluation and publication of scientific and technological work in Paris. The abolition of privileges and censorship had eliminated this at a stroke. The appearance of a free press and the right of association opened up other arenas for scientific activity and debate, however.

Of course, this was not entirely new. Clubs and societies of all kinds had proliferated in Paris before the Revolution, often based on the Masonic model. The Palais-Royal and the area around it had housed many of them. Private establishments like the Salon of Correspondence, the Musée de Paris, and the

Musée de Monsieur had already had as their mission the dissemination of the sciences, arts, and letters among the educated public, and the Academy of Sciences had willingly granted them its approval. But after 1783 it had shown itself more concerned about its authority. Supported by the government, it had wanted to control the production and dissemination of scientific work more closely by firmly condemning "charlatans." This belated rigidity ran against the tide. While the Lycée had been respectful of the savants, journalists and amateurs had rejected their claim to govern the sciences. Writers and inventors who had been turned away now did not hesitate to attack the Academy in public. It was in this climate that voluntary societies for the study of the sciences were formed in Paris just before the Revolution.

The first was the Linnaean Society of Paris, founded in December 1787 by several naturalists: Broussonnet, Louis-Augustin Bosc d'Antic, Aubin-Louis Millin de Grandmaison, and Pierre Willemet. Broussonnet, permanent secretary of the Society of Agriculture, was a proponent of the Linnaean system for the classification of plants, which was not in favor among the savants of the Royal Botanical Garden but was used by many amateurs. He wanted to join with other botanists to promote these methods in France, taking inspiration from the Linnean Society in London, which he knew well. The meetings of the new society were held at his home on the rue des Blancs-Manteaux. Bosc presided, and Millin was secretary; the former, an official in the postal service, was a young naturalist full of personal skills and a close friend of the Rolands; the latter, his exact contemporary, made a living by doing translations and was well connected in the publishing world. A few other naturalists sometimes joined the group: Buffon's gardener, André Thouin; the editor of the *Journal de Physique* Delamétherie; and the entomologist Guillaume-Antoine Olivier, who at the time worked for Gigot d'Orcy. On May 24, 1788, the society organized an elaborate celebration of Linnaeus's birthday, but a year later it was forced to stop meeting because the Academy of Sciences had taken umbrage.[5]

When the Linnaean Society suspended its work, another society had already been born in Paris. The Philomathic Society was founded on December 10, 1788, by six young amateurs who were virtually unknown at the time: Augustin François Silvestre, who served as librarian to Monsieur (the Count de Provence); the naturalist Claude Antoine Riche, who also frequented the Linnaean Society; the physicians Jacques Joseph Audirac and Marie Antoine Petit; the mathematician Charles de Broval; and finally Alexandre Brongniart, nephew of the chemist Antoine Brongniart, who was barely eighteen years old. Begun modestly as a society for mutual support under the motto "Study and friendship," the Philomathic Society became more important and more institutionalized in the first years of the Revolution. By 1791, it had eighteen members and eighteen correspondents.

At its weekly meetings the members read papers and reported on experiments they had conducted. The society's *Bulletin*, first distributed in manuscript, then in printed form after September 1792, kept the correspondents connected to the society. After the suppression of the Academy of Sciences in August 1793, the Philomathic Society served as a refuge for many savants.

Another voluntary society, the Society of Natural History, whose membership overlapped with that of the Philomathic Society, was formed in Paris in the summer of 1790. Founded by two veterans of the Linnaean Society, Bosc and Millin, it was heir to that society but had greater ambitions. Its mission was to bring all naturalists together, whatever their interests and status. The savants of the Royal Botanical Garden like Thouin, Lamarck, Fourcroy, and Faujas could be found there, as well as amateurs like Cels, Lermina, and L'Héritier, supporters of the Linnaean system, as well as those who continued to uphold Jussieu's "natural method." Its meetings took place every Wednesday, first, it seems, at the home of Bosc and then starting in March 1791 in a townhouse on the rue d'Anjou-Dauphine, at the foot of the Pont-Neuf. There they shared quarters with the Philomathic Society, and for more than a century the building was the seat of several Parisian learned societies. The Society of Natural History grew quickly under the leadership of its secretary, Millin. By 1792, it had some sixty members, in addition to ninety correspondents in the provinces and abroad. It organized excursions around the Paris region to study natural history and set up a reading room, a library, and soon a natural history cabinet. It also received authorization from the National Assembly to organize an expedition led by Antoine Bruny d'Entrecasteaux to find Lapérouse, who had not been heard from since 1787. Finally, from the start it was committed to publishing its *Actes*, of which the first and only volume appeared in December 1792.[6]

Indeed, publishing a bulletin or a journal was one of the essential features of the new learned societies. These functioned as extensions of their meetings and allowed the external correspondents to become part of the associations. They also offered members the opportunity to publish at a time when the traditional circuits of printing and distribution were shattered. Some societies were even organized around the publication of a periodical. In the scientific domain, this was the case of the Society for the *Annales de Chimie*, which served as the editorial committee for the journal and met on at least two Wednesdays each month. Before it could be published in the journal, each paper had to be approved by this committee, and the experiment on which it was based had to be repeated by it. In 1791 Fourcroy adopted the same organization for his journal *La Médecine éclairée par les sciences* (Medicine enlightened by the sciences). A voluntary society made up of "distinguished savants" was supposed to meet twice a month in his laboratory to read and discuss the papers to be included in the journal.[7]

Alongside these specialized learned societies, there were general interest societies, often of Masonic origin: the National Society of the Nine Sisters, founded by Cordier in January 1790, and the Cercle Social, founded by Nicolas de Bonneville and Abbé Fauchet in October of the same year. Several men of science, including Lalande and Jussieu, were members of the National Society of the Nine Sisters, an extension of the lodge of the same name discussed in chapter 5. It held its regular meetings every Sunday in the Hôtel de Clermont-Tonnerre on the Quai des Miramiones and organized a few public meetings in its garden. In July 1790, it launched the first issue of its monthly journal, the *Tribut*, and set up a print shop under the control of a committee presided over by Jussieu. Moderate in its views, the society ceased all public functions, including the printing of the journal, after August 10, 1792, and disappeared a year later.[8]

The Cercle Social was a very different kind of society or editorial board. Its highly ambitious plan was to create a space for debate, the Universal Confederation of Friends of Truth, which would be a sort of democratic version of Pahin's Salon of Correspondence, placed under the direct control of public opinion. The confederation began its "federative assemblies" on October 13, 1790. The meetings were always public and were held twice a week in the amphitheater of the Palais-Royal before an immense crowd. The "friends of truth" submitted their ideas, which they could then publish in the journal of the Cercle Social, the *Bouche de Fer* (The iron mouth). Condorcet himself participated in this new type of experience, which lasted less than a year. The confederation, which had become a crucible of republicanism, ceased operations shortly after the fusillade on the Champ de Mars in July 1791.

Meanwhile the Cercle Social had acquired a printing shop that was effectively run by Jean-Louis Reynier, a young naturalist from Lausanne. This excellent botanist, an anti-Newtonian and a convinced phlogisticist, had settled in Paris in 1788, in the entourage of the superintendent Bertier. Like Millin, he had contributed to the French translation of the *Philosophical Transactions*, focusing on papers having to do with physics, then in 1790 he launched a *Journal d'agriculture*, before becoming the printer-bookseller for the Cercle Social on the rue du Théâtre Français. Under the aegis of the Cercle Social, Reynier printed and distributed many books and journals, including some dealing with the sciences and their applications. He reissued his own book *Du feu* (On fire), which had appeared in 1787, and entrusted the editorship of his *Journal d'agriculture* to the academician Teissier. In 1792, he published a *Journal d'histoire naturelle* edited by Lamarck, Bruguière, Olivier, Haüy, and Pelletier, which included an important article in which he expressed his transformist views. Lamarck, who admired him, had him elected to the Society of Natural History, whose *Actes* he was soon publishing. Reynier also signed a contract with Jean-Marie Roland for a *Journal des arts utiles* that never saw the light of day. After August 10, the Cercle Social benefited from government press subsides, thanks again to Roland. The fall of the Girondins in June 1793 put a temporary end to its activities, which did not resume until after Thermidor.[9]

The Elite Parisian Craftsmen Take Action

Among the many voluntary societies that were formed during the first years of the Revolution, we should also mention the societies of inventors, which intervened very directly in the debate about the right of industrial property. In the summer of 1790, a Society for Inventions and Discoveries (Société des inventions et découvertes) was set up on the initiative of Baron de Servières, an aristocrat who had long been interested in the sciences and agronomy. It had fifty-seven members when it was founded, among whom were makers of scientific instruments, artisan-chemists, architects, and building engineers. They met weekly on Thursday evenings, first at the Archbishop's Palace, near Notre Dame, and then at the Louvre, in the old hall of peers that the government had put at the disposal of learned societies.[10] The society, which favored the adoption of a patent system comparable to that in Britain, had been created as a pressure group when the Constituent Assembly was preparing a law on inventions. Its lobbying was successful: with passage of the law on patents and the Allarde law that suppressed the guilds, the society got everything it wanted during the winter of 1791. From now on, an inventor owned his discovery. The assembly instituted a Patents Board to register them under the direction of Servières, and the system for registration was fully instituted in September 1791 with the creation of an Advisory Board for Arts and Trades (Bureau de consultation des arts et métiers), which assumed the prerogatives of the Academy of Sciences for everything having to do with the evaluation of inventions and the dispensing of monetary awards to inventors.[11]

But the savants had not lost everything. Out of the thirty seats on the Advisory Board, only half, appointed by the minister of the interior, went to representatives of the various communities and societies of elite craftsmen, while the other half went to the Academy of Sciences. While this compromise seemed to satisfy the Society for Inventions and Discoveries, which held four seats on the board, it soon aroused criticism from other elite craftsmen. Indeed, why should the Academy of Sciences, thanks to its representatives at the Bureau, retain its privilege of evaluating their work?

The recriminations came primarily from the Point Central des Arts et Métiers (Center of the arts and trades), a new society of elite craftsmen that had been founded on the initiative of Benoît André Houard, a public works contractor. Launched on May 13, 1791, during a Federative Assembly of the Friends of Truth

in the amphitheater of the Palais-Royal, the Point Central held its meetings at
the office of the Cercle Social on the rue du Théâtre-français. Its first public demonstration was in support of a petition from workers in public works demanding
that the Constituent Assembly reestablish the relief workshops that had just been
closed down. The Point Central was in favor of opening several workshops according to plans furnished by its members, who obviously had an interest in the
matter. A second petition prepared by the Point Central to demand the opening
of public work projects was also signed by fraternal societies of workers and
the Cordeliers Club. The Constituent Assembly, which had just passed the Le
Chapelier law banning workers' coalitions, was hostile to the petitioners. The
affair occurred in an atmosphere of political crisis, just after the king's arrest at
Varennes, and it contributed to the crystallization of the sans-culotte movement
during the summer of 1791.[12]

The Point Central did enjoy some success. At the end of 1791, it had two
hundred members, many more than the Society for Inventions and Discoveries.
Associated with the Cercle Social and the Cordeliers Club, it developed a democratic and anti-elitist rhetoric in its journal. It sharply criticized the composition
of the Advisory Board for Arts and Trades, in which the academicians were given
the lead roles. It addressed a new petition to the Legislative Assembly in which it
denounced the machinations of the Academy of Sciences, managing to tar with
the same brush the Society for Inventions and Discoveries, the National Society
of the Nine Sisters, and the Commune of the Arts, a society of artists created by
David. According to the historian Charles Coulston Gillispie, almost a thousand
artisans and elite craftsmen rose up against academic authority. In March 1792,
the Point Central presented the Legislative Assembly with a plan for a decree
to replace the legislation on patents with a democratic organization of arts and
trades. Its author was the engineer Desaudray, an enterprising man whom we have
already met, and who now represented the Point Central on the Advisory Board
for Arts and Trades.[13]

Of great complexity, Desaudray's plan imagined primary assemblies of savants
and artisans in each department (with that of Paris becoming the Point Central
des Arts et Métiers), as well as a General Board of Sciences and Arts divided
into six committees, charged with granting both patents and monetary awards
to inventors. The adoption of such a system presupposed the outright suppression of the Academy of Sciences, the Patents Board, and the Advisory Board for
Arts and Trades. In June 1792, Desaudray revived his scheme of a Point Central
des Arts et Métiers. He imagined in the amphitheater of the Palais-Royal, now
abandoned by the Cercle Social, a free assembly of savants and elite craftsmen,
exhibitions of machines and objects, free public courses, a journal, festivals, and
ceremonies. A board composed of its own professors along with representatives

of the major scientific, artistic, and literary institutions of the capital would ex-
amine inventions and awards. The plan was inspired by the experience of both the
Universal Confederation of Friends of Truth and the Lycée (the former Musée de
Monsieur), which continued to operate on the rue de Valois.[14]

For his plan to succeed it would be necessary to raise an enormous amount
of capital. To this end, Desaudray would spend four hundred thousand livres of
his own money in the following months. He had to put down a hundred thousand livres just for the rental of the amphitheater, whose cost rose to sixty thousand a year. With all this money, according to Lavoisier, he engaged in the "costly
constructions of shops, assembly halls, and spectacles," paid professors, and
financed prizes. He hoped to cover the operating costs by renting out the shops.
Lavoisier did not hide his skepticism: "Such an establishment, which has nothing
but expenditures and promises only meager sources of income, is puzzling, to say
the least." But Desaudray could count on the support of the authorities.[15]

When the new establishment opened its doors on April 7, 1793, it was called
the Lycée des Arts (Lyceum of the arts). The crowds returned to the amphitheater of the Palais-Royal, which had been renamed Palais-Égalité. Five thousand
people attended the inaugural meeting. Fourcroy, who presided over the board,
gave a speech on the state of the sciences and arts in the French Republic. Prizes
were awarded to the first laureates. The event ended brilliantly with a concert. The
regular activities of the Lycée des Arts included board meetings every Thursday
and public meetings in the amphitheater on the first Sunday of each month.
Public courses began on April 15 and were all free of change. Here the sciences
and their useful applications were taught together. Fourcroy taught the physics of
plants, Millin and Brongniart taught natural history, Sue taught physiology, and
Hassenfratz taught technology. In a single bound the Lycée des Arts had jumped
to the first rank of scientific establishments in Paris—at least on paper.[16]

Plans for Public Education

What happened to the official scientific institutions at this time? Their situation was difficult, not because of hostility to the sciences and the savants (at
least not in the beginning) but because, starting in 1790, they found themselves
caught up in the challenge to the Ancien Régime as a whole. The ideas of privilege, corps, and hierarchy on which their organization and its functioning were
founded were being replaced by those of equality, citizenship, and national sov-
ereignty. Taking up a proposal from La Rochefoucauld d'Enville, the Academy of
Sciences undertook the writing of new regulations that placed all its members on
an equal footing and suppressed honorary membership, but it remained divided
over the question of its relationship with the legislative power. Other institutions

also had to adapt. With varying degrees of success, the different savant academies and societies wrote up new regulations. At the Paris Observatory, Cassini tried against all odds to maintain a hierarchy between the director (himself) and the three assistants charged with making observations. At the Royal Botanical Garden, where Buffon, who had died in 1788, was succeeded by the nondescript Count de La Billardière, elder brother of the Count d'Angiviller, the professors and officers presented a united front, agreeing fairly well on both the principle of a collective and egalitarian organization and the need to preserve their budget, which was threatened with cuts.

Assailed by numerous projects, the Constituent Assembly finally decided to leave things as they were until a general plan was adopted. The obvious solution seemed to be to unite all the great cultural institutions of the capital into a single organization that would itself be integrated into the general system of public education. At least this was the plan proposed by Talleyrand in his *Rapport sur l'instruction publique*, read before the assembly in September 1791. In it he envisioned the establishment in Paris of a great "National Institute" dedicated to the improvement of literature, the sciences, and the arts. The project was ambitious: "To bring all men of superior talent into a single and respectable family by means of ever-widening correspondences and of course local institutions, [it will link] all literary establishments, all laboratories, all public libraries, all collections . . . to a central point." The National Institute would hold its meetings in the Louvre. It would have two major divisions: "the philosophical sciences, literature, and fine arts," and "the mathematical and physical sciences and the arts," each subdivided into ten "classes" or categories.

Concretely, the new institute would essentially be constituted out of the old academies, whose functions and missions it would take over: for example, the first six classes of the second section would correspond to the Academy of Sciences, the seventh to the Royal Society of Agriculture, the eighth to the Royal Society of Medicine and the Royal Academy of Surgery combined, and the ninth to the Academy of Architecture, while the tenth, on the arts, would be created entirely from scratch. Moreover, chairs "for teaching that which is most transcendent and most elevated in human knowledge" would be annexed to the National Institute, making this organization "a sort of encyclopedia, always studying and always teaching." This would be accomplished by transferring to the new institution the educational missions of the Collège de France, the Royal Botanical Garden, the Louvre, and the School of Mines. The collections of natural history, physics, and machines, the botanical gardens, the National library (formerly the King's Library), a print shop, and a translation bureau would be also attached to the National Institute.[17]

We know nothing about the conditions under which Talleyrand prepared and wrote his plan, but he seems to have worked on it for more than a year. According to his own account, he consulted several savants at the Academy of Sciences. Lavoisier, whom Talleyrand had asked to write up his advice before he published his report, endorsed the creation of the National Institute enthusiastically.[18] Indeed, the project responded to the wishes of the academic élite. It maintained the network of great Parisian institutions as it had existed before the Revolution, with its posts, its chairs, its endowments, and its resources, all while strengthening its hegemony over the Republic of Letters. The new institute in fact would dominate the organization of public education at the expense of the University of Paris, which was shut down permanently, and would crush the provincial academies, reduced to the rank of satellites of the Parisian body.

The deputies of the Constituent Assembly preferred not to make a decision on Talleyrand's plan, leaving the organization of public education up to the Legislative Assembly, which began to meet in October 1791. It soon created a committee for public instruction, which in turn entrusted the preparation of a general plan to a few of its members who came from the Republic of Letters: Pastoret from the Academy of Inscriptions and Belles-Lettres, the mathematicians Gilbert Romme and Arbogast, the naturalist Lacépède, and especially Condorcet, the permanent secretary of the Academy of Sciences. It was the last of these who wrote the plan and presented it to the Legislative Assembly on behalf of the committee in April 1792. Starting from the same premises as Talleyrand, Condorcet proposed a vast plan of free education at all levels, from the first degree (which would be universal) to the fourth degree, the latter for the training of savants and professors and advanced study in the sciences.[19] The pyramid would be crowned at the fifth degree by a National Society of the Arts and Sciences, "established to supervise and direct the teaching institutions, to be concerned with the advancement of the sciences and arts, and to gather, encourage, apply, and spread useful discoveries."[20]

Despite the overlap with Talleyrand's plan, Condorcet's plan was distinctive in several crucial respects. In particular, the National Society of the Arts and Sciences would be very different from Talleyrand's National Institute. It would have neither chairs nor major equipment, and from this perspective it would be more like the old academies. Condorcet reserved higher education for nine "lycées" that would comprise the fourth degree of education, of which only one would be in Paris, which would correct the excessive centralization of the earlier plan. In exchange, the National Society would have authority over the whole system of public education, which it would supervise and manage, and this was a major innovation. The internal organization of the National Society was also completely different from Talleyrand's National Institute. Condorcet envisioned four classes. The first, mathematical and physical sciences, was a sort of continuation of the Academy of Sciences; the second, moral and political sciences, was entirely new; the third, applications of science to the arts, assumed the missions of societies like the Royal Society of Medicine, the Academy of Surgery, the Society of Agriculture, and the Academy of Architecture, extending them also to the mechanical and chemical arts; finally, the fourth and final class, literature and the fine arts, grouped together what until then had been divided up among the Académie Française, the Academy of Inscriptions and Belles Lettres, the Academy of Painting and Sculpture, and the Academy of Music.

Within the National Society, the sciences would clearly occupy by far the most important position. Writers found themselves relegated to the last rank, lumped together with painters, sculptors, and musicians. The "useful arts" were considered only as the beneficiaries of the sciences applied to them. The primacy given to the sciences was one of the most original things about the plan at all levels of instruction. Condorcet justified this primacy on intellectual, moral, and practical grounds. Studying the sciences, he said, would be the best means of developing the intellectual faculties, the best antidote to all prejudices, and the best preparation for life. But his vision was even grander; Condorcet entrusted universal education to the care and supervision of savants, because he ascribed to the sciences a historic task: to enable the human species to constantly improve itself. This task, which he described in a gripping manner in his celebrated *Esquisse d'un tableau historique des progrès de l'esprit humain* (Sketch for a historical picture of the progress of the human mind), written in 1793, was the grandiose and optimistic vision of a man being hunted and on the edge of the grave, which would later be read as a testament of the Enlightenment.

Condorcet had presented his report the same day as the king came in person to ask the Legislative Assembly to declare war on Austria. The examination of a plan for public education was postponed until better days. But the Legislative Assembly was dissolved after the king's fall in September 1792, without having been able to address the issue; Condorcet's project was not discussed by the Convention until December 1792. Of all the proposals that it contained, the creation of a National Society of Arts and Sciences was by far the least popular. Its adversaries saw it as a means of withdrawing control over education from the nation and entrusting it to a caste of savants, in effect a new clergy. In the face of these attacks, the Convention's Committee of Public Instruction preferred to withdraw this part of the project, and the representatives postponed the discussion without having made any decision.

In the course of the debate, the power of savants had been the prime target, but on a deeper level it was the very idea of an education based on the sciences that had been attacked. It was one thing to require it for higher education, but as Rabaut Saint-Étienne made clear, a national education, one that was designed for the people, "should train the heart. . . . It requires circuses, gymnasia, arms, public games, national festivals, the fraternal competition of people of all ages and all sexes, and the imposing and gentle spectacle of human society gathered together." This Rousseauist conception was radically opposed to Condorcet's cerebral rationalism.

Condorcet responded by denouncing the dangers of enthusiasm: "Once excited, it serves error as much as truth; and hence it actually serves only error, because without it, the truth would still triumph by its own strength. A cold and severe examination, which listens only to reason, must precede the moment of enthusiasm." And he added: "It is no doubt necessary to speak to the imagination of children; for it is good to exercise this faculty just like all the others; but it would be a sin to get carried away with it, even for what we believe, at the bottom of our hearts, to be the truth." Condorcet was thus taking up again the arguments used so many times by the savants of the Academy against the charlatans, but this time to denounce the dangers of political indoctrination. However, he had no chance at all of convincing revolutionaries for whom the stakes were nothing less than the regeneration of the nation, which they were convinced should come about through the establishment of a civic religion that would speak as much to the heart as to the mind. And in this debate, Condorcet was abandoned by everyone, including his political friends.[21]

Invention of the Metric System

While the projects of Talleyrand and Condorcet did grant academic institutions great intellectual and moral authority, they also reduced their role in supplying expertise, which had been so important under the Ancien Régime. Far from having renounced their vocation as advisors to power, however, the savants of the Academy won from the Constituent Assembly a scientific and technical mission of vast economic and cultural (but also political) scope: the reform of weights and measures. The invention of the metric system is without question the prime example (and perhaps the most remarkable one) of a major reform of social practices launched and conceived by men of science. By proposing to base the new system of weights and measures on a universal definition of the unit of length, the meter, the Academy of Sciences was falling directly in line with both the Enlightenment project and the revolutionary enterprise and thus demonstrating its usefulness even as its very existence was being threatened. Less directly, it was placing itself (and with it Paris and the French nation) at the center of a plan for an institution among all men and all nations of something that was supposed to be based on nature. This posture was in perfect accord with the long-standing pretentions of the Academy, but also with a revolutionary discourse that was always prompt to dress national ambitions in the rags of universality.

On March 9, 1790, after having consulted with the savants, Talleyrand proposed to the Constituent National Assembly the creation of a new system of weights and measures based on the length of a simple pendulum beating one second at sea level at 45° latitude. The Assembly adopted the idea two months later. By a first decree, the authorities were charged with collecting standard examples of the different units of measure used in the provinces and depositing them with the Academy of Sciences, as well as making arrangements with the British government to have the new unit of length be determined jointly by the Academy of Sciences and the Royal Society. By a second decree, the Academy of Sciences was also asked to study the question of how the units of measure and currency would be divided into smaller units.

In choosing as the basis of the new system a physical measurement, the length of a pendulum swing, the National Assembly decided a question that had pitted the advocates of a natural system of weights and measures against those in favor of a conventional system. The latter had simply proposed adopting as units of measurement throughout the kingdom those that were used in Paris. Note that on the eve of the Revolution there were about eight hundred different units of measure used in France. This extraordinary Babel of measurement, which had been criticized in the *cahiers de doléances*, was not only a major obstacle to trade but also a headache for the various government agencies. It was the source of delays, misunderstandings, and errors. By ensuring a uniform standard of weights and measures, the centralizers believed that the transformation of Parisian units into national units would solve the problem at the least cost.

Moreover, for a long time the monarchy had been proposing projects of uni- fication that would make Parisian units of measure obligatory throughout the country, such as the *toise de Paris* (*toise royal* or *toise de France*) for length and the *livre de Paris* (or *livre de poids de marc*) for weight, but it ran up against resistance from the provinces and the trades every time. The Academy of Sciences itself had engaged in the promotion of Parisian units of measure. The *toise de Paris* was de- fined by an iron ruler sealed into the wall of the Châtelet accessible to the public, called the *toise du Châtelet*, which served as the standard prototype. In 1735, the Academy of Sciences had two new *toises* for measuring meridian arcs made based on this standard *toise*: the one used in Peru was called the *toise du Pérou*, and the other one, used in Lapland, the *toise du Nord*. The *toise du Pérou* was deposited in the cabinet of the Academy of Sciences after its return to France, and was adopted as the new standard prototype of the *toise de Paris* in 1766 and dubbed the *toise de l'Académie*. In order to make sure that the Parisian unit was put in use everywhere, the Academy of Sciences then asked Tillet, the inspector of the Mint, to make eighty copies, which were then sent to the provinces and abroad.

A strong advocate of a conventional and centralized system of weights and measures, this same Tillet proposed the adoption of Parisian units of measure as national units again in 1790 in the *Observations* he and Abeille submitted to the Constituent National Assembly's Committee on Agriculture and Commerce. However, supporters of a conventional system such as Tillet and Lalande were in the minority in the Academy of Sciences. Most of the savants leaned toward the adoption of a natural basis, which had the advantage of giving the system of weights and measures a truly universal character. Was this not the precondition for its adoption by all nations? Since the end of the seventeenth century, two solutions had been proposed for the natural definition of the basic unit of length: either the length of the pendulum beating one second or else the length of the earth's meridian (a great circle passing through the poles) or one of its divisions.

The first definition was the simplest, since it was relatively easy to measure the length of the pendulum swing, but it also had drawbacks. First, one had to specify the geographical position of the pendulum, since the oscillation period depends on the altitude and latitude. Some people proposed putting the pendulum at sea level at the equator, others at the midpoint between the pole and the equator, meaning at 45° latitude, which had the advantage (for the northern parallel) of crossing France. This was the solution adopted by Talleyrand. The choice of the pendulum presented another problem, of a theoretical order: it made the unit of length depend arbitrarily on a unit of time, the second, and of weight, both susceptible to slight variations. The length of the meridian presented other difficulties: its determination required very complicated geodesic calculations, especially since the earth is not a perfect sphere but a spheroid flattened at the poles, which reduces the precision of the result. The margin of error was thirty-four *toises* per degree, according to Talleyrand. This was no doubt why he preferred to adopt the length of the pendulum swing as the natural base.[22]

Charged with preparing the reform, the Academy of Sciences went to work. In fact, it had not waited for the assembly to take up the problem. Not only had it discussed many measurement proposals since its foundation, but in June 1789 it named a commission to work on preparing a plan for uniform weights and measures that prefigured the great commission of 1791. Nothing came of it, at least officially, but it is probable that Talleyrand had consulted it to prepare his own project. After the vote of the National Assembly, the Academy began by examining the scale of division, opting for decimalization. As for creating a system of weights and measures, it waited until the king had endorsed the assembly's decree before carrying out the tasks entrusted to it: a more precise determination of the length of the pendulum and a collection of standard measures then in use. During this period, negotiations with the British government to set up international cooperation on weights and measures began.

When the British rejected the French proposal to cooperate at the end of 1790, Talleyrand's project was called into question altogether. The length of the pendulum beating one second had been adopted as the basis for the standardization of weights and measures because this choice had seemed acceptable to foreign nations. Now that the prospect of an international metrological system was remote, the savants of Paris felt free to reconsider their choice. This time, the Academy set aside the pendulum solution in favor of the one based on the meridian, proposing as the basis for the system one ten-millionth of a quadrant of the meridian circle. The Constituent Assembly adopted this definition of the unit of length, which would soon be called the "meter," without discussion, annulling the previous definition de facto. At the same time, it charged the Academy of Sciences with carrying out a new measurement of the meridian arc between Dunkirk and Barcelona. To apply the decree, the company appointed five sub-commissions a few days later to accelerate the work: the first to do the astronomical calculations and triangulations, the second to measure the bases, the third to establish the length of the pendulum, the fourth to measure the weight of the distilled water, and the fifth to make the comparisons with the old units of measures. In total, eighteen ordinary academicians out of sixty (almost all convinced proponents of the severe science) would participate closely or from afar in the enterprise.[23]

Paris and the Meter

Much ink has been spilled over the choice of the meter as the basic unit of the metric system. Some people wonder why the academicians finally abandoned the length of the pendulum for that of the meridian. The reason the commission gave in its report was that "it is much more natural to relate the distance from one place to another to the quarter of the earth's meridian than to the length of a pendulum." This official explanation, which has a certain logic, was developed by Laplace in the lectures he gave at the École Normale in Year III. While nothing casts doubt on it, other reasons certainly entered into the final decision. In fact, according to Lagrange and Delambre, the members of the commission hoped to profit from the opportunity to determine more precisely the length of the meridian of France (the meridian arc between Dunkirk and Perpignan) by extending it to Barcelona. In short, it was a matter of resuming the geodesic work that the Academy of Sciences had engaged in since its foundation and that it had pursued throughout the eighteenth century.[24]

The 1787 joint survey between the observatories of Paris and Greenwich, which was part of this long-term project, had made possible considerable improvement in the precision of measurements and calculations. The temptation was now great to extend to the meridian the same successful techniques used in the previous operation. And Borda, who had come up with the idea for the commission, had a personal interested in making use of his repeating circle. Such geodesic and metrological work was long and costly, however—a price tag of several million livres was bandied about. In the end, the commissioners promised to terminate operations in two years for a total cost of three hundred thousand livres. On August 8, 1791, the National Assembly granted them a hundred thousand livres for the first stage of the project, which was more than the Academy's annual budget. The company's adversaries protested. In *Les Charlatans modernes*, Marat vehemently criticized the greed of the savants.[25]

Obviously, by returning the determination of the basic unit of the system of weights and measures to the measurement of the meridian of France on which it had been laboring for more than a century, the Academy of Sciences would reap the benefits and honor of a prestigious enterprise with a universal aim. In order to justify its choice, the commission claimed that the meridian, by a sort of miracle, was the only arc that could yield the meter: its position at the midpoint between the pole and equator, its two extremities (Dunkirk and Barcelona) situated at sea level, its length (neither too long nor too short), and finally the fact that the arc had already been measured, all promised easy and precise results. "Therefore, there is nothing here that could give the slightest pretext for the reproach of having wanted to achieve some kind of preeminence," it concluded disingenuously. As we have already noted, the commissioners had decided to extend the meridian of France southward to Barcelona. But the meridian of France was itself merely an extension of the meridian of the Paris Observatory traced on the ground since the construction of the building on June 21, 1667, and which Jean Picard had measured as far as Amiens. So once again, Paris found itself at the center of operations.

The preliminary work took more than a year (from April 1791 to June 1792), because the instruments had to be made. The commission had first planned to entrust the triangulation to the 1787 team, Cassini, Legendre, and Méchain. When Legendre resigned, it decided to divide the work between Cassini in the north (from Dunkirk to Rodez) and Méchain in the south (from Rodez to Barcelona). But while Cassini, director of the Paris Observatory, fully intended to command the operation, he refused to go into the field himself, and so in May 1792 the Academy called on Delambre, Lalande's protégé and Laplace's calculator and the last person to be elected to the company. Operations began at the end of June before all the instruments had even been delivered. And these were extraordinary.

As we saw in chapter 9, the savants had stimulated the creation of a scientific instrument industry in Paris in the last years of the Ancien Régime. A corps of engineer–instrument makers had even been established under the protection of the Academy of Sciences, and high-precision instruments for astronomy and geodesy had been ordered from the most elite craftsmen. The determination of the meter presented an opportunity—a pretext, some would say—to support them once again, especially since they were hard hit by the collapse of the market.

To make the geodesic instruments, the commission naturally turned to Étienne Lenoir, who had already made the repeating circle used in 1787. He was asked for four circles of the same kind for measuring angles, as well as high-precision double rulers, in platinum and in copper, for measuring the bases. This was an enormous commission for a small workshop on the Île de la Cité. After a year, despite Lenoir's efforts, nothing had been delivered and the government was starting to get impatient. At the beginning of April, Minister Roland suggested going back, at least provisionally, to the Paris units for the new measurements. When Delambre and Méchain finally began operations in the early summer of 1792, they were still missing one of the repeating circles. The preparatory work mobilized other Parisian manufacturers as well: Fortin supplied the platinum cylinders, the balances, and a comparator of his own invention to determine the unit of weight (the future kilogram), equal to the weight of a cubic decimeter of distilled water in a vacuum at the temperature of melting ice; the optician Carrochez received the commission for the achromatic lenses for the repeating circles; the jeweler Janety worked the platinum needed for the instruments; and, finally, the clockmaker Louis Berthoud delivered several astronomical clocks.

With remarkable precision, Borda and Coulomb (the latter soon replaced by Cassini) determined (with the aid of a special ruler fabricated by Lenoir) the length of the pendulum beating one second at the Paris Observatory. Note that this was not at all what had been envisaged at the start. The objective, according to the law, was in fact to determine the period of a simple pendulum, one meter long, at sea level and at 45° latitude, which in the future would provide a simple if indirect means of finding the length of the meter again. The experiments could have been conducted in Bordeaux. In the end the commissioners thought it would be sufficient to measure the length of the pendulum beating one second in Paris and to calculate the results required from that measurement. A little later, at the start of 1793, Lavoisier and Haüy reported on the first results of their experiments on the determination of the unit of weight, work that unfortunately could not be completed. Finally, in May and June 1793, in the garden of Lavoisier's new residence on the boulevard de la Madeleine, Borda and Lavoisier, with the assistance of Lenoir, proceeded to calibrate the bimetallic rulers that were to serve as base measures for the triangulation. This was the last scientific work in which Lavoisier participated.[26]

By this time, almost a year had passed since the geodesic operations in the field had begun. We will not follow Delambre and Méchain in their tribulations along the meridian of France. The two astronomers went far away from the capital, Delambre to Dunkirk and Méchain to Barcelona. However, it was indeed in Paris and around Paris that the adventure began. As Méchain left for Barcelona on June 28, 1792, Delambre and his assistants had already started work around Paris, finding the triangulation points that Cassini had used in 1740: the bell tower of the Saint-Pierre Church in Montmartre, the Montlhéry tower, the Church of Brie Comte-Robert, the Malvoisine farm, the Montjay tower, the Collegial of Dammartin. He quickly realized that most of these points were unusable in their current state. In Montmartre, which had served as the triangulation point for Paris, the bell tower had been leveled, and Delambre had to settle for the dome of the Invalides. At Montjay, the hostility of the inhabitants obliged him to find another point.

Political tensions could not have been higher. The monarchy had fallen on August 10, and the Prussians were marching on Paris. On September 3, while massacres were being carried out in the prisons, Delambre and his companions were arrested by the National Guard of Lagny, twenty-one miles from Paris, while they were working. The astronomer protested and showed his passports: he was an academician. This annoyed a sans-culotte: "There is no longer a Cademy," he repeated, "no more Cademy; everyone is equal, you come with us." After a night under guard, Delambre and his assistants were freed with apologies. Two days later, in Saint-Denis, new troubles: a suspicious crowd gathered around the two carriages laden with instruments. They had to explain themselves. A few voices could already be heard demanding their heads. Delambre owed his life to the authorities who lodged him at the abbey. An emergency decree by the National Assembly on September 7 allowed him to resume his work in relative calm. Alas, in October, after weeks of observations, he had to face the fact that the dome of the Invalides, chosen in August as the signal for Paris, would also not work. Delambre resorted to the Pantheon, but everything had to be redone. After revisiting every triangulation point, he returned to the city in January 1793 and took the last measurements from the summit of the national monument. At the beginning of March, the work was done. Delambre finally left the capital for the Department of the Nord in May 1793.[27]

Naturalists and the Creation of the Museum

While Delambre was trying to triangulate the environs of Paris in the face of all these obstacles, Bernardin de Saint-Pierre was taking the measure of the former Royal Botanical Garden (now simply the Botanical Garden (Jardin des plantes)), where he had just been appointed superintendent to replace La Billarderie, who had emigrated. Known for his *Études de la Nature*, published in 1784, and his novel *Paul et Virginie* three years later, the writer prided himself on his knowledge of both the sciences and literature. A former student at the School of Bridges and Roads, an engineer and a traveler, he had defended a resolutely anti-Newtonian theory of the shape of the Earth. He maintained that the globe was elongated at the poles, and from this conclusion he deduced a similarly unorthodox theory of currents and tides, which resulted, he believed, from the diurnal and annual melting of the polar ice caps. Bernardin's ideas, like those of so many other system makers, were rewarded only with silence from the savants.

By the time the second edition of his *Études de la Nature* appeared in 1788, Bernardin had become famous and could complain about this treatment in his preface, to which Lalande responded somewhat haughtily in the *Journal des savants*: "He lives rather close to the College Royal, where there is a professor of astronomy, so he might have easily learned that the earth is flattened, that the tides are produced by the moon, and this is all so well demonstrated that there cannot be the least doubt for those who have just studied the subject a bit."[28] Furious and humiliated, Bernardin resumed the attack in a long footnote appended to the preface of his *Chaumière indienne* (the Indian cottage), a charming Rousseauist tale that appeared in 1791. In it he depicted Lalande as a hypocrite, one who went so far as to come to his home to convince him of his presumed error. Delambre then tried to bring Bernardin back to more reasonable ideas. He wrote to him anonymously (under the name of "Le Franc," or "Candid") to explain once more the astronomers' arguments. Although he succeeded in winning the writer's trust, he also failed, since the inflexible Bernardin remained until the end besotted with his own theories. We may assume that he was hardly a supporter of the meter.[29]

This is the man the king had appointed a few weeks before his fall to the post formerly occupied by Buffon. His nomination displeased the savants at the Botanical Garden. Since 1790 they had been demanding the suppression of the superintendent's post, but since somebody had to be in charge, they suggested Daubenton, who was their dean. However, there was nothing shocking for them about Bernardin's strange theories and his attacks on the Newtonian astronomers. The chasm between the naturalists on one side and the mathematicians and physicists on the other had considerably widened in the decade of the 1780s. The latter dominated the Academy at the expense of the former. Buffon himself had been marginalized by the company in his last years. Some like Haüy and even Daubenton had indeed been attracted by the methods of the geometers and new chemists, but the great majority of naturalists resisted the hegemonic claims of the partisans of the severe science.

Could not the precision of physical measurement, of which the determination of the meter offered the model, be matched by the descriptions in Jussieu's *Genera plantarum* and Olivier's *Entomologie*? And against Newtonian physics, limited to inert bodies, did they not offer a general physics bearing on the animate as well as the inanimate? Against the action of forces at a distance, they had fluids circulating among all bodies; and against a nature governed by universal and eternal laws, they had a historicized nature that could be seen in its specificities and its singularities. For the naturalists, because man himself was not an external mind that observed and calculated but a being in the world that lived and felt, the path was open not only to a natural history of humankind but also to an ethics based on human nature. Of course, the stakes were not only scientific but also philosophical and even political. In general, the naturalists felt much closer to the ideas of Rousseau, who had been a recreational botanist and a friend of the Royal Botanical Garden, than to those of the philosophes. Even the proud Buffon had liked and admired Jean-Jacques. "Observe nature and follow the route she lays out for you," the author of *Émile* had written. How could these modest savants, seeking the truth with a simple heart, not have felt some sympathy for Bernardin, the champion of nature and friend of Rousseau, who preferred the cottage of a pariah lost in the forest to the temple of Puri and his pundits? In any case, the new superintendent, who proved to be a good administrator, made an effort to win them over.

It was with their agreement that he undertook to round out the establishment with a menagerie. The opportunity to do so arose with the closing of the one at Versailles. Accompanied by Thouin and Desfontaines, Bernardin went to Versailles to examine what remained of the king's animals: a quagga, a hartebeest, a crested pigeon from Banda, a rhinoceros, and a lion who had a dog as a companion. The superintendent wrote an eloquent paper in which he stressed the need to complement the three kingdoms of dead nature with those of living nature, since, he noted, while the garden contained "fertile soil and plants that grew, there are no animals that feel, that love, that know."[30] The idea would succeed, but without the writer: unfortunately for him, the naturalists of the Botanical Garden also shared with Rousseau the conviction that men are naturally and politically equal; therefore they did not want a superintendent, whether Bernardin or anyone else.

Because natural history developed in relation to collections, the naturalists had been very dependent on rich patrons who financed their voyages and their publications and opened the doors of their cabinets to them. Under the Ancien Régime they maintained close relations, and sometimes friendships, with the upper echelons of the Parisian elite: aristocrats, financiers, and diplomats. However, the milieu of the Parisian naturalists was definitely on the side of the Revolution, soon expressing its desire to be liberated from the patronage of the powerful and to organize itself freely and in an egalitarian way. It was from within this milieu that the voluntary societies were launched, a movement that was at the very least an expression of their distrust of the academic model. As for the savants at the Botanical Garden, since 1790 they had been demanding the elimination of the post of superintendent and the collective management of an institution they wanted to be national. At the same time, the newly created Society of Natural History organized a ceremony for the unveiling of a bust of Linnaeus at the garden, a way of taking power symbolically (Fig. 10.1).

Even as they were thus freeing themselves from old forms of power and patronage, the naturalists were making connections with political figures, both the leaders of the revolutionary section in which the Botanical Garden was located and the Parisian municipality, as well as the National Assembly, and even engaging directly in the battle for public opinion. Millin, who led the Society of Natural History, had just created a new daily newspaper, the *Chronique de Paris*, to compete with the *Journal de Paris*, and Reynier was getting started in the book trade. Patriotic and Jacobin leanings quickly predominated among the naturalists, and their Rousseauism now had political overtones. At the Society of Natural History, moderates like Broussonet kept their distance, or else (like Ramond) were expelled and replaced by supporters of Roland and Brissot. At the Botanical Garden, everyone was a patriot. André Thouin and his circle of friends and relations did not hide their Jacobin sympathies. In such an environment, the political star of Bernardin de Saint-Pierre, who had been compromised by royal power, seemed quite dim. Despite his skill, he could not withstand the movement that was marginalizing him. After having overseen the transfer of the collections of the château de Chantilly to the cabinet of the Botanical Garden, representative Joseph Lakanal of the Committee of Public Instruction made direct contact with Daubenton without going through the superintendent and, taking up the plan of 1790, got the Convention to adopt (without any discussion) in June 1793 the transformation of the Botanical Garden into the National Museum of Natural History.

The organization of the new institution was republican. The post of superintendent was definitively eliminated, and twelve professorial chairs were instituted for those who had been employed by the Botanical Garden. All of them, whether they had been professors, demonstrators, guardians of collections, or, like Thouin, gardeners, were now on an equal footing. Above all, the reform allowed for the share allocated to the study of the animal kingdom to be expanded. Alongside Mertrud, who occupied the chair in comparative anatomy, two new arrivals entered the field: Geoffroy Saint-Hilaire for the natural history of quadrupeds, cetaceans, and fish, and Lamarck for the natural history of insects, worms, and microscopic animals. The former, still quite young, had published only a few articles on mineralogy, while the latter, a member of the Academy of Sciences, was known only as a botanist. The management of the institution and the nomination

FIGURE 10.1: Bust of Linnaeus at the Royal Botanical Garden. Engraving by François Allix after a drawing by Pierre-Henri de Valenciennes, frontispiece of *Actes de la Société d'histoire naturelle*, 1791.

of its members were entrusted to an assembly of the professors who were equal in rights and duties. It would elect the museum's director, appointed for a one-year term, renewable only once, as well as filling vacant chairs. In short, the new Museum of Natural History was entirely controlled by the savants.

Lavoisier, One Last Time

The contrast between the fate of the former Royal Botanical Garden and that re-served, two months later, for the beacon of the Parisian Republic of the Sciences has often been noted: on one side the survival of the institution, reformed ac-cording to the wishes of the naturalists, and on the other the brutal suppression of the Academy of Sciences. It is tempting, as several historians of science have done, to relate these contrasting destinies to the differences between the universe of the mathematical and physical sciences and the world of the naturalists: on one side austere methods reserved for savants alone and aimed at determining precisely the laws of nature through experimentation and reason in order to make them serve the common good, and on the other activities that put savants and amateurs on the same footing, in which curiosity followed as much from the pleasure of feeling as from the desire to understand, and whose ultimate motive was less the knowledge of things, perhaps, than personal advancement and collective utility. In other words, an élite science, hermetic and severe, versus a science for everyone, open and pleasant. This opposition itself refers back to the great split that spanned the Enlightenment between the abstract rationalism of the philosophes with its Newtonianism, and sentimental empiricism with its cult of Nature, a split dra-matically replayed during the Revolution between the Constituants of 1789, heirs of the philosophes and close to the savants of the Academy, and the Jacobins of 1793, admirers of Rousseau and friends of the naturalists of the Botanical Garden.

This schema, which the history of ideas has elaborated considerably, supplies an explanatory framework for the analysis of the relationship between the savant world, political figures, and public opinion during the Revolution. It makes it possible to account for the representations at play among actors involved in the politics of science, whether savants, elite craftsmen, journalists, or members of assemblies. But by displacing the analysis from the field of history to the theater of ideas, it also tends to confine the protagonists to roles written in advance, re-ducing the sphere of possibilities in which they actually acted. We can see this by following Lavoisier, the prototype of the severe savant, during the twelve months between the fall of the king and that of the Academy. For far from playing the part that his positions and dispositions should have imposed on him, he revealed a remarkable capacity to adapt: evolving, negotiating, compromising, and com-bining and shifting strategic lines in order to save the Academy of Sciences— a lost cause, since in the end he failed, but which at the very least shows that nothing was predetermined.

At the start of 1793, the Parisian savant world was holding its breath. The reor-ganization of public education was being debated. The academies were attacked on all sides, as was the University of Paris, whose faculties and colleges had been suspended. At the Collège de France, courses continued as before under the direction of Lalande, who had put astronomy in the lead and all the sciences ahead of the humanities, but for how long? At the Paris Observatory, things were not going well between Director Cassini and his assistants Nicolas-Antoine Nouet, Jean Perny, and Alexandre Ruelle, who were demanding to be treated as equals. At the Botanical Garden, as we have seen, the savants were also challenging the authority of the superintendent. As for the Academy of Sciences, it was pursuing its work as usual and waiting to see how its fate would be decided. Although the savants still attended meetings and the commission of weights and measures was working away, people were very worried.

In November 1792, the Convention decided to stop replacing members in the academies. It had also been necessary to purge émigrés—or those presumed to be—discreetly, which the authorities were demanding if pensions were to be paid: four honorary members, four academicians, and one associate (Cornette) were struck off the lists. However, the National Treasury still refused to pay out the sums due, this time on the pretext that the law forbade dual appointments. Lavoisier had to play the role of banker, paying advances to his colleagues out of his own funds. The Treasury's unwillingness to pay revealed how much credit the institution had lost: the Convention's unfavorable reception of the plan for a Society of Arts and Sciences, followed by its rejection, had weakened the Academy greatly. Even within the administration, the Academy of Sciences seemed to have lost all legitimacy.

With Condorcet no longer performing the functions of permanent secretary, the company lacked direction. Only one person could replace him: Lavoisier. After being named treasurer and secretary of the powerful commission of weights and measures in August 1791, he succeeded Tillet as the Academy's treasurer in December, which made him its sole permanent officer. It was in this capacity that he gradually took in hand the destiny of the Academy of Sciences. Lavoisier believed in the utility of academic institutions. Indeed, he considered them necessary. He had already emphasized this point in 1790, when they were first threatened,.[31] However, aware of the hostility to the academies, he was ready to adapt to the new situation, to limit their power and take into account the demands of their adversaries. His experience as an administrator and great patron of science prepared him, unlike Condorcet, for the necessary compromises. He understood very well the elite craftsmen who rejected academic tutelage, and he had contacts with all the actors engaged in the scientific enterprise, including those who did not share his ideas. Finally, his large fortune and many connections offered him the means to take action.

Before the Revolution, Lavoisier had gathered around him a team that met at the Arsenal. These men fought together to make the new chemistry triumphant, and the *Annales de Chimie* continued to bind them together. Now several had rallied to the Jacobins. At the beginning of 1793, Monge was in the government,

Meusnier was serving in the War Ministry, and Guyton-Morveau was a member of the Convention. Notably, Lavoisier remained on good terms with two activists who had been his protégés: Fourcroy and Hassenfratz. The chemist Fourcroy, a member of both the Academy of Sciences and the Society of Medicine, professor at the Botanical Garden, brilliant propagandist for Lavoisier's ideas, and a major authority on plant and animal chemistry, was passionately engaged in the Revolution. Since 1790, he had been in favor of a radical reform of academic institutions. At the Botanical Garden, he was one of the first to demand the elimination of the post of superintendent and equality among the professors. A physician by training, he was also very hostile to the Faculty of Medicine and wanted it to be abolished, pure and simple. At the Society of Medicine and at the Academy he had demanded the purge of émigrés after the fall of the king. Because he was a partisan of liberty in all its forms, he condemned the monopolies enjoyed by the academies and defended the voluntary societies. He became president of the Lycée des Arts in April 1793. Elected to the Convention that July to replace the assassinated Marat, he immediately began to take part in the work of the Committee of Public Instruction. Even though he then rallied to the radical solution of closing the academies, he never broke with Lavoisier.

As for Hassenfratz, he was just one of the elite craftsmen, a carpenter who had become a mining engineer. Before the Revolution, he had been a frequent visitor at the Arsenal, where he had become friends with the Lavoisiers. After following his patron into the Society of 1789, he (like Monge, Meusnier, and Fourcroy) had gone over to the Jacobins. He had a major role in the Commune on August 10. He then served Pache in the Ministry of War. In the spring, still close to Pache, he was among the radical sans-culottes who engaged in the struggle against the more moderate Girondins. Although by this time Hassenfratz was taken up with politics and no longer doing science, he had not lost contact with Lavoisier, who paid for his apartment in the rue des Bourdonnais, in the same building where Fourcroy lived, and he saw him regularly at meetings of the Advisory Board for Arts and Trades.[32]

As we have seen, this office had been created in September 1791 after the passage of the law on patents. Its mission was to encourage and reward inventors. Half the members were elite craftsmen selected by the voluntary societies, and the other half were savants from the Academy. Hassenfratz represented the Society of the *Annales de Chimie*. Lavoisier himself sat on it as an academician. His interest in the arts was long-standing and sincere. He had written countless academic reports on inventions and, using both his fortune and his social position, consistently supported elite craftsmen, chemists, and instrument makers. He now collaborated with them on the Advisory Board. There he met again the engineer Desaudray, now representing the Point Central, whom he had known before the

Revolution and who now professed anti-academism. Handling this influential man whom he hoped to win over carefully, in 1793 Lavoisier agreed to participate with Fourcroy and Hassenfratz in the venture of the Lycée des Arts. He had long been engaged in the voluntary society movement anyway. The Society of the *Annales de Chimie* was basically his creation. Lavoisier had also been a member of the Lycée since its founding in 1790 and had joined the Society of Natural History in 1791. A liberal in politics as in economics, Lavoisier viewed with sympathy the birth of all these associations. His defense of the Academy of Sciences did not include a defense of its monopoly.

At the Convention's Committee of Public Instruction, Lakanal looked after the affairs of the Academy of Sciences. After meeting with Lavoisier he became a zealous defender of the institution. The attack on the academies was launched at the Convention on July 1, 1793. At first it considered only the Academy of Painting and Sculpture, but the Committee of Public Instruction decided to expand the scope to include all the academic societies. Urgently, Lakanal was charged with making a report. Convinced that all the literary academies would be closed, he sought at least to preserve the Academy of Sciences. His strategy was to separate the fate of the savants from that of artists and men of letters. Lavoisier supplied him with the argument: he insisted on the important role the Academy of Sciences played in supplying expertise, on its ties with the Advisory Board for Arts and Trades, and on the importance of its work on the reform of weights and measures. He rejected the idea of bringing savants and artists together in the same society, as seemed to be the plan. In the end, Lakanal, no doubt in a minority on the committee, did not submit his report. It was Abbé Grégoire who presented the Convention with a draft decree on the academies on August 8, 1793. It called for their suppression, even as it asked the Academy of Sciences, which would remain open provisionally, to continue its work. The Convention refused to make this exception, and the Academy of Sciences was disbanded like the others.

Lavoisier, who had remained in constant contact with the members of the committee—at least those who were favorable to keeping the Academy open—refused to admit defeat. On August 10 he wrote a long official letter to the Committee of Public Instruction in the name of his colleagues and had it printed. In it he announced the formation of a club, a voluntary society composed of the former academicians. In a second letter written the same day in his capacity as the former treasurer of the Academy, he noted that it "exercised in some fashion the duties of a Ministry of Sciences of the Republic; this ministry should by no means be abandoned." He therefore suggested that if the Academy was not maintained, "the former academicians, meeting together as a Free and Fraternal Society for the Advancement of the Sciences, might at least continue to occupy the premises at the Louvre, to conduct their work, particularly concerning weights and measures, and to receive the sums allocated to the old Academy of Sciences." Despite the support of Lakanal, this final attempt failed. Meanwhile, Lavoisier, always pragmatic, had already taken up another solution, this one promoted by Romme and Fourcroy, who had both joined the Committee of Public Instruction in July. It consisted of transforming the Academy's commission of weights and measures into a temporary commission answering directly to the executive council, with the same functions and the same means. This way, at least the work on the meter would be saved.[33]

In his fight to save the Academy of Sciences, Lavoisier pulled out all the stops. Very active on the Advisory Board for Arts and Trades, he approached the elite craftsmen like Desaudray who were hostile to academic institutions and tried to win them over to his cause. To this end, he integrated a plan for reorganizing the academies into the Plan for Public Instruction that he had already written for the board. After having proposed two completely separate national societies, one for the sciences, the other for the arts, then four societies, as in Condorcet's plan, he now submitted a plan for a single national society of the sciences and the arts. In doing so he anticipated the wishes of the elite artisans, who always wanted to be placed on the same level as the savants. The maneuver succeeded, and the Advisory Board adopted the plan on September 11. But the success was illusory, since two weeks later the Board decided to defer presenting this plan. In its place, it sent to the Convention Desaudray's grand plan that the Point central had already submitted to the Legislative Assembly in March 1792. And so ended Lavoisier's vain attempts to save the Academy of Sciences.[34]

Notes

1. J. TREY and A. DE BAECQUE, *Le Serment du Jeu de Paume: Quand David réécrit l'histoire*, Versailles, Éditions Artys, 2008. On Bailly during the Revolution, see SMITH, 1954, 509–518.
2. GILLISPIE, 2004, 7–100.
3. On the Society of 1789, see BAKER, 1975, 272–285.
4. On the savants at the Legislative Assembly, see GILLISPIE, 2004, 101–110.
5. DURIS, 1993, 69–87.
6. DURIS, 1993, 92–99, and CHAPPEY, 2010.
7. CHAPPEY, 2007.
8. J.-L. CHAPPEY, "La Société nationale des Neuf Sœurs (1790–1793): Héritages et innovations d'une sociabilité littéraire et politique," in Ph. BOURDIN and J.-L. CHAPPEY, dirs., *Réseaux et sociabilité littéraire en Révolution*, Clermont-Ferrand, Presses Universitaires Blaise Pascal, 2007, 51–86.

9. G. KATES, *The Cercle Social, the Girondins and the French Revolution*, Princeton, NJ, Princeton University Press, 1985), and M. DORIGNY, "Le Cercle social ou les écrivains au cirque," in BONNET, 1988, 49–66.

10. C. DEMEULENAERE-DOUYÈRE, "L'itinéraire d'un aristocrate au service des arts utiles: Servières, alias Reth (1755–1804)," *Documents pour l'histoire des techniques* 15 (2008), 64–76, and "Inventeurs en Révolution: la Société des inventions et découvertes," *Documents pour l'histoire des techniques* 17 (2009), 19–56.

11. DE PLACE, 1988, and GILLISPIE, 2004, 195–200.

12. S. LACROIX, ed., *Actes de la Commune pendant la Révolution*, 2nd ser., 5, 1907, 235–241.

13. *Nouvelle constitution des arts et métiers . . . rédigée par la Société du Point central des arts et métiers*, March 1792; GILLISPIE, 2004, 200–205.

14. See GILLISPIE, 2004, 205–206 and 214–215.

15. A.-L. LAVOISIER, "Compte rendu à l'administration du Lycée de la rue de Valois de l'établissement formé au cirque du Palais-Égalité sous le nom de Lycée des Arts," in *Œuvres*, vol. 6, 559–569.

16. H. Guénot, "Une nouvelle sociabilité savante: le Lycée des arts," in BONNET, 1988, 67–78. On the beginnings of the establishment, see W. SMEATON, "The Early Years of the Lycée and the Lycée des arts: A Chapter in the Lives of A. L. Lavoisier and A. F. de Fourcroy," *Annals of Science* 11 (1955), 309–319.

17. TALLEYRAND, *Rapport sur l'Instruction publique fait au nom du Comité de constitution à l'Assemblée nationale les 10, 11 et 19 septembre 1791*, Paris, 1791.

18. Talleyrand mentions having consulted the savants in his *Mémoires*. The report prepared for him by Lavoisier was published by J. GUILLAUME, "Lavoisier anticlérical et révolutionnaire," *Études révolutionnaires* 1, 1908, 354–379.

19. CONDORCET, *Rapport et projet de décret sur l'organisation générale de l'instruction publique, présentés à l'Assemblée nationale, au nom du Comité d'instruction publique, les 20 et 21 April 1792*, Paris, Imprimerie nationale, 1792.

20. Ibid., 35.

21. J.-P. RABAUT SAINT-ÉTIENNE, *Projet d'Éducation nationale*, Paris, Imprimerie nationale, Year I (1793). Condorcet responded by criticizing enthusiasm in note E of the second edition of his report (printed in 1793 on the orders of the Convention).

22. BIGOURDAN, 1901, 1–12.

23. Y. NOËL and R. TATON, "La réforme des poids et mesures. 1. Origines et premières étapes (1789–1791)," in LAVOISIER, *Correspondance*, vol. 6 (1789–1791), 1997, 439–465, and GILLISPIE, 2004, 223–249.

24. "Rapport fait à l'Académie des sciences sur le choix d'une unité des mesures," *HMAS*, 1788, 9–10, and Laplace's lesson at the École normale in J. DHOMBRES, ed. *Leçons de l'École normale, Mathématiques*, Dunod, 1992, 121. On the hidden reason, see J.-B. DELAMBRE, *Grandeur et figure de la Terre*, Paris, 1912, 203.

25. J. P. MARAT, *Les Charlatans modernes*, Paris, 1791, 40.

26. On the work of various sub-commissions, see Bigourdan, 1901, 83–89 and 94–108.

27. Bigourdan, 1901, 114–130, and Alder, 2005, 29–55.

28. *Journal des savants*, August 1788, 540–542.

29. Delambre recounted his arguments with Bernardin in Delambre, 1827, 560–561.

30. J.-H. Bernardin de Saint-Pierre, *Mémoire sur la nécessité de joindre une ménagerie au Jardin des plantes de Paris*, Paris, 1792.

31. A.-L. Lavoisier, with Séguin, "Premier Mémoire sur la transpiration des animaux lu le 14 avril 1790," *HMAS*, 1789, 569–570 (*Œuvres de Lavoisier*, vol. 2, 704–714, 707).

32. Grison, 1996, 123–173.

33. B. Belhoste, "L'Académie des sciences et la Révolution: 2° Lavoisier et la fin de l'Académie des sciences," in Lavoisier, *Correspondance*, vol. 7.

34. Lavoisier, *Œuvres*, vol. 4, 649–668, and vol. 6, 516–558, and Th. Charmasson, "Lavoisier et le plan d'éducation du Bureau de consultation des arts et métiers," in C. Demeulenaere, ed., *Il y a 200 ans Lavoisier*, Paris, Tec & Doc, 201–216.

Epilogue

After the closing of the Academy of Sciences, the savants quickly dispersed. Some, like Lavoisier, Vicq d'Azyr, Berthollet, Fourcroy, Monge, Laplace, and Lamarck, joined the Philomathic Society, but it proved to be only a short-term substitute for the Academy. In truth, political and material conditions made any scientific activity in Paris beyond official commissions difficult. Many former members of the Academy of Sciences preferred to retreat to the provinces. Others survived in Paris but in obscurity and poverty. Some who had become suspect could not even do this. Bailly, guillotined in November 1793, was the first savant to be executed. Dietrich, Dionis du Séjour, and above all Lavoisier also mounted the scaffold, victims of the Terror not because they were savants but because they were accused of plotting and betraying the fatherland. When he was arrested, Condorcet chose suicide. The old system of patronage was dragged down within the fall of the Academy. Certain patrons had fled; others died, some, like Malesherbes and Bochard de Saron, victims of the Revolution and others, like Joubert and Gigot d'Orcy, in their beds. But many also served the Republic, pursuing the measurement of the meter or participating in the war effort, such as Vandermonde, Monge, Fourcroy, Berthollet, Haüy, Delambre, and many others. For several months, Paris resounded with Jacobin and sans-culotte rhetoric.

For the Republic of Letters, the fall of Robespierre provided some consolation. Even those who had gone along with the Montagnards now denounced his tyranny; Condorcet and Lavoisier were the new martyrs. The Convention elected after Thermidor wanted to reestablish the power of the savants, and thus before it disbanded it created major educational institutions in Paris as well as the National Institute. This "living encyclopedia," conceived on the model proposed by Talleyrand and Condorcet, was supposed to form the intellectual foundation of the new régime, which was republican, liberal, and moderate. The Academy of Sciences was reborn from the ashes as the First Class of the new Institute. With the terrible interlude over, scientific life seemed to pick up where it had left off in the 1780s, but in reality it emerged from the Revolution profoundly transformed. The institutional framework had been shattered and aristocratic patronage had vanished. The constitution of a national political sphere and the rise of the press, especially after 1815, completely changed its relationship with the general public. The Republic of Letters itself had ceased to exist as an organic whole despite the vain attempts of the Institute, because the chasm between the savants and the writers had grown considerably: as the figure of the inspired writer was born, the man of science became more and more a specialist and a professor.

Yet nothing would be more misleading than to superimpose on Paris Savant before 1840 the character of science at the end of the nineteenth century, ensconced in its laboratories and palatial universities. Our savants were important public figures. Their work was followed closely in the salons, clubs, and societies. Scientific discussions reverberated through the press. Something yet remained of that passion for the sciences that had aroused Parisians before the Revolution. Paris was more than ever dazzling in its scientific production, but from now on, the heirs of Lavoisier dominated the field. The first years of the nineteenth century saw the triumph of the severe science, of which the young graduates of the École Polytechnique were enthusiastic supporters. Mathematics and the physical sciences witnessed extraordinary progress. The makers of scientific instruments finally rivaled those of London. At the Museum of Natural History, Cuvier, an admirer of Laplace, rejected the tradition of Buffon and tried, in the style of the physicists, to impose laws on natural history. The reformation of medical education based on the clinical and experimental method had made its reputation. In all domains, the level of science was higher than ever. In short, for a few decades more, Paris could claim the title of capital of the sciences in Europe.

Such success was in some ways a realization of the promises of the 1780s. From this perspective, continuity triumphed over rupture. Moreover, the sites had changed very little. The old structures of the Latin Quarter still housed schools and bookshops; at the Palais-Royal and on the boulevards, spectacles still offered as they used to special effects borrowed from chemistry and physics; and industrial activity and urban renewal remained for Parisian savants the terrain of experiments and applications. However, the geography of Paris Savant had been drastically modified. In 1808 the Academy of Sciences left the Louvre for the Left Bank, taking up residence with the Institute in the former Collège des Quatre-Nations. Far from being an isolated move, this transfer, imposed by the creation of the art museum in the Louvre, was part of a wholesale reorganization of Parisian savant space.

In the new geography of science, the Latin Quarter reclaimed the central position it had formerly occupied. Without deserting the rest of the city entirely, the sciences tended to retreat to their own territory with its concentration of learned societies, private courses, laboratories, bookshops specializing in the sciences and medicine, makers of physics supplies, and especially the main public institutions dedicated to scientific study. The oldest of these, such as the Collège de France, the Sorbonne, and the School of Medicine, took on a new brilliance. The Paris Observatory and the National Museum of Natural History assembled exceptional research tools, the first in the physical sciences, the second in the natural sciences. There were in addition new institutions like the École polytechnique, the École normale, and the School of Mines, whole new faculties including a faculty of sciences, and the great *lycées*. In short, a new Paris Savant was being invented.

Cast of Characters

Abeille, Louis-Paul (1719–1807). Lawyer, inspector of manufactures and trade; secretary of the Bureau of Commerce in 1768, agronomist and economist, elected a member of the Institut in 1799.

Adanson, Michel (1727–1806). Naturalist, traveler, and collector. After spending five years in Senegal, he entered the Royal Academy of Sciences as deputy botanist in 1757. He was promoted to associate in 1773 and pensioner in 1785. He was elected a member in the first class of the Institut in 1795.

Alban, Léonard (1740–1803). Chemist and industrialist. In 1776 he created with an associate a manufactory of sulfuric acid that two years later moved to Javel, on the banks of the Seine outside Paris. He directed the Javel factory, which produced all sorts of salts and acids, until his death.

Argand, Ami (1750–1803). Genevan inventor and industrialist. After moving to Paris, where he studied chemistry, he went to Montpellier, where he improved distillation processes and invented the draft lamp that bears his name in England. He tried to establish his invention by associating with Quinquet and Lange in Paris and with Boulton and Parker in London. He directed a lamp factory at Versoix near Geneva, where he died utterly ruined.

Arlandes (Marquis d'), François-Laurent (1742–1809). Aeronaut. A childhood friend of Joseph Montgolfier, with Pilâtre de Rozier he performed the first untethered flight of the Montgolfier balloon on November 21, 1783, in Paris.

Bailly, Jean-Sylvain (1736–1793). Astronomer. Entered the Academy of Sciences in 1763, the Académie Française in 1783, and the Academy of Inscriptions and Belles-Lettres in 1785, and was a political leader of the early part of the French Revolution. He served as mayor of Paris from 1789 to 1791 and was ultimately guillotined for his role in the repression of the sans-culotte movement.

Baumé, Antoine (1727–1804). Apothecary and chemist. Admitted to the Academy of Sciences in 1773; supporter of phlogistics.

Bernardin de Saint-Pierre, Jacques-Henri (1737–1814). Writer. An engineer who traveled around the world, he was closely tied to Jean-Jacques Rousseau upon his return to France in 1771. In 1784 he published *Études de la Nature*, which made him famous, then in 1788 his masterpiece, the novel *Paul et Virginie*. He was named superintendent of the Royal Botanical Garden in 1792 and professor at the École Normale in Year III (1795), when he also became a member of the Institut.

Berthollet, Claude-Louis (1748–1822). Physician (in the service of the Duke d'Orléans) and chemist. He was elected to the Academy of Sciences in 1780. He rallied to Lavoisier's new chemistry and helped reform chemical terminology. A partisan of the Revolution, he advised the government under the Convention, taught at the École Normale in Year III, and participated in Napoleon's Egypt expedition.

Berthoud, Ferdinand (1727–1807). Clockmaker in Paris. After having made a series of marine watches that were partly inspired by Harrison's work in England, he was appointed the king's mechanical horologist in 1770. He contributed to the *Encyclopédie* and published several horology treatises. He was admitted to the Institut in 1795. In his workshop he trained his nephew Louis Berthoud (1754–1813), who continued his work on precision watchmaking.

Bertier de Sauvigny, Louis-Bénigne-François (1737–1789). Superintendent of the Généralité de Paris, succeeding his father in this post in 1776 after being his deputy for eight years. An enlightened reformer, he established a land registry and in 1783 revived the Royal Society of Agriculture. He was killed by rioters a few days after July 14, 1789.

Bochart de Saron, Jean-Baptiste-Gaspard (1730–1794). Magistrate and astronomer. President of the Parlement de Paris, then first president in 1789, he was an amateur astronomer, appointed honorary member of the Academy of Sciences in 1781. The protector of Messier and Laplace, he was above all a calculator. He was arrested as a former high court judge and guillotined during the Terror.

Borda (Chevalier de), Jean-Charles (1733–1799). Engineer and physicist. Trained as a military engineer, he entered the naval service in 1767. He participated in many naval cartography missions. He tested the naval chronometers and perfected the astronomical circle, a precision goniometer utilized in geodesic work. He became a member of the Academy of

Sciences in 1764. During the Revolution, he participated in the establishment of the metric system.

Bossut, Charles (1730–1814). Mathematician. He was a cleric (but only tonsured), hence his title of *abbé*. Protected by d'Alembert, he contributed to the *Encyclopédie*. He was professor at the engineering school in Mézières in 1752, then an engineering examiner in 1768, the year he became a member of the Academy of Sciences. His work bore principally on hydraulics.

Bralle, François-Jean (1750–1831). Engineer and entrepreneur; went into the service of the Count d'Artois. He was first an inspector of canals, then in 1786 he created a clockmaking factory Paris. He occupied various posts during the Revolution and empire, including as director of hydraulic works for the city of Paris.

Breguet, Abraham-Louis (1747–1823). Parisian clockmaker. From Neufchâtel in Switzerland, in 1775 he founded a clockmaking workshop in Paris, where he became a master artisan in 1784. He supplied both the court and the nobility and his inventions were responsible for many improvements in timekeeping. He became a member of the Bureau of Longitudes in 1806 and of the Academy of Sciences in 1816.

Breteuil (Le Tonnelier, Baron de), Louis-Auguste (1730–1807). Diplomat and minister under Louis XVI. He was ambassador to Vienna, then minister and secretary of state to the King's Household from 1783 to 1788. He supported savants and was received as an honorary member of the Academy of Sciences in 1785. He came back to government as principal minister on July 11, 1789. He emigrated after the fall of the Bastille.

Brisson, Mathurin-Jacques (1723–1806). Naturalist. A protégé of Réaumur, he taught physics at the Collège de Navarre. He was elected to the Academy of Sciences in 1759 and the following year published an *Ornithologie*, his principal scientific contribution.

Brissot, Jean-Pierre (1754–1793). Journalist and politician from Chartres. In 1774 he came to Paris, where he launched a career as a journalist and man of letters. He was a disciple of Rousseau and close to the patriot party and to Lafayette. He went to England several times and on his return to Paris in 1788 created the Society of Friends of the Blacks. As its president he then traveled to the United States to meet with abolitionists there. Back in France, he was elected deputy to the Legislative Assembly, then to the Convention, where he assumed the leadership of the party of the Girondins. Arrested in June 1793, he was guillotined a few months later.

Broussonet, Auguste (1761–1807). Naturalist from Montpellier. He first settled in London, where he was connected to Joseph Banks. Returning to Paris in 1782, he taught at the Veterinary School in Alfort and was elected

to the Academy of Sciences in 1785. The same year, he became permanent
secretary of the Society of Agriculture of Paris. He was one of the founders
in 1789 of the Linnaean Society of Paris. Under the Revolution, he was
a deputy to the Legislative Assembly. Pursued for his Girondin political
views, he managed to emigrate in 1793. He returned to France in 1797 and
ended his life in Montpellier.

Bucquet, Jean-Baptiste (1746–1780). Physician and chemist. He gave a famous public course on chemistry in Paris. He became a member of the
Royal Society of Medicine in 1777 and of the Academy of Sciences the
following year. He collaborated actively with Lavoisier.

Buffon (Count de), Georges-Louis Leclerc (1707–1788). Naturalist
and writer. He was elected to the Academy of Sciences in 1734 as a mechanic for his mathematical work. In 1739, he was named superintendent
of the Royal Botanical Garden, which he directed for almost fifty years,
transforming and enlarging it. He was elected a member of the Académie
Française in 1753. In 1749 he published in the first volumes of his *Histoire
Naturelle*, which would include thirty-six volumes by 1789. The work made
him famous.

Cadet de Vaux, Antoine-Alexis (1743–1828). Apothecary, chemist, and journalist. He was one of the founders and directors of the *Journal de Paris*,
where he conducted a campaign for public hygiene. He was appointed
inspector of hygiene by Lieutenant of Police Lenoir in 1779 and the following year founded a baking school with Parmentier. He was the younger
brother of **Charles-Louis Cadet de Gassicourt** (1731–1789), an apothecary and chemist.

Cassini, Jean Dominique, called **Cassini IV** (1748–1845): Astronomer. He
belonged to the Cassini dynasty that had directed the Paris Observatory
since 1669. He was elected to the Academy of Sciences in 1770 and assisted
his father, **César François Cassini de Thury** (1714–1784), in directing
the Observatory before replacing him in 1784. He undertook to reform
that establishment with the support of Minister Breteuil. Hostile to the
Revolution, he resigned in 1793 and was arrested the following year.

Charles, Jacques-Alexandre-César (1746–1823). Physicist. He opened a
physics cabinet in Paris in 1780. After launching the first hydrogen balloon
on August 27, 1783, he became famous by performing the second human
flight in the company of Noël Marie Robert, on December 1, 1783. He was
named a member of the Institut in 1795.

Condorcet (marquis de), Marie-Jean-Antoine Nicolas de Caritat (1743–
1794). Mathematician, politician, and philosopher. A protégé of d'Alembert,
he was elected to the Academy of Sciences in 1769. His mathematical work

mainly concerned the analysis and calculation of probabilities. He was one of the advisors to Minister Turgot, who made him currency inspector in 1775. He became permanent secretary of the Academy of Sciences in 1776 and entered the Académie Française in 1782. During the Revolution he was a deputy in the Legislative Assembly, where he prepared a major plan for public instruction, and in the Convention. Close to the Girondins, he was arrested and died before being tried, probably as a result of suicide.

Cornette, Claude-Melchior (1744–1794). Physician and chemist. Protected by the first physician to the king, François de Lassone, he was elected to the Academy of Sciences in 1778 and named inspector of dyeing at the Gobelins the same year. Physician to the king's aunts, he emigrated with them in 1791.

Coulomb, Charles-Augustin (1736–1806). Engineer and physicist. He had a career in military engineering as an army officer, while doing research in mechanics and on electricity. He was elected to the Academy of Sciences in 1781. He participated in the work on the meter during the Revolution, but retired to the provinces after the Academy was suppressed. He was named a member of the Institut in 1795.

D'Alembert, Jean Le Rond (1717–1783). Mathematician and philosopher. His mathematical work bears principally on analysis. He was elected a member of the Academy of Sciences in 1741. He was one of the two general editors of the *Encyclopédie*, with Denis Diderot, from 1747 to 1757. Elected to the Académie Française in 1754, he became its permanent secretary in 1772.

Darcet, Jean (1724–1801). Physician and chemist. Starting out as Montesquieu's secretary, in 1774 he was appointed professor of chemistry at the Collège de France. He was elected a member of the Academy of Sciences in 1784 and became director of the Sèvres manufactory and inspector of dyeing at the Gobelins.

Daubenton, Louis-Jean-Marie (1716–1799). Physician and naturalist. Summoned by Buffon to the Royal Botanical Garden to be keeper and demonstrator at the Cabinet of Natural History in 1742, he assisted him in preparing his *Histoire naturelle*. In 1744 he was elected to the Academy of Sciences. He was named professor of natural history at the Collège de France in 1778 and professor of rural economics at the Veterinary School in Alfort in 1783. He taught natural history at the École Normale in Year III (1795).

David, Jacques-Louis (1748–1825). Painter. Awarded the Prix de Rome in 1774 and elected a member of the Royal Academy of Painting and Sculpture in 1783, he became leader of the new neoclassical painting style.

Very involved in the Revolution, he was elected deputy to the Convention, played a major role in the suppression of the academies in August 1793 and rallied to the Montagnards. Later he served Napoléon. He ended his life in exile as a regicide.

Desaudray (Gaullard-), Charles-Emmanuel (1740–1832). Engineer and industrialist. He imported hardware techniques for the small metal goods industry from Birmingham to France in the 1770s and created several firms in Normandy and in Paris. During the Revolution, he created a republican society of artisans, the Point Central des Arts et Métiers, then the Lycée des Arts.

Desault, Pierre-Joseph (1738–1795). Surgeon. Member of the College of Surgery, surgeon in chief at the Hôpital de la Charité, then at the Hôtel-Dieu.

Desfontaines, René (1750–1833). Botanist. A student of Bernard de Jussieu, he was elected to the Academy of Sciences in 1783. He traveled in North Africa in 1785 and 1786 and succeeded Le Monnier in the chair of botany at the Royal Botanical Garden in 1788.

Desmarest, Nicolas (1725–1815). Naturalist, known in particular for his work on the Auvergne volcanos. A protégé of the Duc de La Rochefoucauld d'Enville, he was elected to the Academy of Sciences in 1771. He was named inspector of manufactures in 1788.

Diderot, Denis (1713–1784). Writer and philosophe from Langres. The author of a wide variety of writings, including plays, philosophical dialogues, novels, and art criticism, he is best known for having directed the *Encyclopédie* from 1748 to 1765 and contributed to the *Correspondance littéraire, philosophique et critique* from 1769 to 1773. In his philosophical writings, he professed materialism.

Duhamel du Monceau, Henri-Louis (1700–1782). Chemist, botanist, and agronomist. He was elected to the Academy of Sciences in 1728, and in 1739 appointed inspector general of the navy. In 1741 he established a naval college in Paris to train officers. In 1759 he became editor of the *Description des arts et métiers*, published by the Academy of Sciences.

Faujas de Saint-Fond, Barthélémy (1741–1819). Geologist. A protégé of Buffon, he became an assistant naturalist at the Royal Botanical Garden in 1776. He conducted research on volcanos and in 1788 was named commissioner for coal mines at the Bureau of Commerce. He was named professor of geology at the National Museum of Natural History in 1793.

Fourcroy, Antoine-François (1755–1809). Physician and chemist. He was elected a member de la Royal Society of Physicians in 1778, named professor of chemistry at the Royal Botanical Garden in 1784, and received into the Academy of Sciences in 1787. Close to Lavoisier, he was an adept of the new chemistry. A partisan of the Revolution, he was a Jacobin and deputy in the Convention. He was director of public instruction under the Consulate.

Franklin, Benjamin (1706–1790). Writer, physicist, and politician. Starting out as a printer and journalist in Boston, then Philadelphia, he was named postmaster in 1737. In 1743 he founded the American Philosophical Society, and by 1748, he was devoting himself to scientific research, and politics. In 1750 he invented the lightning rod. From 1764 to 1775, he was the representative of the American colonies in London, and in 1776 he settled in Paris as official ambassador of the United States. In 1783 he signed the Paris Treaty with England that ended the American Revolution and in 1785 returned to America.

Geoffrin, Marie-Thérèse (1699–1777). Parisian *salonnière*. Married to the director of the mirror manufactory of Saint-Gobain, from 1749 until 1776 she kept a salon on rue Saint-Honoré, where she received artists, men of letters, savants, and aristocrats.

Gigot d'Orcy, Jean-Baptiste (1737–1793). Administrator and entomologist. He was a tax collector for the Champagne region. He possessed a significant cabinet of natural history in Paris, renowned especially for its butterflies. He supported the naturalist Guillaume-Antoine Olivier and had a description of his collection published by Engremelle.

Guettard, Jean-Étienne (1715–1786). Physician and naturalist. A protégé of Réaumur, he was admitted to the Academy of Sciences in 1743. In 1747 he became the personal physician to the Duke d'Orléans and keeper of his natural history cabinet.

Hallé, Jean-Noël (1754–1822). Physician. He was a doctor on the Faculty of Medicine in Paris and member of the Royal Society of Medicine in 1778. He was appointed a professor of hygiene at the School of Health in Paris in 1794.

Haüy, René-Just (1743–1822). Mineralogist. He was an ordained priest and taught Latin in a college in the Latin Quarter. He devoted himself to mineralogy and analyzed the structure of crystals. He was admitted to the Academy of Sciences in 1783. He taught physics at the École Normale in Year III (1795) and after 1800 mineralogy at the National Museum of Natural History.

Helvétius, Claude-Adrien (1715–1771): Financier and philosophe. A farmer-general (tax farmer), he held a salon in rue Saint-Anne with his wife, Anne-Catherine, and in 1766 created with the astronomer Lalande the Lodge of the Nine Sisters. His book *De l'Esprit*, published in 1758, caused a scandal for its radical materialism.

Holbach (Thiry, Baron d'), Paul-Henri (1723–1789): Philosophe. Of German origin, he was a lawyer in Paris. He hosted men of letters at his home in rue Royale Saint-Roch and actively supported the Encyclopedists. He wrote many scientific articles, in particular on mineralogy, for the *Encyclopédie*. His *Système de la Nature*, published in 1770, was openly materialistic and atheistic.

Jussieu, Antoine-Laurent (1748–1836). Physician and botanist. A nephew of the naturalists Bernard and Joseph de Jussieu, in 1770 he studied medicine and became a botany demonstrator at the Royal Botanical Garden. He was admitted to the Academy of Sciences in 1773 and became a member of the Royal Society of Medicine in 1776. He proposed a natural system of plant classification in his *Genera plantorum*, published in 1789.

La Rochefoucauld d'Enville (Duke de), Louis-Alexandre (1743–1792). Member of the liberal nobility; a science amateur and a protector of savants. He was a partisan of the American cause and translated the Declaration of Independence into French in 1778. He was admitted to the Academy of Sciences as an honorary member in 1781. He was a deputy in the Constituent Assembly in 1789, but he was killed a few days after the fall of the monarchy.

Lacépède (Count de), Bernard-Germain-Étienne (1756–1825). Naturalist. Originally from Agen, he came to Paris in 1776 and thanks to Buffon's protection, became the keeper of the natural history collections at the Royal Botanical Garden in 1785. A Freemason, he belonged to the Lodge of the Nine Sisters. In 1794 he was named professor of ichthyology at the National Museum of Natural History. Meanwhile, he had a political career after 1789: deputy in the Legislative Assembly in 1791, senator in 1799, and grand chancellor of the Légion d'Honneur in 1803.

Lagrange, Joseph-Louis (1736–1813). Mathematician. Originally from Turin, he was summoned to the court of the Prussian king Frederick II in 1766 on the advice of d'Alembert and became a member of the Academy of Berlin. He settled in Paris in 1788, where he became a member of the Academy of Sciences. During the Revolution, he was professor at the École Normale of Year III and at the École Polytechnique.

Lalande (Lefrançois de), Joseph-Jérôme (1732–1807). Astronomer. He entered the Academy of Sciences in 1753 and became professor of astronomy at the Collège de France in 1762. An eminent Freemason, along with Helvétius, he was the co-founder of the Lodge of the Nine Sisters. He was named Director of the Paris Observatory in 1795.

Lange, Ambroise-Bonaventure. Parisian shopkeeper and distiller. He was one of the inventors of the air draft lamp, along with Argand and Quinquet.

Laplace, Pierre-Simon (1749–1827). Mathematician, astronomer, and physicist. A protégé of d'Alembert, he was admitted to the Academy of Sciences in 1773 and became an examiner of artillery in 1783. During the Revolution, he was a professor at the École Normale of Year III. After Bonaparte's coup d'état, he was for several months interior minister, then chancellor of the Senate.

Lassone (de), François (1717–1788). Physician. He entered the Academy of Sciences in 1743. He was appointed the queen's physician in 1751 and became first physician to King Louis XVI in 1774. In 1776 he founded the Royal Society of Medicine, over which he presided.

Lavoisier (de), Antoine-Laurent (1743–1794). Chemist and tax farmer. He was admitted to the Academy of Sciences in 1768 and became a tax farmer in 1770. He was appointed by Turgot as to head the gunpowder administration in 1775 and moved to the Arsenal. He laid the bases for his new pneumatic chemistry (theory of combustion) over the following years and become leader of a school. He was also a member of the Royal Society of Medicine. During the Revolution he participated in the creation of the metric system and defended the Academy of Sciences. During the Terror he was arrested along with other tax farmers and guillotined.

Le Breton, André-François (1708–1779). Parisian bookseller and printer. He was the principal publisher of the *Encyclopédie*, in association with the booksellers **Antoine-Claude Briasson, Michel-Antoine David, and Laurent Durand.**

Le Monnier, Louis-Guillaume (1717–1799). Physician. He was admitted to the Academy of Sciences in 1735 and named professor of botany at the Royal Botanical Garden in 1759. He became first physician at the court in 1770 and first physician to the king in 1788, replacing Lassone.

Le Roy, Jean-Baptiste (1720–1800). Physicist. Known for his work on electricity, he was admitted to the Academy of Sciences in 1751. He contributed to the *Encyclopédie* and maintained close ties with Franklin.

Macquer, Pierre-Joseph (1718–1784). Physician and chemist. He entered the Academy of Sciences in 1746 and in 1757 was named chemist at the Sèvres manufactory, where he rediscovered the processes for making porcelain; in 1766 he became commissioner for dyeing at the Bureau of Commerce. Until his death he remained a true believer in the theory of phlogistics.

Malesherbes (de Lamoignon de), Chrétien-Guillaume (1721–1794). Magistrate and politician. As the director of the book trade from 1750 to 1763, he protected the *Encyclopédie*. An amateur savant and botanist, he became an honorary member of the Academy of Sciences in 1750 and a member of the Académie Française in 1775. He served as Louis XVI's defense lawyer at his trial in 1793, for which he was condemned and guillotined during the Terror.

Marat, Jean-Paul (1743–1793). Physician, journalist, and politician. After a long stay in Great Britain, he established himself as a physician in Paris in 1775. His research on light and electricity was rejected by the Academy of Sciences. In 1789, he launched into journalism, supporting the Jacobins and sans-culottes. He was assassinated in August 1793.

Mercier, Louis-Sébastien (1740–1814). Man of letters. A prolific writer, between 1781 and 1788 he published the *Tableau de Paris*, in which he described in detail the life of the capital. During the Revolution, he was a deputy in the Convention.

Mesmer, Franz-Anton (1734–1815). Physician. He first practiced in Vienna, where he discovered animal magnetism in 1774. He went to Paris in 1778 and became celebrated. His success was considerable, but the scientific and medical authorities who advised the government condemned his doctrine in 1784. He left France the following year and died almost forgotten.

Meusnier, Jean-Baptiste-Marie (1754–1793). Engineer and mathematician. An officer in the engineering corps known for his work on geometry, he was admitted to the Academy of Sciences in 1784. He collaborated with Lavoisier on ballooning and the great water experiment. He was killed in combat during the Revolution.

Monge, Gaspard (1746–1818). Mathematician and politician. He began to teach geometry in 1766 at the Engineering School at Mézières and entered the Academy of Sciences in 1780. In 1783, he settled definitively in Paris and was named a naval examiner. During the Revolution, he was minister of the Navy in 1792, then founded the École Polytechnique in 1794 and taught at the École Normale of Year III (1795). Close to Bonaparte, he participated in the Egypt expedition in 1797. He was elected a senator in 1799.

Montgolfier, Étienne (1745–1799). Industrialist and inventor. The Montgolfier brothers were papermakers in the town of Annonay. With his elder brother Joseph (1740–1810), he invented the hot-air balloon, or "Montgolfière," in 1782. The first flights took place in 1783, first in Annonay and then in Paris and Versailles.

Montucla, Jean-Étienne (1725–1799). Administrator and mathematician. He was a high official in the King's Building Administration and royal censor for mathematics. He published the first *Histoire des mathématiques* in 1758.

Necker, Jacques (1732–1804). Financier and politician. He worked in banking, first in Geneva, then after 1748 in Paris. Although a foreigner and a Protestant, he was appointed director general of finance (i.e., minister) in 1775 and undertook important reforms. He was dismissed in 1781. He returned to the ministry in 1788, but was dismissed a second time on July 11, 1789. Recalled again after the fall of the Bastille three days later, he remained in office another year before leaving France definitively for Switzerland with his wife, the salonnière Suzanne Necker.

Nollet (Abbé), Jean-Antoine (1700–1770). Physicist. He opened a course in experimental physics in Paris in 1735 and was admitted to the Academy of Sciences in 1735. A chair in experimental physics was established for him at the Collège de Navarre in 1753. In 1758, he was named master of physics at court.

Panckoucke, Charles-Joseph (1736–1798). Parisian printer-bookseller. Originally from Lille, he settled in Paris in 1768 and became one of the principal Parisian publishers, specializing in the sciences. He published a supplement to the *Encyclopédie* and then (starting in 1782) the *Encyclopédie méthodique.*

Parmentier, Antoine-Augustin (1737–1813). Pharmacist and agronomist. He made a career as a military pharmacist and became known for his work on the nutritional value of the potato.

Périer, Jacques-Constantin (1742–1818). Engineer and inventor. In 1781, under the patronage of the Duke d'Orléans, he and his brother Auguste-Charles Périer, set up the first of Watt's steam engines in Paris, designed to pump water from the Seine. They created a joint-stock company for water distribution, the Compagnie des Eaux, as well as a foundry in Chaillot to manufacture machinery.

Pilâtre de Rozier, Jean-François (1754–1785). Pharmacist and physicist. He became known for his physics experiments and in 1781 created the *Musée de Monsieur,* a club dedicated to science and technology. He became (with the Marquis d'Arlandes) the first aeronaut on November 21, 1783. He died two years later while trying to cross the Channel in a balloon.

Romé de l'Isle, Jean-Baptiste (1736–1790). Mineralogist. After a career as an army officer in India, he settled in Paris, where in 1772 he started giving public lectures on mineralogy. With Abbé Haüy, he was one of the founders of crystallography.

Rousseau, Jean-Jacques (1712–1778). Writer and philosophe. Born in Geneva, he came to Paris in 1740, gradually became known in the Republic of Letters, and became famous in the 1750s after publishing his first essays. He was forced to leave France in 1762 after his book *Émile* was condemned by the church. He returned to Paris in 1770, where he lived a retired life, devoting himself in part to botany, until his death.

Sage, Balthazar-Georges (1740–1824). Apothecary and chemist. He was admitted to the Academy of Sciences in 1770 thanks to the patronage of Louis XV. Starting in 1778, he taught mineralogy at the Mint, where he founded the School of Mines in 1783. A believer in phlogistics, he was an adversary of Lavoisier.

Tenon, Jacques-René (1724–1816). Surgeon. A professor at the College of Surgery, he was admitted to the Academy of Sciences in 1759. In 1785 he proposed a reform of Parisian hospitals. He was elected a deputy to the Legislative Assembly in 1791.

Thouin, André (1747–1824). Botanist and agronomist. In 1764 he succeeded his father as head gardener at the Royal Botanical Garden. He was admitted to the Academy of Sciences in 1784; in 1793, he became professor of botany at the National Museum of Natural history.

Trudaine de Montigny, Philibert (1733–1777). Administrator and chemist. In 1757 he succeeded his father as director of the corps of bridges and roads, then became head of the Bureau of Commerce. He entered the Academy of Sciences in 1764 as an honorary member and collaborated with Lavoisier.

Turgot, Anne-Robert-Jacques (1727–1781). Administrator, economist, and politician. He was a royal intendent, in charge of the Limousin region, then minister of finance at the start of Louis XVI's reign in 1774. He undertook economic reforms and a reorganization of the trades but was forced to resign after two years.

Vandermonde, Alexandre-Théophile (1735–1796). Mathematician. He was admitted to the Academy of Sciences in 1771. In 1783 he was named a commissioner at the Bureau of Commerce and administrator of the Machine Depository at the Hôtel de Mortagne. In 1795 he taught political economy at the École Normale of Year III.

Vicq d'Azyr, Félix (1748–1794). Physician. He gave a public course in anatomy, entered the Academy of Sciences in 1774, and became permanent secretary of the Royal Society of Medicine, when it was created in 1776. He was professor of comparative anatomy at the Veterinary School of Alfort from 1782 to 1788. In 1790 he proposed a complete reform of medical education in France.

Voltaire (François-Marie Arouet de) (1694–1778). Writer and philosophe. The premier man of letters of his age, he was one of the first to introduce Newton into France and the acknowledged leader of the party of philosophes. After 1750 he lived far from Paris, in Prussia and then Geneva, and finally at Ferney, just across the border from Geneva in France. He returned to Paris in triumph in 1778 but died soon after his arrival.

Paris Savant: locations

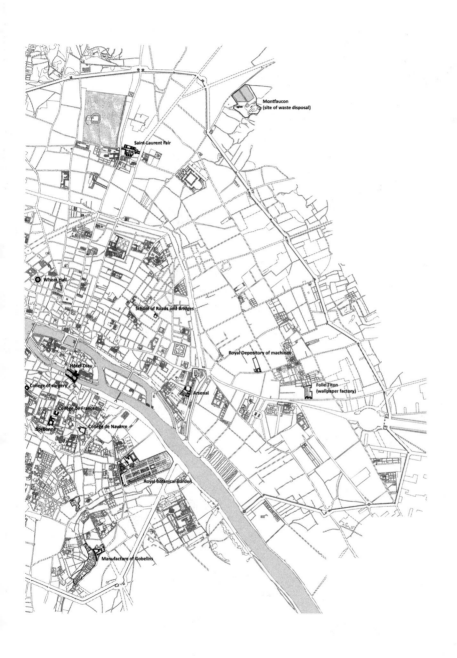

Montfaucon
(site of waste disposal)

Saint-Laurent Fair

Wheat Hall

School of Roads and Bridges

Royal Depository of machines

Hôtel-Dieu

College of surgery

Collège de France

Arsenal

Folie Titon
(wallpaper factory)

Sorbonne

Collège de Navarre

Royal Botanical Garden

Manufacture of Gobelins

Bibliography

ARCHIVAL SOURCES

Almanach Royal (annual, 1770–1792), Paris.

AN = Archives nationales de France.

ARTFL Encyclopédie, https://encyclopedie.uchicago.edu/.

HMAS = *Histoire et Mémoires de l'Académie royale des sciences* (annual, 1666–1792), Paris.

HMSM = *Histoire et Mémoires de la Société royale de médecine* (annual 1776–1789), Paris.

Index biographique de l'Académie des sciences, 1666–1978, Paris: Gauthier-Villars, 1979.

Journal de Paris (daily starting January 1, 1777).

P.-V.= Minutes of the Meetings, *Séances de l'Académie royale des sciences* (annual manuscript volume 1699–1792, online on Gallica).

Turgot map of Paris, Wikipedia, https://en.wikipedia.org/wiki/Turgot_map_of_Paris.

BOOKS AND ARTICLES

Aguillon, L. 1889. *L'École des mines de Paris, notice historique*. Paris: Vve C. Dunod.

Alder, K. 2002. *The Measure of All Things: The Seven-Year Odyssey and Hidden Error that Transformed the World*. New York: Free Press.

Amiable, L. 1989. *Une loge maçonnique d'avant 1789: La loge des Neuf soeurs*. Paris, 1897, 2nd ed. Ch. Porset. Paris: Edimaf, 1989.

Bachaumont, L. Petit de. 1777–1789. *Mémoires secrets pour servir à l'histoire de la République des Lettres en France, depuis 1767 jusqu'à nos jours*. London: John Adamson.

Badinter, É. 1999–2007. *Les Passions intellectuelles*. 3 vols. Paris: Fayard.

Badinter, É., and R. Badinter. 1990. *Condorcet: Un intellectuel en politique*. Paris: Le Livre de Poche.

Baker, K. M. 1975. *From Natural Philosophy to Social Mathematics*. Chicago: University of Chicago Press.

Balayé, S. 1988. *La Bibliothèque nationale des origines à 1800*. Geneva: Droz.

Belhoste, B. 2012. "A Parisian Craftsman among the Savants: The Joiner André-Jacob Roubo (1739–1791) and his Works." *Annals of Science* 69, no. 3, 395–411.

Belhoste, B. 2015. "Dyeing at the Gobelins in the Eighteenth Century—The Challenge of Quemizet." In *Colour Histories: Science, Art, and Technology in the 17th and 18th centuries*, ed. Magdalena Bushart and Friedrich Steinle, 67–91. Berlin: De Gruyter.

Belin, J.-P. 1913. *Le Mouvement philosophique de 1748 à 1789: Etude sur la diffusion des idées des philosophes à Paris d'après les documents concernant l'histoire de la librairie.* Paris: Belin.

Bertucci, P. 2017. *Artisanal Enlightenment: Science and the Mechanical Arts in Old Regime France.* New Haven, CT: Yale University Press.

Bigourdan, G. 1901. *Le Système des poids et mesures.* Paris: Gauthier-Villars.

Bigourdan, G. 1930. *Histoire de l'astronomie d'observation et des observatoires en France.* 2 vols. Paris: Gauthier-Villars.

Birembaut, A. 1957. "L'Académie royale des Sciences en 1780 vue par l'astronome suédois Lexell (1740–1784)." *Revue d'histoire des sciences* 10, 148–166.

Bonnet, J.-C., ed. 1988. *La Carmagnole des Muses: L'homme de lettres et l'artiste sous la Révolution.* Paris: Armand Colin.

Bouchary, J. 1942. *Les Compagnies financières à Paris à la fin du XVIIIᵉ siècle.* 3 vols. Paris: Marcel Rivière.

Bouchary, J. 1946. *L'eau à Paris à la fin du XVIIIᵉ siècle: La Compagnie des eaux de Paris et l'entreprise de l'Yvette.* Paris: Marcel Rivière.

Bourdier, F. 1962. "Le cabinet d'histoire naturelle au muséum, 1635–1935." *Sciences* 18, 35–50.

Bret, P. 2002. *L'État, l'armée, la science: L'invention de la recherche publique en France (1763–1830).* Rennes: PUR.

Brian, É. 1994. *La Mesure de l'État: Administrateurs et géomètres au XVIIIᵉ siècle.* Paris: Albin Michel.

Brian, É., and C. Demeulenaere-Douyère, eds. 1996. *Histoire et mémoire de l'Académie des sciences: Guide de recherches.* Paris: Lavoisier Tec & Doc.

Britsch, A. 1926. *La Maison d'Orléans à la fin de l'Ancien Régime: La jeunesse de Philippe-Égalité (1747–1785).* Paris: Payot.

Brockliss, L. W. B. 1987. *French Higher Education in the Seventeenth and Eighteenth Centuries.* Oxford: Clarendon.

Brockliss, L. W. B., and C. Jones. 1997. *The Medical World of Early Modern France.* Oxford: Oxford University Press.

Campardon, É. 1877. *Les Spectacles de la foire.* 2 vols. Paris: Berger-Levrault.

Chabaud, G. 1996. "La physique amusante et les jeux expérimentaux en France au XVIIIᵉ siècle." *Ludica, annali di storia e civiltà del gioco* 2, 61–73.

Chagniot, J. 1988. *Paris au XVIIIᵉ siècle.* Nouvelle Histoire de Paris. Paris: Hachette.

Chappey, J.-L. 2004. "Enjeux sociaux et politiques de la vulgarisation scientifique en révolution (1780–1810)." *Annales historiques de la Révolution française* 338, 11–51.

Chappey, J.-L. 2007. "Sociabilités intellectuelles et librairie révolutionnaire." *Revue de Synthèse*, 128, 71–96.

Chappey, J.-L. 2010. *Un lieu du savoir naturaliste sous la Révolution française: Les procès-verbaux de la Société d'histoire naturelle de Paris (1790–1798)*. Paris: CTHS.

Chartier, R. 1990. *Les Origines culturelles de la Révolution française*. Paris: Le Seuil.

Chevreul, M. 1854, "Les tapisseries et les tapis des Manufactures nationales." In *Travaux de la commission française sur l'industrie des nations*, vol. 5, 1–98. Paris.

Compère, M.-M. 2002. *Les Collèges français*. Vol. 3. Paris: INRP.

Coquard, O. 1993. *Marat*. Paris: Fayard.

Coquery, N. 2000. *L'Espace du pouvoir: De la demeure privée à l'édifice public; Paris 1700–1790*. Paris: Seli Arslan.

Corbin, A. 1986. *The Foul and the Fragrant: Odor and the French Social Imagination*. Cambridge, MA: Harvard University Press.

Corlieu, A. 1877. *L'ancienne Faculté de médecine de Paris*. Paris: A. Delahaye.

Crow, T. E. 1985. *Painters and Public Life in Eighteenth-Century Paris*. New Haven, CT: Yale University Press.

Cuvier, G. 1819–1827. *Recueil des éloges historiques*. 3 vols. Strasbourg: Levrault.

Dalbyan, D. 1983. *Le Comte de Cagliostro*. Paris: Robert Laffont.

Darnis, J.-M. 1988. *La Monnaie de Paris: Sa création et son histoire du Consulat et de l'Empire à la Restauration (1795–1826)*. Levallois: Centre d'Études Napoléoniennes.

Darnton, R. 1968. *Mesmerism and the End of the Enlightenment in France*. Cambridge, MA: Harvard University Press.

Darnton, R. 1979. *The Business of Enlightenment: A Publishing History of the Encyclopédie: 1775–1800*. Cambridge, MA: Harvard University Press.

Darnton, R. 1982. *The Literary Underground of the Old Regime*. Cambridge, MA: Harvard University Press.

Dartein, F. de. 1906. "Notice sur le régime de l'ancienne École des ponts et chaussées et sur sa transformation à partir de la Révolution." *Annales des ponts et chaussées*, 8th ser., 22, 5–143.

Daumas, M. 1955. *Lavoisier théoricien et expérimentateur*. Paris: PUF.

Daumas, M. 1972. *Scientific Instruments of the 17th and 18th Centuries and their Makers*. London: Praeger.

Davy, R. 1955. *Contribution à l'étude des origines de la droguerie pharmaceutique et de l'industrie du sel ammoniac en France: l'apothicaire Antoine Baumé (1728–1804)*, Cahors.

Delambre J. B. J., 1827. *Histoire de l'astronomie au dix-huitième siècle*. Paris: Bachelier.

Delaunay, P. 1905. *Le Monde médical parisien au dix-huitième siècle*. Paris: Rousset.

Demeulenaere-Douyère, C. 2009. "Inventeurs en Révolution: La Société des inventions et découvertes." *Documents pour l'histoire des techniques* 17, 19–56.

Demeulenaere-Douyère, C. and Brian, É., eds. 2002. *Règlement, usages et science dans la France de l'Absolutisme*. Paris: Éd. Lavoisier Tech & Doc.

Deming, M. K. 1984. *La Halle au blé de Paris, 1762–1813: 'Cheval de Troie' de l'abondance dans la capitale des Lumières*. Brussels: Archives d'Architecture Moderne.

De Place, D. 1988. "Bureau de consultation pour les arts, Paris, 1791–1796." *History and Technology* 5, 139–178.

Dion, M.-P. 1987. *Emmanuel de Croÿ (1718–1784), Itinéraire intellectuel et réussite nobiliaire au siècle des Lumières.* Brussels: Editions de l'Université de Bruxelles.

Doyon, A. 1963. "L'Hôtel de Mortagne après la mort de Vaucanson (1782–1837)." *Histoire des entreprises* 11, 5–23.

Doyon, A., and Liaigre, L. 1966. *Jacques Vaucanson mécanicien de génie.* Paris: PUF.

Dulaure, J.-A. 1787. *Nouvelle description des curiosités de Paris.* 2nd ed. Paris: Lejay.

Duprat, C. 1993. *Le Temps des philanthropes, La philanthropie parisienne des Lumières à la Monarchie de Juillet,* vol. 1. Paris: CTHS.

Duris, P. 1993. *Linné et la France (1780–1850).* Geneva: Droz.

Easterby-Smith, S. 2017. *Cultivating Commerce: Cultures of Botany in Britain and France, 1760–1815.* Cambridge: Cambridge University Press.

Falls, W. F. 1933. "Buffon et l'agrandissement du Jardin du Roi à Paris." *Archives du Muséum d'Histoire naturelle,* 6th ser. 10, 131–200.

Fosseyeux, M. 1912. *L'Hôtel-Dieu de Paris au XVIIe et au XVIIIe siècle.* Paris: Berger-Levrault.

Frijhoff, W. 1990. "L'École de chirurgie de Paris et les Pays-Bas analyse d'un recrutement, 1752–1791." *Lias* 17, 185–239.

Garrioch, D. 2002. *The Making of Revolutionary Paris.* Berkeley: University of California Press.

Gelfand, T. 1980. *Professionalizing Modern Medicine: Paris Surgeons and Medical Science and Institutions in the 18th Century.* London: Greenwood Press.

Gillispie, C. C. 1980. *Science and Polity in France: the End of the Old Regime.* Princeton, NJ: Princeton University Press.

Gillispie, C. C. 1983. *The Montgolfier Brothers and the Invention of Aviation, 1783–1784, with a Word on the Importance of Ballooning for the Science of Heat and the Art of Building Railroads.* Princeton, NJ: Princeton University Press.

Gillispie, C. C. 1997. *Pierre-Simon Laplace, 1749–1827: A Life in Exact Science.* Princeton, NJ: Princeton University Press.

Gillispie, C. C. 2004. *Science and Polity in France: The Revolutionary and Napoleonic Years.* Princeton, NJ: Princeton University Press.

Gillmor, C. S. 1971. *Coulomb and the Evolution of Physics and Engineering in Eighteenth Century France.* Princeton, NJ: Princeton University Press.

Goodman, D. 1994. *The Republic of Letters: A Cultural History of the French Enlightenment.* Ithaca, NY: Cornell University Press.

Goodman, D. 2009. *Becoming a Woman in the Age of Letters.* Ithaca, NY: Cornell University Press.

Greenbaum, L. S. 1973. "Jean-Sylvain Bailly, the Baron de Breteuil and the 'Four New Hospitals' of Paris." *Clio Medica* 8, 261–284.

Greenbaum, L. S. 1975. "Scientists and Politicians: Hospital Reform in Paris on the Eve of the French Revolution." In *The Consortium on Revolutionary Europe, 1750–1850, Proceedings, 1974*, ed. D. D. Horward, 168–191. Gainesville: University Press of Florida.

Grimaux, É. 1899. *Lavoisier: 1743–1794*. Paris: Alcan.

Grimm, F. 1829–1830. *Correspondance littéraire, philosophique et critique de Grimm et de Diderot depuis 1753 jusqu'en 1790*. Paris: Furne.

Grison, É. 1996. *L'Étonnant parcours du républicain Hassenfratz (1755–1827)*. Paris: Les Presses de l'École des mines.

Guénot, H. 1986. "Musées et lycées parisiens (1780–1830)." *Dix-Huitième Siècle* 18, 149–167.

Guerlac, H. 1976. "Chemistry as a Branch of Physics: Laplace's Collaboration with Lavoisier." *Historical Studies in Physical Sciences* 7, 193–276.

Guillerme, A. 2007. *La Naissance de l'industrie à Paris entre sueurs et vapeurs, 1780–1830*. Seyssel: Champ Vallon.

Hahn, R. 1964. "The Chair of Hydrodynamics in Paris 1775–1791: A Creation of Turgot." *Actes du dixième Congrès international d'histoire des sciences* 2, 751–754.

Hahn, R. 1971. *The Anatomy of a Scientific Institution: The Paris Academy of Sciences, 1666–1803*. Berkeley: University of California Press.

Hahn, R. 2004. *Pierre Simon Laplace, 1749–1827: A Determined Scientist*. Cambridge, MA: Harvard University Press, 2005.

Hamon, M. 2010. *Madame Geoffrin: Femme d'influences, femme d'affaires au temps des Lumières*. Paris: Fayard.

Hannaway, C. 1972. "The Société Royale de médecine and Epidemics in the Ancien Régime." *Bulletin of the History of Medicine* 46, 257–273.

Hannaway, O., and C. Hannaway. 1977. "La fermeture du cimetière des Innocents." *Dix-Huitième Siècle* 9, 181–191.

Hazard, P. 2013. *The Crisis of the European Mind: 1680–1715*, 1952, repr. New York: New York Review Books, 2013.

Hazard, P. 1954. *European Thought in the Eighteenth Century from Montesquieu to Lessing*. London: Hollis & Carter.

Hesse, C. 1986. *Publishing and Cultural Politics in Revolutionary Paris, 1789–1810*. Berkeley: University of California Press.

Hilaire Pérez, L. 2000. *L'Invention technique au siècle des Lumières*. Paris: Albin Michel.

Holmes, F. L. 1985. *Lavoisier and the Chemistry of Life: An Exploration of Scientific Creativity*. Madison: University of Wisconsin Press.

Isherwood, R. M. 1986. *Farce and Fantasy: Popular Entertainment in Eighteenth Century Paris*. Oxford: Oxford University Press.

Israel, J. 2011. *Democratic Enlightenment: Philosophy, Revolution, and Human Rights, 1750–1790*. Oxford: Oxford University Press.

Jones, C. 1990. "The *Médecins du Roi* at the End of the Ancien Régime and in the French Revolution." In *Medicine at the Courts of Europe, 1500–1837*, ed. V. Nutton, 214–267. London: Routledge.

Jones, C. 2014. *The Smile Revolution in Eighteenth Century Paris.* Oxford: Oxford University Press.

Juratic, S. 1997. "Le commerce du livre à la veille de la Révolution." In *Le Commerce de la librairie en France au xixe siècle, 1789–1914*, ed. J.-Y. Mollier, 19–26. Paris: IMEC éditions.

Kaplan, S. 1984. *Provisioning Paris: Merchants and Millers in the Grain and Flour Trade during the Eighteenth Century.* Ithaca, NY: Cornell University Press.

Kaplan, S. 1996. *The Bakers of Paris and the Bread Question, 1700–1775.* Duke University Press.

Kaplan, S. 2001. *La Fin des corporations.* Paris: Fayard.

Kates, G. 1985. *The Cercle Social, the Girondins and the French Revolution.* Princeton, NJ: Princeton University Press.

Lacordaire, A.-L. 1853. *Notice historique sur les Manufactures impériales de tapisseries des Gobelins et de tapis de la Savonnerie, suivie du catalogue des tapisseries exposées et en cours d'exécution.* Paris: À la Manufacture des Gobelins.

Langlois, G.-A. 1991. *Folies, Tivolis et attractions: Les premiers parcs de loisirs parisiens.* Paris: Délégation à l'action artistique de la ville de Paris.

Lanoë, C. 2008. *La Poudre et le fard: Une histoire des cosmétiques de la Renaissance aux Lumières.* Seyssel: Champ-Vallon.

Lavoisier, A.-L. 1955–2011. *Correspondance.* Edited by R. Fric et. al. Paris.

Lavoisier, A.-L. 1862–1893. *Œuvres.* Edited by J.-B. Dumas and E. Grimaux. Paris.

Lefranc, A. 1893. *Histoire du collège de France depuis ses origines jusqu'à la fin du Premier Empire.* Paris: Hachette.

Le Roux, Th. 2011. *Le laboratoire des pollutions industrielles, Paris, 1770–1830.* Paris: Albin Michel.

Letouzey, Y. 1989. *Le Jardin des plantes à la croisée des chemins avec André Thouin, 1747–1824.* Paris: Muséum national d'histoire naturelle.

Lilti, A. 2015. *The Society of Salons: Sociability and Worldliness.* New York: Oxford University Press.

Lough, J. 1971. *The Encyclopédie.* New York: David McKay.

Lopez, Cl.-A. 1966. *Mon Cher Papa: Franklin and the Ladies of Paris.* New Haven, CT: Yale University Press.

Lynn, M. R. 2006. *Popular Science and Public Opinion in Eighteenth Century France.* Manchester: Manchester University Press.

Maindron, É. 1888. *L'Académie des sciences: Histoire de l'Académie; Fondation de l'Institut national; Bonaparte membre de l'Institut national.* Paris: Alcan.

Maury, A. 1864. *L'ancienne Académie des Inscriptions et Belles-Lettres.* Paris: Didier.

McClellan, J. E. 1993. "L'Europe des académies." *Dix-Huitième Siècle* 25, 153–165.

Mercier, L.-S. 1782–1788. *Tableau de Paris*. Rev. ed. Amsterdam.

Mi Gyung Kim. 2006. "'Public' Science: Hydrogen balloons and Lavoisier's Decomposition of Water." *Annals of Science* 63, 291–318.

Mi Gyung Kim. 2017. *The Imagined Empire: Balloon Enlightenments in Revolutionary Europe*. Pittsburgh: University of Pittsburgh Press

Noël, Y. and R. Taton. 1997. "La réforme des poids et mesures. 1. Origines et premières étapes (1789–1791)." In Lavoisier 1955–2011, vol. 6, *1789–1791*, 439–465.

Passeron, I., ed. 2008. "La République des sciences." Special issue, *Dix-Huitième Siècle 40*.

Pattie, F. A. 1994. *Mesmer and Animal Magnetism: A Chapter in the History of Medicine*. Hamilton, NY: Edmonston.

Payen, J. 1969. *Capital et machine à vapeur au xviiie siècle: les frères Périer et l'introduction en France de la machine à vapeur de Watt*. Paris: Mouton.

Pellisson, M. 1911. *Les Hommes de lettres au XVIIIe siècle*. Paris: Armand Colin.

Perkins, J. 2010. "Chemistry Courses, the Parisian Chemical World and the Chemical Revolution, 1770–1790" *Ambix* 57, 27–47.

Picon, A. 1992. *L'Invention de l'ingénieur moderne: l'École des ponts et chaussées, 1747–1851*. Paris: Presses de l'École nationales des ponts et chaussées.

Pinard, A., et al. 1903. *Commentaires de la Faculté de médecine: 1777 à 1786*. 2 vols. Paris: G. Steinheil.

Pomeau, R. 1995. *Voltaire en son temps*. 2 vols. Paris: Fayard.

Proust, J. 1957. "La Documentation technique de Diderot dans l'Encyclopédie." *Revue d'Histoire littéraire de la France* 57, 335–352.

Proust, J. 1995. *Diderot et l'Encyclopédie*. 3rd ed. Paris: Albin Michel.

Pujoulx, J.-B. 1801. *Paris à la fin du XVIIIe siècle*. Paris: B. Mathé.

Railliet, A., and L. Moulé. 1908. *Histoire de l'École d'Alfort*. Paris: Asselin et Houzeau.

Reinhardt, M. 1971. *Nouvelle histoire de Paris: La Révolution, 1789–1799*. Paris: Hachette.

Regourd, F. 2008. "Capitale savante, capitale coloniale: sciences et savoirs coloniaux à Paris aux XVIIe et XVIIIe siècles." *Revue d'histoire moderne et contemporaine*, no. 55, 121–151.

Reverd, L. 1946. "La Manufacture des Gobelins et les colorants naturels." *Hyphé* 1, no. 2, 91–104, and no. 3, 141–147.

Riskin, J. 2002. *Science in the Age of Sensibility: The Sentimental Empiricists of the French Enlightenment*. Chicago: University of Chicago Press.

Roche, D. 1988. *Les Républicains des lettres: Gens de culture et Lumières au xviiie siècle*. Paris: Fayard.

Roche, D., ed. 2000. *La ville promise: Mobilité et accueil à Paris (fin XVIIe–début XIXe siècle)*. Paris: Fayard.

Roche, D. 1993. *La France des Lumières*. Paris: Fayard.

Roger, J. 1989. *Buffon, un philosophe au Jardin du Roi*. Paris: Fayard.

Salomon-Bayet, C. 1978. *L'institution de la science et l'expérience du vivant: Méthode et expérience à l'Académie royale des sciences: 1666–1793*. Paris: Flammarion.

Scheurrer, H. 1996. "La 'roskopf' de Breguet, ou comment produire une montre." *Chronométrophilia* 40, 83–96.

Schmitt, S. 2009. "From physiology to classification: comparative anatomy and Vicq d'Azyr's plan of reform of life sciences and medicine." *Science in Context* 22, 145–193.

Ségur, P. de. 1905. *Julie de Lespinasse*. Paris: Calmann-Lévy.

Sgard, J. 1991. *Dictionnaire des journaux (1600–1789)*. 2 vols. Paris: Universitas.

Shank, J. B. 2008. *The Newton Wars and the Beginning of the French Enlightenment*. Chicago: University of Chicago Press, 1994.

Simon, J. 2005. *Chemistry, Pharmacy and Revolution in France, 1777–1809*. Ashgate: Aldershot, 2005.

Smeaton, W. 1955. "The Early Years of the Lycée and the Lycée des Arts: A Chapter in the Lives of A. L. Lavoisier and A. F. de Fourcroy." *Annals of Science* 11, 309–319.

Smith, E. B. 1954. "Jean-Sylvain Bailly: Astronomer, Mystic, Revolutionary, 1736–1793." *Transactions of the American Philosophical Society*, n.s., 44, 427–538.

Smith, J. G. 1979. *The Origins and Early Development of the Heavy Chemical Industry in France*. Oxford: Clarendon.

Sonenscher, M. 1989. *Work and Wages: Natural law, Politics and the Eighteenth-Century French Trades*. Cambridge: Cambridge University Press.

Spary, E. C. 2000. *Utopia's Garden: French Natural History from Old Regime to Revolution*. Chicago: University of Chicago Press.

Sutton, G. 1995. *Science for a Polite Society: Gender, Culture and Demonstration of Enlightenment*. Boulder CO: Westview.

Taton, R., ed. 1964. *Enseignement et diffusion des sciences en France au XVIIIe siècle*. Paris: Hermann.

Terrall, M. 2002. *The Man Who Flattened the Earth: Maupertuis and the Sciences in the Enlightenment*. Chicago: University of Chicago Press.

Terrall, M. 2014. *Catching Nature in the Act: Réaumur and the Practice of Natural History in the Eighteenth Century*. Chicago: University of Chicago Press.

Thébaud-Sorger, M. 2009. *L'aérostation au temps des Lumières*. Rennes: PUR.

Thiéry, L.-V. 1783–1787. *Almanach du voyageur à Paris*. Paris: anonymous, several editions.

Thiéry, L.-V. 1787. *Guide des amateurs et des étrangers à Paris, ou Description raisonnée de cette ville, de sa banlieue et de tout ce qu'elles contiennent de remarquables*. 2 vols. Paris: Hardouin & Gattey.

Torlais, J. 1953. "Un prestidigitateur célèbre, chef de service d'électrothérapie au XVIIIᵉ siècle: Ledru dit Comus (1731–1807)." *Histoire de la médecine* 5, 13–25.

Torlais, J. 1954. *Un physicien au siècle des Lumières, l'abbé Nollet: 1700–1770*. Paris: Editions Jonas.

Trousson, R. 2005. *Denis Diderot ou le vrai Prométhée*. Paris: Taillandier.

Tucoo-Chala, S. 1977. *Charles-Joseph Panckoucke et la librairie française de 1736 à 1798.* Paris: Touzot.

Turner, A. 1989. *From Pleasure and Profit to Science and Security: Étienne Lenoir and the Transformation of Precision Instrument-Making in France, 1760–1830.* Cambridge: Whipple Museum of the History of Science.

Van Marum, M. 1970. "Journal physique de mon séjour à Paris—1785." (with an English translation), in *Martinus van Marum Life and Work*, vol. 2, ed. R. J. Forbes, 31–52. Haarlem. English translation, 220–239.

Vidal, M. 1995. "David among the Moderns: Art, Science and the Lavoisier." *Journal of the History of Ideas* 56, 595–623.

Weiner, D., and M. J. Sauter. 2003. "The City of Paris and the Rise of Clinical Medicine." *Osiris*, 2nd ser., 18, 23–42.

Wilson, C. 1985. "The Great Inequality of Jupiter and Saturn: From Kepler to Laplace." *Archive for the History of Exact Sciences* 33, 15–290.

Wolf, C. 1902. *L'Histoire de l'Observatoire de Paris, de sa fondation à 1793.* Paris: Gauthier-Villars.

Wolfe, J. J. 1999. *Brandy, Balloons, and Lamps: Ami Argand, 1750–1803.* Cardondale: Southern Illinois University Press.

Young, A. 1792. *Travels during the Years 1787, 1788 and 1789.* Dublin: J. Rackham for W. Richardson.

Index